THE IMPOSSIBLE BOMB

Also by Gareth Williams

***Angel of Death: The Story of Smallpox* (2010)**
'Wonderfully researched, vividly written ... medical history at its absolute best.' Michael Neve, Wellcome Book Prize

'Williams recounts the history of smallpox in a breezy, accessible style. And what a history it is.' Clive Anderson, *New Scientist Big Read*

'A meticulously researched story with pace and flair ... Both the history and the science are terrific.' Medical Journalists' Association

***Paralysed with Fear: The Story of Polio* (2013)**
'His splendid book, riveting from beginning to end, is a model of its kind ... a consistently fascinating account.' *Literary Review*

'An incredible story told by a great storyteller.' *The Lancet*

'Wonderful ... a revealing story of the development of 20th-century medicine, warts and all.' *BBC History*

***A Monstrous Commotion: The Mysteries of Loch Ness* (2015)**
'Sheds intriguing light on the origins of our obsession ... Williams does a fine job of weaving all these threads into a coherent narrative ... The final revelation is a good one.' *Sunday Telegraph*

'Lively and entertaining ... surely the best and sanest of the recent books on the Loch Ness Monster.' *Literary Review*

'Gareth Williams is the first to present a non-partisan account of events at the loch ... [He] brings a dry wit and a scientist's illuminating perspective.' *TLS*

***Unravelling the Double Helix: The Lost Heroes of DNA* (2019)**
'Truly superb ... Williams tells the story of science's greatest discovery with fluency and a real feel for narrative.' *The Times*

'Williams refreshes a familiar chronicle ... looking beyond giants to the many scientists, now half-forgotten, whose contributions paved the way to an icon of science.' *Nature*

'Williams brings a fresh look and detail ... a "rattling good yarn" which I thoroughly recommend.' *Institute of Physics News*

THE IMPOSSIBLE BOMB

The Hidden History of British Scientists and the Race to Create an Atomic Weapon

GARETH WILLIAMS

YALE UNIVERSITY PRESS
NEW HAVEN AND LONDON

Copyright © 2025 Gareth Williams

All rights reserved. This book may not be reproduced in whole or in part, in any form (beyond that copying permitted by Sections 107 and 108 of the U.S. Copyright Law and except by reviewers for the public press) without written permission from the publishers.

All reasonable efforts have been made to provide accurate sources for all images that appear in this book. Any discrepancies or omissions will be rectified in future editions.

For information about this and other Yale University Press publications, please contact:
U.S. Office: sales.press@yale.edu yalebooks.com
Europe Office: sales@yaleup.co.uk yalebooks.co.uk

Set in Minion Pro by IDSUK (DataConnection) Ltd
Printed and bound in the UK using 100% renewable electricity at CPI Group (UK) Ltd

Library of Congress Control Number: 2025935745
A catalogue record for this book is available from the British Library.
Authorized Representative in the EU: Easy Access System Europe, Mustamäe tee 50, 10621 Tallinn, Estonia. gpsr.requests@easproject.com

ISBN 978-0-300-28488-1

10 9 8 7 6 5 4 3 2 1

*With love and thanks to Alison, Tim, Jo, Tessa, Emily and Alice,
and in memory of Caroline*

*Special thanks to Robin for getting me started, and to Jane and Alan,
without whom I might never have finished*

CONTENTS

	List of Plates, Figures and Maps	*ix*
	Timeline, 1939–45	*xiii*
	Who's Who	*xx*
	Preface: By Way of Explanation	*xxviii*
1	Setting the Scene	1
2	War Games	11
3	Pure Physics	19
4	Years of Wonder	28
5	Beyond Nature?	40
6	To the Brink	51
7	Warm-Up	67
8	Liaisons Dangereuses	80
9	Memorandum of Understanding	93
10	Separation Anxieties	107
11	Men with Missions	119
12	In the Dark	130
13	Minority Reports	141

vii

CONTENTS

14	Tube Alloys	155
15	Über Alles	169
16	Double Dealing	179
17	Critical Masses	192
18	Breakdown and Repair	203
19	Over There	215
20	Missionaries	223
21	Liberation	230
22	The Giving of All Help	243
23	Countdown	254
24	On the Run	267
25	Proof of Concept	274
26	Instrument of War	286
27	Aftermaths	298
28	Winners and Losers	306
29	Whodunnit	316
30	Follow-Up	330
	Postscript: Full Circle	348

Appendix I Frisch–Peierls Memorandum	*351*
Appendix II Members of the British Mission to the Manhattan Project	*358*
Glossary	*360*
Acknowledgements	*369*
Notes	*372*
Select Bibliography	*410*
Index	*414*

PLATES, FIGURES AND MAPS

Plates

1. Otto Hahn and Lise Meitner in their laboratory, Berlin. CBW / Alamy.
2. James Chadwick. Photograph by Burrell and Hardman, Liverpool. Reproduced by courtesy of AIP Emilio Segrè Visual Archives, gift of Lawrence Cranberg.
3. Werner Heisenberg and Niels Bohr, Copenhagen. Photograph by Paul Ehrenfest, Jr. Reproduced by courtesy of AIP Emilio Segrè Visual Archives, Weisskopf Collection.
4. Marcus (Mark) Oliphant. Historic Collection / Alamy.
5. John Cockcroft in the Cavendish Laboratory, Cambridge. Chronicle / Alamy.
6. Otto Robert Frisch. Reproduced by courtesy of AIP Emilio Segrè Visual Archives, *Physics Today* Collection.
7. Attendees at Niels Bohr's annual symposium, Copenhagen, July 1937. Photograph by GF Hund. GFHund / CC BY 3.0.
8. Frédéric Joliot with Hans von Halban and Lew Kowarski, Paris. Photograph by Charles Baulard. Reproduced by kind permission of the Musée Curie and Agence Di.
9. Rudolf (Rudi) Peierls. Reproduced from *Science Monthly*, 12 (1945), by kind permission of the British Council. Courtesy of Churchill Archives Centre, Papers of Sir Michael Perrin, PERR 3/3.

PLATES, FIGURES AND MAPS

10. Sir Henry Tizard. SuperStock / Alamy.
11. Jacques Allier with Frédéric Joliot, Paris. Photograph by Charles Baulard. Reproduced by kind permission of the Musée Curie and Agence Di.
12. George Thomson. piemags / ww2archive / Alamy.
13. Franz (Francis) Simon. World History Archive / Alamy.
14. Paul Rosbaud. Photograph by Lotte Meitner-Graf (1898–1973). © The Lotte Meitner-Graf Archive.
15. Klaus Fuchs. Photograph by Los Alamos National Laboratory.
16. Ernest Lawrence, Arthur Compton, Vannevar Bush and James Conant. U.S. Department of Energy, HD.1A.018.
17. Wallace Akers. Photograph by Walter Stoneman. © National Portrait Gallery, London.
18. General Leslie R. Groves and J. Robert Oppenheimer. Science History Images / Alamy.
19. The 'calutron girls' in the Y-12 electromagnetic U-235 separation plant at Oak Ridge, 1944. U.S. Department of Energy, HD.30.844.
20. Mark Oliphant's research group at Berkeley. U.S. National Archives.
21. Otto Frisch playing the piano for the Los Alamos radio station. Science History Images / Alamy.
22. The Alsos team dismantling the *Uranverein*'s reactor shell. Photograph by Sergeant Malcolm Thurgood. Reproduced by courtesy of AIP Emilio Segrè Visual Archives, Goudsmit Collection.
23. The Alsos team digging up uranium cubes from the Haigerloch reactor. Photograph by Samuel Goudsmit. Reproduced by courtesy of AIP Emilio Segrè Visual Archives, Goudsmit Collection.
24. A US Army sergeant delivers the plutonium core for the Trinity bomb. Federal Government of the United States.
25. The Trinity fireball, photographed by an ultra-high-speed camera. U.S. Department of Energy, HD.4G.042, HD.4G.043, HD.4G.045, HD.4G.046.

26. Little Boy in the loading pit on Tinian Island. U.S. National Archives, NAID: 519394.
27. The centre of Hiroshima, before and after the detonation of Little Boy, 6 August 1945. ibiblio.org.
28. Under the mushroom cloud in Nagasaki, 9 August 1945. Photograph by Hiromichi Matsuda. Shawshots / Alamy.
29. The British Mission's farewell to Los Alamos. Atomic / Alamy.
30. Sir James Chadwick and General Leslie Groves, with Richard Tolman and Henry Smyth. United States Army Signal Corps.
31. William Penney, Otto Frisch, Rudi Peierls and John Cockcroft in 1946. Los Alamos National Laboratory.
32. Albert Einstein and Leo Szilard re-enacting the drafting of Einstein's letter to Roosevelt.

Figures

All figures were drawn by Ray Loadman, except for Figure 7.

1.	Components of ionising radiation: alpha- and beta-particles.	6
2.	The atom, as envisaged by Ernest Rutherford and modified by Niels Bohr.	9
3.	Structure of the nucleus, as envisaged after Chadwick's discovery of the neutron in 1932.	29
4.	Nuclear fission in uranium (U-235).	57
5.	Nuclear chain reaction in uranium, confined to U-235.	60
6.	Creation of plutonium by irradiating uranium (U-238) with neutrons.	83
7.	The 'calutron', used for the electromagnetic separation of U-235 from U-238. Original drawing by Robert Hile, Westinghouse.	236
8.	Little Boy, the U-235 bomb destined for Hiroshima, showing its internal structure.	256
9.	Making nuclear explosive for Little Boy: the U-235 enrichment cascade at Oak Ridge.	257

10. Fat Man, the plutonium bomb destined for Nagasaki, showing its internal structure. 260
11. Diagram of the *Uranverein*'s B-VIII reactor at Haigerloch. 272

Maps

Maps drawn by Martin Brown.

1. Northern Europe before the Second World War. xxxi
2. Research and supply centres for the Tube Alloys programme, 1943–4. xxxii
3. Research and supply centres for the German *Uranverein* programme, 1944. xxxiii
4. The three 'Atomic Cities' of the Manhattan Project, 1943–5. xxxiv
5. Los Alamos and its surroundings, New Mexico, 1944–5. xxxv
6. The villages of Haigerloch, Hechingen and Tailfingen, close to the Swiss border in southern Germany. xxxvi

TIMELINE, 1939–45

Military and political landmarks are italicised; detonations of atomic bombs are in bold.

1939

Jan Hahn and Strassmann show that neutrons split uranium ('fission')
Meitner and Frisch show that fission releases a huge burst of energy
Joliot, Szilard and others report that fission generates neutrons which could trigger an explosive nuclear chain reaction
Feb Bohr proposes that only the rare isotope U-235 undergoes fission
Mar Rosbaud briefs Cockcroft about German nuclear research
Hitler invades Czechoslovakia, breaking Munich Agreement
Apr Tizard's Committee (UK) considers the possibility of a uranium fission bomb
Uranverein ('Uranium Club') of German nuclear researchers meets to develop atomic bombs and reactors
May Joliot files secret patents for nuclear reactors and bombs
Rosbaud updates Cockcroft about the *Uranverein*
British government drafts secret report on possibility of an atomic bomb
July Tizard sets up two secret research projects into uranium fission

TIMELINE, 1939–45

Aug Einstein signs a letter alerting Roosevelt to the risk of a German atomic bomb
Sep *Hitler invades Poland; Britain and France declare war on Germany*
     Uranverein conference on uranium reactors and bombs, led by Heisenberg
Oct New *Uranverein* fission research centres begin construction in Berlin
     Uranium Advisory Committee (USA) funds atomic reactors but not bombs
Dec Chadwick begins experiments to develop an atomic bomb
     Heisenberg reports positively on atomic bomb to German Army
     Germany orders huge quantity of heavy water from Norsk Hydro plant in Vemork

1940

Feb Nier and Dunning confirm that only U-235 undergoes fission
Mar Frisch and Peierls write a secret memorandum, proposing that a few kilos of pure U-235 will make a devastating 'super-bomb'
     Allier steals entire heavy-water stock from Vemork
Apr *Germany invades Denmark and Norway*
     Thomson's M.A.U.D. Committee (UK) meets to develop the Frisch–Peierls U-235 super-bomb
     USA's Uranium Advisory Committee funds Fermi's uranium pile but not atomic bombs
May *Belgium and Holland fall to the Nazis*
     Winston Churchill becomes British Prime Minister
June *Fall of Paris*
     USA's revamped Uranium Committee dismisses possibility of atomic bomb
     McMillan and Abelson create element 94 by bombarding uranium with neutrons
     Howard rescues Halban, Kowarski and French heavy-water stocks from Bordeaux
     France surrenders to Germany

TIMELINE, 1939–45

July *Battle of Britain begins*
Aug   Tizard's scientific and technological mission to the USA begins
Sep   British send Americans all information about the U-235 super-bomb
Dec   Halban and Kowarski's heavy-water-moderated reactor (UK) generates excess of neutrons, suggesting incipient chain reaction
      Kemmer names element 94 'plutonium'
      Bretscher identifies plutonium as suitable for an atomic bomb

1941

Jan   M.A.U.D. Committee abandons plutonium as a nuclear explosive
Feb   Seaborg isolates plutonium
Apr   *Germany invades Greece and Yugoslavia and now occupies continental Europe*
May   Seaborg shows that plutonium is more fissionable than U-235
      USA's National Academy of Science (NAS) experts dismiss atomic bomb
June *Germany invades Russia*
July M.A.U.D. Committee concludes that the U-235 super-bomb will work but must be built in collaboration with the Americans; final report sent to USA
Aug   Oliphant discovers that British information about the U-235 super-bomb was withheld from the USA's Uranium Committee; persuades Lawrence that the bomb is viable
Sep   Churchill decides that Britain will build the U-235 super-bomb alone and covertly obstructs collaboration with USA
      Heisenberg meets Bohr in Copenhagen to discuss atomic bombs
Oct   *German Army reaches Moscow*
      Roosevelt urges Churchill to collaborate on the U-235 super-bomb
      A third NAS review concludes that the atomic bomb is viable
Nov   Tube Alloys Directorate (UK) supersedes M.A.U.D. Committee
Dec   *Pearl Harbor; US declares war on Japan and then Germany*

1942

Jan — *German Army retreats after defeat in Battle of Moscow*
Roosevelt instructs Bush, USA's scientific supremo, to 'build the bomb and build it fast'
Tube Alloys delegation finds no coherent atomic bomb planning in USA

June — Nazi military chiefs support *Uranverein*'s reactors but axe the atomic bomb
Oppenheimer appointed to lead American atomic bomb programme

July — Churchill accepts that Britain can only build the atomic bomb in collaboration with the Americans and in the USA

Aug — US Army sets up Manhattan District Engineer Project to build atomic bombs

Sep — Groves appointed Director of Manhattan Project
Churchill signs Anglo-Russian Treaty, potentially sharing British and American nuclear secrets with the Soviets

Oct — Bush cuts off British access to key aspects of American atomic research

Nov — Fermi begins building CP-1, a graphite-moderated 'pile' (reactor) in Chicago
Sintered-nickel gaseous diffusion membranes (UK) successfully separate U-235
Operation Freshman (UK), sent by Special Operations Executive (SOE) to destroy the heavy-water plant at Vemork, fails catastrophically

Dec — Fermi's CP-1 pile in Chicago sustains a nuclear chain reaction

1943

Feb — Oppenheimer appointed scientific director at Los Alamos
SOE's Operation Gunnerside (UK) wrecks the heavy-water plant at Vemork
Germans defeated at Stalingrad

Mar — Manhattan Project work begins at Hanford (making plutonium), Oak Ridge (enriching U-235) and Los Alamos (bomb development)

TIMELINE, 1939–45

May   *Germans defeated in North Africa*
Aug   Churchill and Roosevelt sign secret Quebec Agreement for full US–UK collaboration on atomic bombs
British Mission to the Manhattan Project established, led by Chadwick
Sep   Bohr extracted from Denmark and joins British Mission
Allies invade Italian mainland
Nov   US Air Force bombs the renovated heavy-water plant at Vemork
Experimental pile (X-10) at Oak Ridge begins producing plutonium
Dec   Groves sets up Alsos Mission to hunt down *Uranverein* researchers and acquire Nazi atomic knowledge
First members of British Mission arrive in Los Alamos

1944

Feb   Following Allied air raids on Berlin, *Uranverein* scientists relocate to southern Germany
SOE saboteurs blow up ferry carrying last heavy-water stock from Vemork
Apr   Groves commissions Smyth to write official history of the Manhattan Project
June   *D-Day: Allied invasion of Normandy*
July   *Hitler survives briefcase-bomb assassination attempt*
Aug   *Liberation of Paris*
Alsos Mission interrogates Joliot in Paris
Trinity test of plutonium bomb scheduled at Alamogordo, New Mexico
Sep   Roosevelt and Churchill agree potential use of atomic bombs against Japan

1945

Jan   Implosion 'gadget' invented by Christie and Peierls is chosen to detonate 'Fat Man' (plutonium bomb)
Feb   First delivery of plutonium from Hanford to Los Alamos
High-explosive 'lenses' successfully implode a cadmium sphere

TIMELINE, 1939–45

	Roosevelt, Churchill and Stalin meet at Yalta to decide fate of Germany after the war
	US defeats Japan in Battle of Iwo Jima
Mar	Allied troops cross the Rhine into Germany
	Start of regular U-235 deliveries from Oak Ridge to Los Alamos
	Alsos launches Operation Big to seize *Uranverein* personnel
	US Air Force firebombing of Tokyo kills 83,000 people
Apr	Frisch's 'dragon' experiment shows that 'Little Boy' (uranium bomb) needs 52 kg of U-235
	Design of Fat Man finalised, containing 6 kg of plutonium
	Roosevelt dies; Truman succeeds him as President
	Russians capture Vienna and advance on Berlin
	Alsos captures *Uranverein* members and reactor components at Haigerloch
	Hitler commits suicide as Russian troops take Berlin
May	Germany surrenders; Victory in Europe (VE) Day
	Americans capture Okinawa from Japan
	Manufacture of U-235 slugs for Little Boy begins
June	US Chiefs of Staff set 1 November 1945 as date to invade Japan if no surrender
	High-explosive lenses for detonating Fat Man tested successfully
July	Eight *Uranverein* members interned at Farm Hall (UK)
	US–UK Joint Policy Committee approves atomic bombing of Japanese cities
	Truman, Churchill and Stalin meet at Potsdam to divide up postwar Europe
	16 July 1945: successful detonation of Trinity, the test plutonium bomb, at Alamogordo
	Potsdam Declaration threatens Japan with 'utter devastation' unless surrender
Aug	**6 August 1945: Little Boy detonated over Hiroshima, killing at least 80,000 people**
	Russia declares war on Japan
	9 August 1945: Fat Man detonated over Nagasaki, killing at least 40,000 people

TIMELINE, 1939–45

	Publication of Smyth Report, describing only American input to Manhattan
Sep	*Japan surrenders*
Oct	Publication of *Statements Relating to the Atomic Bomb*, the British response to the Smyth Report

WHO'S WHO

Principal Players in the Story

Akers, Wallace (1888–1954)
British industrial chemist and research director at Imperial Chemical Industries in London. Closely involved with the top-secret M.A.U.D. Committee and all aspects of the British nuclear fission programme during the war. Appointed Director of Tube Alloys in late 1941; two years later, the Americans forced him to withdraw from the British Mission to the Manhattan Project.

Bohr, Niels (1885–1962)
Danish theoretical physicist; Nobel Prize in Physics (1922). Director of his world-famous Institute for Theoretical Physics in Copenhagen and father figure for European nuclear physicists. Escaped to England from Nazi-occupied Denmark in 1943; badly disguised as 'Nicholas Baker', joined the British Mission to the Manhattan Project. Actively supported the development of the atomic bombs at Los Alamos (and may have contributed to their design) but steadfastly opposed their use in warfare; failed spectacularly to convince Churchill that Anglo-American nuclear secrets should be shared with the Russians to avoid a nuclear arms race.

Briggs, Lyman (1874–1963)
American soil scientist and scientific administrator; chaired successive secret committees (1939–42) which advised President Roosevelt

about potential military applications of nuclear fission. Briggs knew nothing about nuclear physics and withheld from his committee members (who included top American physicists) crucial British evidence that the U-235 uranium bomb would work. His committees concluded that atomic bombs were neither feasible nor affordable, which helped to harden American opinion against collaborating with the British on the U-235 bomb.

Bush, Vannevar (1890–1974)
American electronics pioneer and President of the Carnegie Institution. America's scientific supremo during the war; chaired the National Defense Research Committee (NDRC) and later the all-powerful Office for Scientific Research and Development (OSRD). Led US policy on the bomb despite his poor understanding of nuclear physics; was highly sceptical of the British U-235 atomic bomb and during 1942–3 actively blocked American collaboration with the British on both the U-235 and plutonium bombs.

Chadwick, James (1891–1974)
Legendarily inscrutable British nuclear physicist; while interned in Berlin during the First World War, continued working for his PhD with experiments on radioactive toothpaste. Won the Nobel Prize in Physics (1932) for discovering the neutron. Leading light in the British nuclear programme from late 1939, including the M.A.U.D. Committee and Tube Alloys. Directed the British Mission to the Manhattan Project; his highly effective working relationship with General Leslie Groves underpinned the Anglo-American collaboration which delivered both the U-235 and plutonium bombs dropped on Japan in August 1945.

Cockcroft, John (1897–1967)
British physicist and 'man of few words'. In 1932, he and Ernest Walton split the lithium atom (Nobel Prize in Physics, 1951). Key member of Sir Henry Tizard's Technical and Scientific Mission to America in 1940, and of the M.A.U.D. Committee. Instrumental in building up the British defensive radar network ('Chain Home')

before the war; later established a successful Anglo-French-Canadian collaboration which built the first functional nuclear reactor outside the USA.

Conant, James (1893–1978)
American chemist and President of Harvard; Vannevar Bush's right-hand man during the war, succeeding him as chair of the National Defense Research Committee. Had little grasp of nuclear physics but, with Bush, determined American policy on the atomic bomb programme and obstructed collaboration with the British during 1942–3.

Fermi, Enrico (1901–1954)
Italian theoretical and experimental physicist; awarded the Nobel Prize in Physics (1938) for creating 'transuranic' elements (later shown to be spurious). His wife was Jewish; they left Fascist Italy for America straight after the Nobel Prize ceremony in December 1938. With Leo Szilard, Fermi built experimental reactors ('piles') of uranium and graphite; his CP-1 pile in Chicago sustained the world's first nuclear chain reaction in December 1943. Subsequent piles were used to generate plutonium for the Trinity test and the Fat Man bomb which devastated Nagasaki. At the Trinity test, Fermi estimated the power of the explosion from watching shreds of paper fall to the ground.

Frisch, Otto Robert (1904–1979)
Austrian-born, Jewish nuclear physicist who fled Nazi Germany in 1933. In late 1938, he and his aunt Lise Meitner deduced that a single neutron could split the uranium nucleus with a massive blaze of energy; he named the process 'fission' and confirmed the energy release experimentally. While working in Birmingham in 1940, Frisch predicted that a few pounds of pure U-235 could make a hugely powerful 'super-bomb' – the basis of the Frisch–Peierls memorandum, co-written with Rudi Peierls. He continued to develop the U-235 bomb through the M.A.U.D. and Tube Alloys committees and at Los Alamos with the British Mission to the Manhattan Project. His hazardous 'dragon' experiment confirmed the feasibility of the Little

Boy bomb dropped on Hiroshima and determined the mass of U-235 required.

Fuchs, Klaus (1911–1963)
German theoretical physicist. Hounded out of Nazi Germany because he was a Communist; fled to Britain where he was arrested as an 'enemy alien' and interned in Canada. Recruited by Rudi Peierls and worked closely with him in Birmingham and at Los Alamos, first on purifying U-235 by gaseous diffusion, and later on modelling the implosion 'gadget' for the plutonium bomb. Was arrested and imprisoned in 1952 for having passed secrets of both bombs to the Russians.

Groves, General Leslie R. (1896–1970)
Megalomaniacal, security-obsessed US Army engineer who had overseen the construction of the Pentagon. As Director of the Manhattan Project, Groves masterminded the creation of the three 'Atomic Cities' at Oak Ridge, Hanford and Los Alamos. Groves is considered one of the three pillars essential to the project's success. Dedicated American patriot and anti-British but worked closely with James Chadwick, ensuring that the American–British collaboration delivered both the bombs dropped on Japan. Groves shaped the Smyth Report, the official American history of Manhattan, which omitted the British contributions and established the misconception that the atomic bombs were exclusively American inventions.

Hahn, Otto (1879–1968)
German chemist whose collaboration with Lise Meitner led in 1918 to the discovery of protactinium (element 91). In late 1938, Hahn and Fritz Strassmann realised that neutrons could split uranium nuclei into smaller fragments including barium (fission). Continued to work on fission during the war; Hahn refused any military involvement but was interned after the war at Farm Hall in Cambridgeshire, together with active members of the *Uranverein*. Awarded the Nobel Prize in Chemistry (1945) for discovering fission, but was prevented from receiving his prize as he was still in internment.

Heisenberg, Werner (1901–1979)
German theoretical physicist, awarded the Nobel Prize in Physics (1932) aged twenty-five. Intellectual leader of the *Uranverein* but lacked leadership and credibility with his peers as he had never done any experiments. Heisenberg effectively killed off the embryonic German atomic bomb programme by giving an inept presentation to Nazi High Command; later revealed his ignorance of atomic bomb physics in a seminar to his fellow internees at Farm Hall, covertly recorded by British Intelligence.

Joliot, Frédéric (1900–1958)
French physicist; son-in-law of Marie Curie; joint winner with his wife Irène of the Nobel Prize in Chemistry (1935) for creating artificial radioisotopes. Pioneer in fission research; with Hans von Halban and Lew Kowarski, filed secret patents in 1939 for nuclear reactors and a uranium atomic bomb (all non-viable). Remained in Nazi-occupied Paris throughout the war; sabotaged his own equipment to prevent *Uranverein* members from using it, and later fought with the Resistance to liberate Paris from the Germans.

Lawrence, Ernest (1901–1958)
American physicist, awarded the Nobel Prize in Physics (1941) for inventing the 'cyclotron' particle accelerator. Initially sceptical about the British U-235 bomb; in late 1941, he was converted by Mark Oliphant into a fervent believer and persuaded Vannevar Bush, James Conant and key American physicists to collaborate with the British in building the U-235 bomb. Helped by Oliphant and his team, Lawrence adapted the cyclotron to separate U-235; the resulting 'calutron' was crucial in producing the nuclear explosive at Oak Ridge for Little Boy.

Meitner, Lise (1878–1968)
Austrian-born, Jewish physicist; aunt of Otto Frisch. Long-term collaborator with Otto Hahn; with him, discovered protactinium (element 91) in 1918. Fled from Berlin to Stockholm in July 1938. Alerted by experimental data sent by Hahn, Meitner and Frisch,

deduced that neutrons split the uranium nucleus and released a huge burst of energy. She remained in Stockholm throughout the war.

Oliphant, Marcus [Mark] (1901–2000)
Volatile, Australian-born Professor of Physics at Birmingham; Frisch and Peierls's boss, who fought tirelessly to have their memorandum about the U-235 bomb taken seriously. Member of the M.A.U.D. Committee; mystified when the Americans failed to respond to M.A.U.D.'s verdict that the U-235 bomb was viable and must be built in collaboration with the USA. On discovering that Lyman Briggs had hidden the M.A.U.D. report from his Uranium Committee, Oliphant confronted Ernest Lawrence and triggered the process of turning American opinion in favour of the bomb. At Berkeley, he led a twenty-five-strong team which played a vital role in enabling the calutrons to enrich U-235 for the Little Boy bomb.

Oppenheimer, J. Robert (1904–1967)
Charismatic, Communist-leaning American theoretical physicist and professor at Berkeley. In early 1943, appointed Director of a 'special laboratory in New Mexico' tasked with producing an 'instrument of war' codenamed 'Projectile S-1-T'. At Los Alamos, Oppenheimer created an extraordinary scientific community and provided the leadership and motivation to design and build two fundamentally different atomic bombs from scratch in just two and a half years. Nicknamed the 'father of the atomic bomb', Oppenheimer is regarded as one of the three indispensable pillars that supported the Manhattan Project.

Peierls, Rudolf [Rudi] (1907–1995)
German-born, Jewish theoretical physicist who left Germany for England in 1932. In 1940, he and Otto Frisch wrote the Frisch–Peierls memorandum, proposing that a devastating super-bomb could be made with a few pounds of pure U-235. Affiliated to the M.A.U.D. Committee, which concluded that the U-235 bomb would work, Peierls became a member of Tube Alloys. He worked with Francis Simon, and later Klaus Fuchs, on using gaseous diffusion to enrich U-235 for the Little Boy uranium bomb; at Los Alamos, Peierls and

Fuchs played crucial roles in devising the implosion shockwave which detonated the Fat Man plutonium bomb.

Rosbaud, Paul (1896–1963)
Austrian-born metallurgist who worked as scientific adviser and talent scout for the Springer publishing company in Berlin, thus winning the confidence of top German scientists including members of the *Uranverein*. In early 1939, he fast-tracked publication of Hahn and Strassmann's paper on nuclear fission. Staunchly but covertly anti-Nazi (his wife was Jewish); in July 1938, Rosbaud helped Lise Meitner to escape from Berlin. Working as an undercover agent (codename 'Griffin') for the British MI6, he sent in secret reports on the V-2 rocket range at Peenemünde, the activities and movements of the *Uranverein*, and the abandonment of the German atomic bomb.

Simon, Franz [Francis] (1893–1956)
German-born, Jewish physicist who left Germany in 1933 and set up a leading laboratory in low-temperature physics in Oxford. From 1940, collaborated with Rudi Peierls to develop a gaseous diffusion enrichment method, using thin metal membranes perforated with millions of tiny channels, for concentrating U-235 from native uranium. A functional plant was never built in Britain during the war, but Simon's work laid the foundations for the huge gaseous diffusion factory at Oak Ridge, which produced U-235 for the Hiroshima bomb.

Szilard, Leo (1898–1964)
Hungarian Jewish physicist and free thinker whose patents for inventions (all unworkable) included an electron microscope and a nuclear chain reaction in beryllium. While living out of two suitcases in a New York hotel, Szilard did early experiments on fission in uranium and predicted a 'grim day for humanity'. Worked with Enrico Fermi on graphite–uranium reactors including the landmark CP-1. While contributing to atomic bomb research in Chicago in spring 1945, Szilard argued that the bombs should be exploded in uninhabited locations to demonstrate their power without loss of life.

Thomson, George (1892–1975)
English physicist, awarded the Nobel Prize in Physics (1937) for discovering the wave properties of the electron. Chaired the M.A.U.D. Committee (1940–1) which concluded that the U-235 super-bomb proposed by Frisch and Peierls would work and could be built in collaboration with the Americans in time to influence the outcome of the war. This laid the ground for Tube Alloys and eventually the British Mission to the Manhattan Project – the third pillar underpinning the project which enabled the uranium and plutonium bombs to be ready for use in August 1945.

PREFACE

By Way of Explanation

Until about 10.30 a.m. on Wednesday, 10 May 2017, I would never have written this book. Pinned to the noticeboard in my study is a black enamel lapel badge stamped with the silver crow's foot of the Campaign for Nuclear Disarmament. When I joined CND in 1980, the Cold War was at its most glacial, and I'd seen enough of the world through the eyes of a young doctor to realise that there could only be losers – hundreds of millions of them – in a nuclear war. Even when a busy career and family life pushed it down the list of things to worry about, the nightmare of watching a mushroom cloud boiling up into the sky above the rooftops still haunted me. Like everyone else, I breathed a sigh of relief when a near-miraculous thaw in East–West relations ended the Cold War in late 1991. I stopped wearing my CND badge but didn't cancel my membership. The planet seemed a safer place, but the visceral fear of nuclear weapons never left me and never will.

On that morning in May 2017, I was dipping into a treasure trove – the index of holdings in the archives of Churchill College, Cambridge. I was researching a book about DNA and had come to sift through the papers of Sir John Randall FRS, who may have thrown away the Nobel Prize for discovering the double helix because he engineered a conflict between Rosalind Franklin and Maurice Wilkins (who, with James Watson and Francis Crick, went on to win the Prize in 1962). If I'd stuck to my brief, you wouldn't be reading this. However, my gaze wandered past 'RANDALL' and was caught by 'TUBE ALLOYS'. These seemingly meaningless words rang a vague bell: something that Wilkins had done during the war? On impulse, I asked to see the material.

A heavy cardboard filing box arrived containing several bound volumes like old-fashioned desk diaries, their covers embossed in gilt capitals – 'MINUTES OF TUBE ALLOYS TECHNICAL COMMITTEE – TOP SECRET' – and each carrying a gummed paper label: *DECLASSIFIED 04/06/2007.* Leafing through the close-typed foolscap pages, I recognised a couple of Nobel Prize winners: James Chadwick, discoverer of the neutron, and John Cockcroft, who split the atom. The other names meant nothing to me. Mark Oliphant. A trio who sounded oddly German: Otto Frisch, Rudolf Peierls and Franz Simon. Two enigmatic chaps – Mr Akers and Mr Perrin – who apparently ran the show. The committee first met on 6 November 1941, at the 'Tube Alloys Directorate Office' in Old Queen Street, London SW. The minutes provided vivid snapshots of frantic activity, excitement, frustration and danger, against the backdrop of Britain at war. Tube Alloys had taken over from an earlier committee cryptically named 'M.A.U.D.', and its purpose was to build an atomic bomb. Five minutes in, the story had become a thriller and I was hooked. I put John Randall on hold and spent all day and the following morning trying and failing to work out where Tube Alloys fitted into the time-honoured saga of how America made the bombs that were dropped on Hiroshima and Nagasaki in August 1945.

The book about DNA stayed on my desk for another couple of years. In August 2019, I dug out my notes to see if Tube Alloys had lost its excitement. It hadn't. Soon after, I found myself listening to a gripping story from an intelligent, articulate man. Before the chance encounter in the Churchill College archives, I would never have talked to him because I'd despised what he did; while working at the Atomic Weapons Establishment in Aldermaston, he'd helped to perfect Britain's hydrogen bombs. He surprised me with occasional flashes of bitterness – which were triggered, I realised, by the same two words that had caught my eye in the archive index. Tube Alloys.

Why was the man from Aldermaston so bitter? Because, he said, everyone knows about the Manhattan Project but very few have even heard of M.A.U.D. or Tube Alloys. He believed that the Americans deliberately airbrushed British scientists and their contributions from accounts of 'the race to build the bomb', leaving the impression – still prevalent today – that the bomb was an all-American invention created from nothing in the USA. Then he told me that America alone couldn't have produced a workable nuclear weapon until after the war had ended; without those forgotten British scientists, the atomic devastation of Hiroshima or Nagasaki could never have happened.

Seeing my surprise, he showed me his own archive of books, articles and cuttings that he'd gathered over fifty years and challenged me to find out if he was right. We shook hands on it, and that's how this book came to be written.

Map 1. Northern Europe before the Second World War.

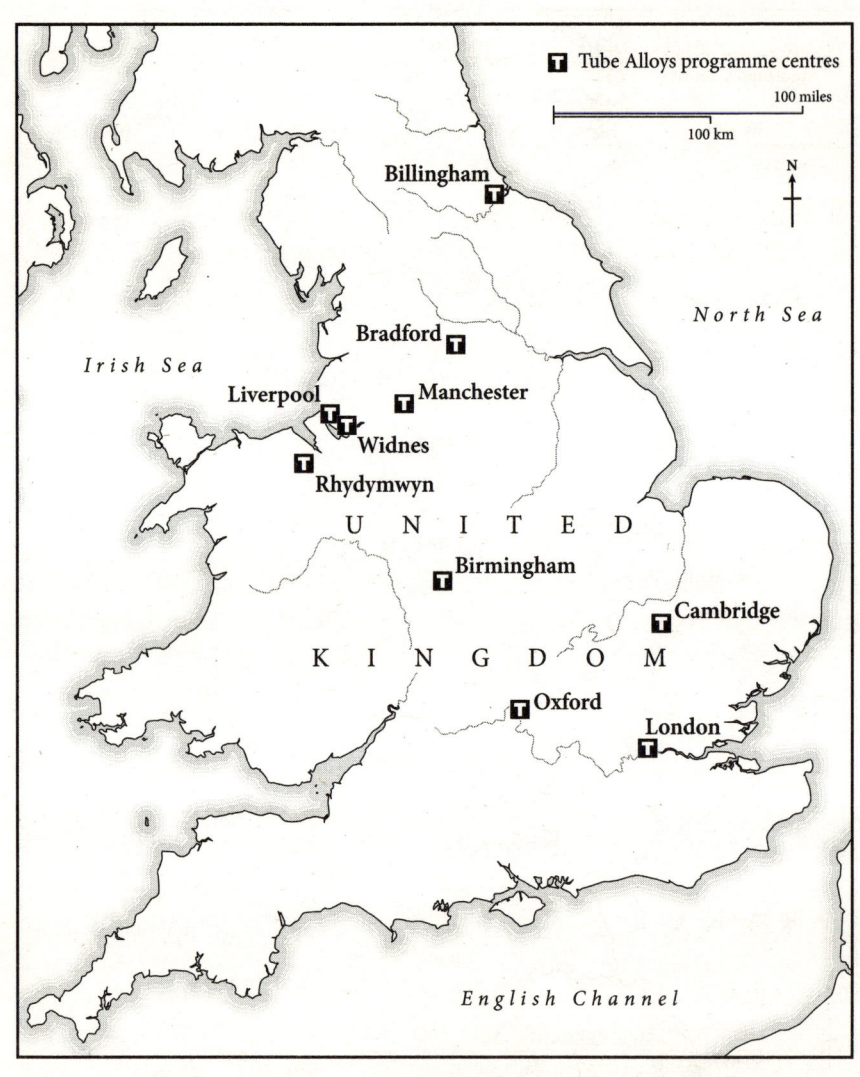

Map 2. Research and supply centres for the Tube Alloys programme, 1943–4.

Map 3. Research and supply centres for the German *Uranverein* programme, 1944.

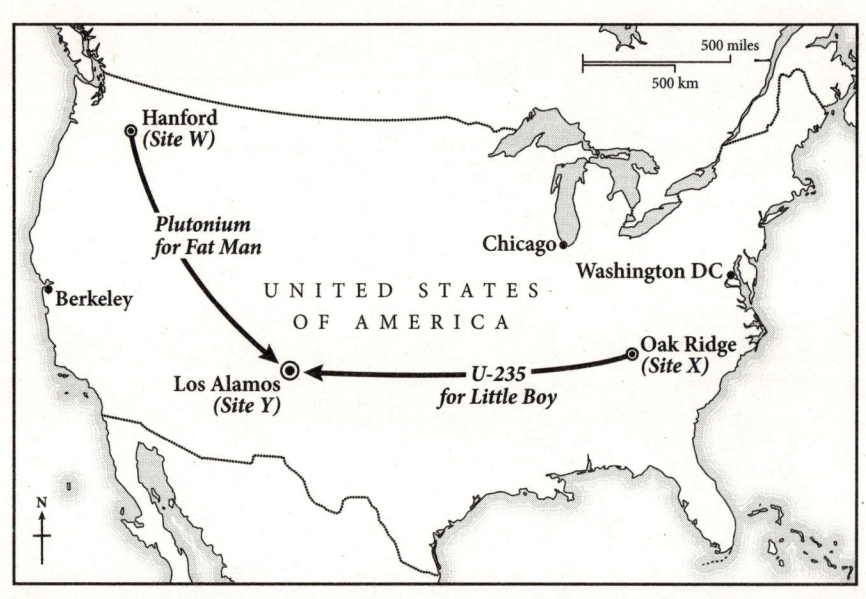

Map 4. Sites W, X and Y: the three 'Atomic Cities' of the Manhattan Project, 1943–5.

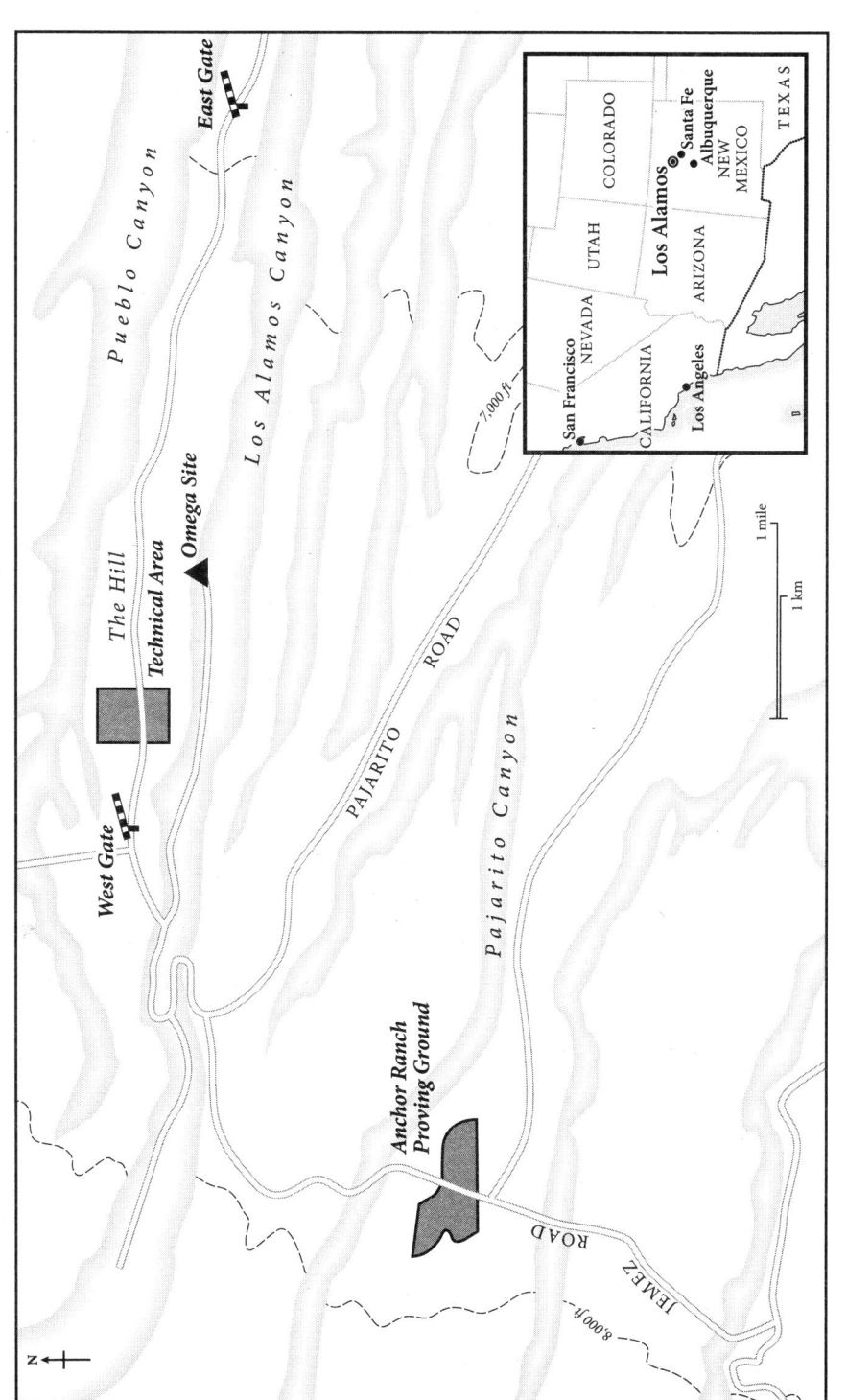

Map 5. Los Alamos and its surroundings, New Mexico, 1944–5.

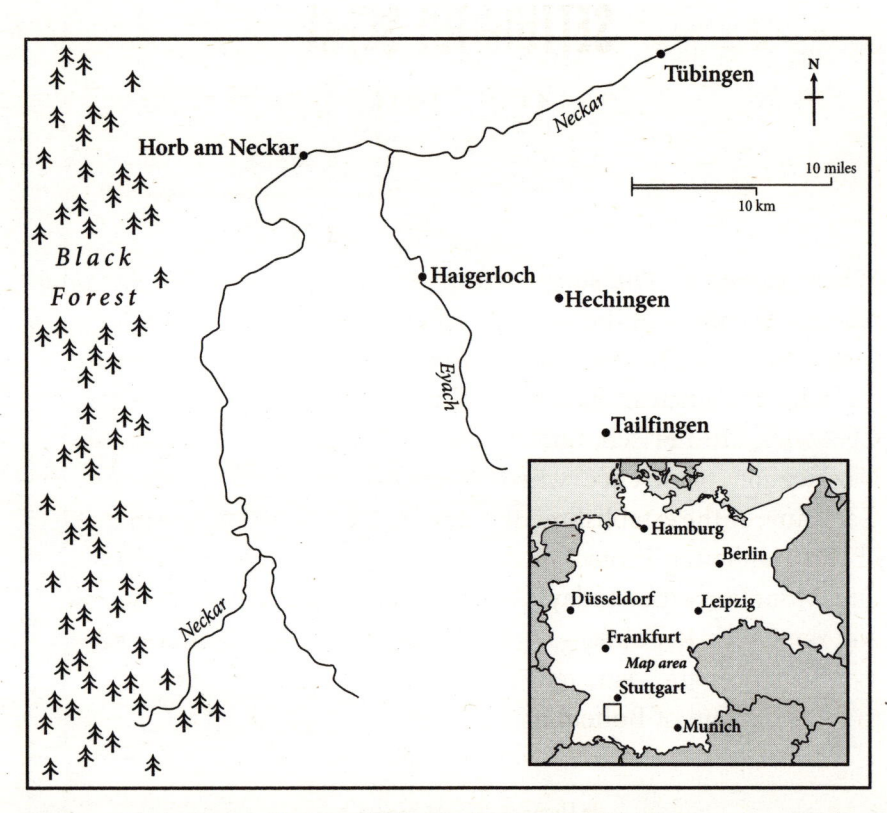

Map 6. The villages of Haigerloch, Hechingen and Tailfingen, close to the Swiss border in southern Germany.

1

SETTING THE SCENE
1595–1918

Where to begin? The story of how the atomic bomb was invented reaches its climax in the summer of 1945, but its roots go back much earlier. We could start with the discoveries of uranium, radioactivity, the neutron, nuclear fission or plutonium; or the births of James Chadwick, Otto Frisch, Rudi Peierls or Robert Oppenheimer; or the Nazification of science which hounded Jewish physicists out of Germany; or the establishment of the M.A.U.D. Committee in Britain or the Manhattan Project in America. In fact, the story opens in the Ore Mountains that today straddle the frontier between Germany and the Czech Republic – and, to begin at the beginning, we have to step back over 500 years to the silver rush which brought vast wealth to the Kingdom of Bohemia.

An Element of Little Importance

The pitch-black mineral discovered in 1595 near the silver-mining town of St Joachimsthal (Map 1) was a disappointment from the start. It was as heavy as the real thing but yielded no silver, so the miners gave it the derisory name *Pechblende* – from the German words for 'pitch' and 'deceiver' – and left it in the ground.[1] *Pechblende* became 'pitchblende' in English but was otherwise ignored for almost two centuries until, in 1789, Martin Heinrich Klaproth showed that pitchblende contained a new element which he named after the recently discovered planet Uranus.[2] It took another half-century to isolate uranium, which turned out to be a silvery grey metal, as heavy as gold and hard as steel. By then, massive deposits of uranium ore had been found in Canada, the Urals and the Congo, but its only commercial use was in colouring the vivid yellow and green 'uranium glass' which was a passing fad during the 1860s.[3] Scientists were similarly unexcited. Dimitri Mendeleyev, the wild-haired Russian architect of the periodic table of the elements, initially gave uranium an atomic weight of 116, less than half the correct value.[4]

The first hint that uranium possessed properties which set it apart from other elements came in 1857, when the French photography pioneer Claude Niepce de Saint-Victor reported that uranium salts poured out invisible rays which exposed a photographic plate. Niepce missed the full significance of his observations, and his work was actively suppressed by a jealous rival and former collaborator. Nonetheless, the 'uranic rays' went on to win a Nobel Prize, albeit for someone else and thirty years after Niepce's death. The lucky recipient of the prize – for Physics, in 1903 – was Henri Becquerel, honoured for having discovered his 'rays of genius'. Becquerel never acknowledged Niepce's prior discovery, even though he must have known all about it: he was the son of Edmond Becquerel, President of the all-powerful Academy of Science in Paris and the collaborator turned jealous rival who had pushed Niepce into oblivion. Niepce died bitter about 'certain people' (Becquerel père) who had robbed him of his reputation, but he wasn't totally forgotten; forty years later, Henri Becquerel's son ranted about the 'bastards' who had resurrected Niepce's claim to the discovery of radioactivity.[5]

Becquerel's uranic rays were feeble – they took days to burn a decent image on a photographic plate – and quickly lost their magic after his flurry of papers in spring 1896. They were eclipsed initially by X-rays, the all-seeing invisible radiation which Konrad Roentgen in Würzburg had discovered a few months earlier. X-rays thrilled everyone, scientific or not; Roentgen's name was effectively stamped on the first Nobel Prize in Physics (as yet unborn) when he passed his hand in front of his X-ray tube and the bones inside his fingers appeared on the detector screen. Becquerel's radiation was just a ten-paper sideshow beside the thousands of articles and numerous books devoted to the miraculous X-rays, which peered inside the living body, cured skin cancer and even drove sales of lead-lined knickers to protect ladies' modesty from the 'naughty Roentgen rays'.[6] Before long, Becquerel himself lost interest. His uranic rays could well have been forgotten – just like Niepce's – but for an extraordinary stroke of luck which assured his place in history.

Becquerel's saviour was a thirty-year-old Polish woman, already displaying academic brilliance and her 'will of iron and maniacal taste for perfection' – and who would later become legendary for weathering personal tragedy ('the saddest eyes I have ever seen,' according to an American benefactress) and for wearing the same simple black dress to accept both her Nobel Prizes. Marya Skłodowska grew up in the Old Quarter of Warsaw under Russian oppression. She watched the Russians sack her father, a physics teacher, and sat through the night with a schoolfriend whose brother was to be hanged at dawn; it was no wonder that she always spat on the monument to the tsar and danced for joy when he was assassinated. Marya excelled at school but Warsaw University only admitted men, so she educated herself from textbooks in Polish, French and German and risked imprisonment by attending the 'Floating University', a fugitive night school banned by the Russians. Aged twenty-four, she bought a fourth-class train ticket to Paris and registered at the Sorbonne as Marie Skłodowska to study physics and mathematics. Paris embodied both heaven and hell: heaven, soaking up knowledge at the feet of France's greatest scientists and in the library until it closed at 10 p.m.; then home to hell, a tiny attic cubicle where water froze in the jug beside her bed.[7]

She came top of the class and was introduced to a tall and eccentric 'scientist of genius' named Pierre Curie. They hit it off immediately, and she began working in his lab on her first research project. In 1895, they took a few days away from the lab bench to get married; their first daughter was born in 1897.[8] Next came a doctorate, the first in France for a woman. Marie Curie chose a high-risk topic – the 'astonishing uranic rays', now abandoned by Becquerel and everyone else. She worked in a derelict storeroom that 'sweated with damp', and scrounged samples of uranium compounds from colleagues and mineral collectors. Pierre built her an exquisitely sensitive instrument which measured the strength of radiation by its ability to make air conduct electricity. With the 'Curie electrometer', Marie quickly proved that the intensity of the radiation emitted by a sample was proportional to its uranium content, confirming that the rays came from the uranium atom itself. Next, she found that thorium was the only other of the eighty known elements to emit radiation. Her tour de force was to show that some pitchblende samples poured out far more radiation than their uranium content would have predicted, and to deduce that these must contain an even more powerful source of radiation. It wasn't thorium, the only other known candidate, and so must be a new element.

Pierre now dropped his own research and became his wife's full-time scientific collaborator. To cut short a long story (hundreds of tons of pitchblende delivered by lorry from St Joachimsthal, thousands of litres of concentrated acid and four years of 'the most desperate and arid effort'), the Curies eventually isolated tiny samples of two new active elements, both present at less than one part in 1,000,000,000 of pitchblende. They named them 'polonium' ('after the country of origin of one of us') and 'radium', which blasted out 'enormous radiation' that lit up solutions of its salts with a beautiful pale blue glow. And Marie invented a new word: *radioactivité*, short for *activité radiante*, which rolled beautifully off the tongue in French and was immediately snapped up by other languages.[9]

Radium made the Curies world famous, not only for their brilliance, dedication and hard work, but also for their oddness and naivety. Marie was regarded with both admiration and suspicion as 'the rarest of animals: a woman physicist'. With radium trading at fifty times the price of gold, they could have been millionaires; instead, they handed all their intellectual property to industrialists, who gave them nothing in return.[10] And while radium cursed Marie and Pierre with the misery of unwanted celebrity, it was remarkably kind to Henri Becquerel. The 1903 Nobel Prize in Physics was shared between him (50 per cent) for discovering radioactivity, and the Curies (25 per cent each) for their research into Becquerel's 'radiation phenomena'. Marie wouldn't have figured at all if Pierre hadn't made a fuss after discovering that the Academy of Science in Paris had only nominated Becquerel and himself.[11] The Curies didn't attend the prize ceremony in Stockholm. They stayed in Paris to honour 'commitments' including the heavy teaching burdens which paid their bills; their Nobel awards were accepted on their behalf by a French minister.[12]

Radium, the spinoff product of uranium research whose existence Becquerel had never even suspected, now pushed uranium into the shade. Everything about radium had charisma: the brilliance, self-sacrifice and modesty of its discoverers, and its own power and potential to transform the future. Radium rays were millions of times stronger than those from uranium, so intense that you could scribble a message on a photographic plate with a radium-tipped stylus. The radiation also created ozone from oxygen and turned diamonds green, while *Curiethérapie* (radium therapy) made skin cancers and even internal tumours shrivel away.[13]

The Curies' new element also turned the spotlight on one of the impenetrable black boxes of physics – the inner working of the atom. Radium ignited a chain reaction of research that threw up cutting-edge questions: the nature of radioactivity and how it was generated; whether uranium, radium and other radioactive elements might be related; and what atoms looked like inside.

And the notion that vast, potentially explosive amounts of energy were somehow locked away inside the atom, waiting to be liberated.

Alpha Male

Henri Becquerel had started to unravel the 'very complicated phenomenon' of radioactivity by firing a narrow beam of radium rays across a photographic plate straddled by a powerful magnet. Switching on the magnet split the beam into two components: one continued straight ahead, while the other was pulled sideways. Becquerel deduced that both components consisted of tiny particles, but with contrasting properties. The ones deflected by the magnetic field were negatively charged, virtually weightless and travelled almost as fast as light, while their undeflected companions were much slower and heavier (Figure 1). He also dissected the radiation from the other radioactive elements. Like radium, thorium sprayed out both types. Uranium emitted only the lightweight, negatively charged particles, and polonium only the heavy, slower ones.[14]

The two particles were fleshed out by Ernest Rutherford, the new Professor of Physics at McGill University in Montreal. Growing up in New Zealand, Rutherford displayed 'unusual powers of concentration' and narrowly survived a swimming accident that killed his two brothers. He won a scholarship to Trinity College, Cambridge, becoming J.J. Thomson's first research student at the famed Cavendish Laboratory. On arrival at McGill, he was twenty-eight years old, tall, blue-eyed, confident and with 'volcanic energy and an immense capacity for work'.[15] He showed that the slow, heavy radiation particles – which he called 'alpha' – travelled at one-tenth of the speed of light and weighed as much as an atom of the second element, helium (atomic weight 4). Alpha-particles were positively charged, but so massive that they were only deflected by an extremely powerful magnetic field; they could be stopped dead by thin aluminium foil or 4 inches of air. Rutherford named the speed-of-light, negatively charged particles 'beta'. These turned out to be electrons, whose discovery J.J. Thomson had recently reported to the Royal Institution in London. Weighing only 1/8,000th of an alpha-particle, beta-particles were high-velocity bullets that could slice through an aluminium block.[16]

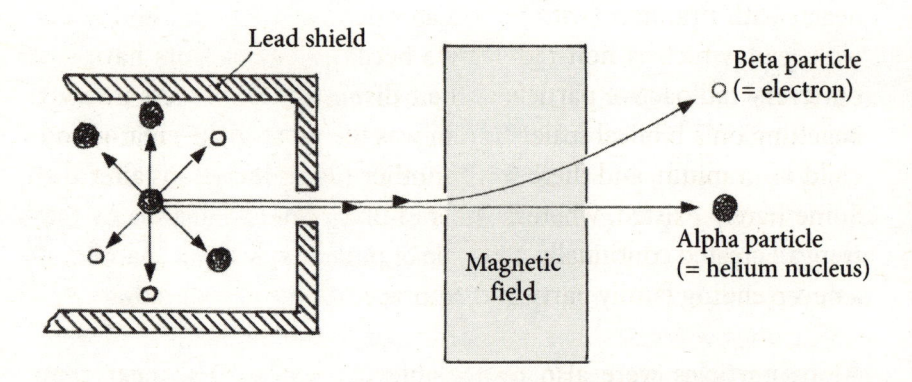

1. Components of ionising radiation: alpha- and beta-particles.

The alpha-particles, as heavy as helium, led Rutherford to make an extraordinary leap in lateral thinking. While investigating a transient radioactivity that sometimes wrecked experiments on radium, he found the culprit to be a dense, intensely radioactive gas that could be boiled off solutions of radium salts and dispersed by a puff of air. This 'emanation' (later renamed 'radon') was released slowly and steadily by radium; it blasted out alpha-particles even more powerfully than its parent and accounted for most of the radioactivity attributed to radium. Its atomic weight turned out to be 222, four less than its parent element radium. Noting that this difference was exactly the weight of an alpha-particle, Rutherford suggested that radium generated an atom of emanation whenever it fired off an alpha-particle.[17] This made him wonder whether uranium might similarly break down into other radioactive products. At this point, Rutherford hired a chemist – a wise decision, as his own knowledge of the subject was notoriously sketchy. Frederick Soddy, twenty-three years old and with a First in Chemistry from Oxford and an 'outstandingly strong personality', joined him in 1900 to begin the four-year crusade to map out the process that they called 'radioactive decay'.[18]

The emanation, the offspring of radium, turned out to be the parent of a new radioactive element (which they called 'Radium A', atomic weight 218), created by emitting an alpha-particle. Radium A wasn't the end of the road, as it popped out an alpha-particle and a beta-particle to yield another new element ('Radium B'). And so it went on. Soddy and Rutherford showed that the cascade of radioactive decay began with uranium (with its top atomic weight of 238) and ended with lead, which is non-radioactive because its ancestors have spat out every radioactive particle at their disposal. The process involved begetting on a biblical scale: radium was the great-great-great-grandchild of uranium, and there were another nine generations after that. Some barely existed, whereas others hung around for ever. As they were all created continually, a sample of pitchblende was a snapshot of a never-ending family party that embraced fourteen generations.[19]

Alpha-particles were also responsible for Rutherford's next coup, after moving from McGill to Manchester University. As a student, he'd been taught that the atom was solid like a billiard ball – 'a nice hard fellow, red or grey in colour, according to taste' – but one afternoon in April 1911 found him standing in the doorway of his lab, looking smug and telling everyone that he knew what the atom really looked like. He had reanalysed experiments done by two students, Hans Geiger and Ernest Marsden, who had shot alpha-particles at gold foil. Most passed straight through as if the foil contained more holes than gold – but about one in 100,000 bounced back. As alpha-particles are high-energy projectiles, Rutherford found these ricochets 'astonishing . . . as if someone had fired a 15-inch shell at a piece of tissue paper and it had come back and hit him'.[20]

Rutherford envisaged the nucleus as unbelievably tiny but carrying essentially all its mass and a number of positive charges; the rest of the atom was a relatively vast emptiness through which whirled the correct number of negatively charged electrons to balance the positive charges in the nucleus. This structure explained why alpha-particles fired at gold foil collided so rarely against something with enough mass and positive charge to sling them back in their tracks. And the billiard ball had shrunk in the wash. Rutherford calculated that the nucleus was 1/100,000th of the diameter of the atom, equivalent to a real billiard ball centred on a circle of green baize twice the size of the Wembley Stadium. Rutherford's revolutionary structure of the atom was published in May 1911.[21] The model has subsequently been tweaked but Rutherford's basic vision has stood the test of time (Figure 2). Like gold foil, everything consists of over 99.9999 per cent nothingness.

Same Outside, Different Inside

In 1904, Frederick Soddy moved to the University of Glasgow and continued to work on radioactive decay. He became famous for his public lectures and jaw-dropping demonstrations on radioactivity. His star turn was to puff the radium emanation over pieces of willemite, a zinc mineral that blazes with jade-green fluorescence when bombarded by alpha-particles. 'One of the most beautiful sights I know,' Soddy told the entranced crowd, as the darkened auditorium filled with the magical glow.[22]

SETTING THE SCENE

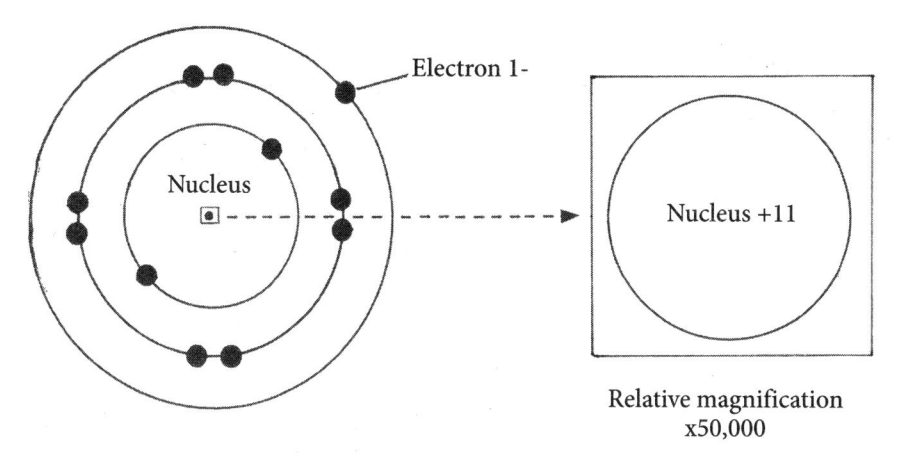

2. The atom, as envisaged by Ernest Rutherford and modified by Niels Bohr.

In 1910, Soddy made a far-reaching discovery that even Rutherford had missed. It had been assumed that each element had sole occupancy of its cell in the periodic table, but Soddy realised that some cells had to accommodate squatters which were chemically identical but had different atomic weights and radioactive signatures. For example, 'mesothorium-1' (atomic weight 228) was chemically indistinguishable from radium (atomic weight 226); the two could stand in for each other in chemical reactions. Soddy summed it up neatly – 'the same on the outside, different inside' – and proposed a snappy new name for chemically identical variants with different atomic weights: 'isotopes', from the Greek meaning 'same place' (the term was actually coined at a dinner party by a friend of Soddy's wife).[23]

The idea was heresy, but Soddy trampled over criticism and laid out the evidence. Many cells in the periodic table were overcrowded: for example, thorium (atomic weight 232) shared its chemistry with four others whose atomic weights ranged from 227 to 234. Soddy pointed out that the squatters' existence would never have been suspected if they hadn't been radioactive, and suggested that non-radioactive elements also had isotopes. Proof soon followed from precise measurements of atomic weight in 'ordinary' lead (207) and 'radio-lead', associated with pitchblende and derived from the decay of uranium (206). Furthermore, ordinary air contained minute amounts of neon with an atomic weight of 22, two more than 'ordinary' neon.[24]

For now, though, there was no hint of isotopes at the beginning or end of the periodic table. Hydrogen (atomic number 1, atomic weight 1) and uranium (atomic number 92, atomic weight 238) each appeared to be alone in its cell.

The year 1913 ended with an accolade for Frederick Soddy that would be noted around the globe. A much-anticipated book, soon to be published, was dedicated to him because it 'owed long passages' to *The Interpretation of Radium*, the anthology of Soddy's public lectures in Glasgow.[25] The author was H.G. Wells, zoology graduate and world-famous father of science fiction. Wells's bestsellers such as *The Time Machine*, *The Invisible Man* and *The War of the Worlds* were all rattling good yarns that implanted futuristic science into Edwardian drawing rooms. His new book was classic Wells: a clairvoyant, credible nightmare.

Wells's imagination had been seized by Soddy's account of how radium had revealed the huge amount of energy stored in the atom. Pierre Curie first observed that radium samples were always warm, and calculated that the element generated enough heat to turn its own weight of ice into steam in just forty-five minutes. On a grander scale, unexpected hot spots along the Simplon Tunnel under the Alps were related to radium-rich rock strata. But how could radium pour out radiation and heat year after year – 'like Aladdin's lamp', Soddy said – when an unbreakable law of physics states that energy cannot be created from nothing? Uranium contained even more energy than radium: Soddy's lecture demonstration jar contained just 1 lb of uranium oxide, but the same energy as 160 tons of coal. Harnessing that energy, if ever achievable, could 'transform a desert continent, thaw the frozen poles and make the whole world one smiling Garden of Eden' – or even enable mankind to 'explore the outer reaches of space and emigrate to more favourable worlds'.[26]

In *The World Set Free*, Wells envisaged civilisation transformed by cheap and unlimited energy from radioactivity – but with a twist that Soddy hadn't predicted. First, humanity had to be purified by fire, also the product of radioactivity. Wells invented a new term for the horrific weapon which engulfed Paris and Berlin in manmade infernos. He called it the atomic bomb.[27]

2

WAR GAMES
1914–1919

During the last few months of peace before the Great War, the fault lines sending tremors across the continent were centred on Germany, heavily armed and aggressive. Germany was also the scientific powerhouse of Europe and had carried off more Nobel Prizes than any other nation. The jewels in its scientific crown were the specialist research institutes recently created by the Kaiser Wilhelm Society for the Advancement of Science. In October 1912, pomp, circumstance and Kaiser Wilhelm II himself had descended on the south Berlin suburb of Dahlem for the inauguration of the Kaiser Wilhelm Institute (KWI) for Physical Chemistry and Electrochemistry and the nearby KWI for Chemistry. The KWI for Physical Chemistry was rightly proud of its first director, Fritz Haber, who had made his name in 1909 by inventing a brilliant catalytic process which fused nitrogen and hydrogen to make ammonia. Haber's discovery revolutionised the production of agricultural fertiliser and boosted global wheat yields; by creating 'bread from air', he had saved hundreds of millions from starvation.[1] Surely Haber deserved a Nobel Prize?

The KWI for Chemistry comprised three Departments, of which Radioactivity was devoted to the discovery and exploitation of 'radiochemicals'. Its director was Otto Hahn, Frankfurt-born, thirty-five years old and one-half of a highly productive chemist–physicist team. Hahn had left Germany with a chemistry degree and gone to rub shoulders with great men in London and Montreal. During his six-month spell at

McGill, Hahn impressed Ernest Rutherford with his 'special nose for discovering new elements'; the two got on so well that Rutherford would borrow Hahn's pipe whenever he mislaid his own. But when Hahn returned to Berlin in 1907, he had to set up his radiochemistry lab in a carpenter's workshop in the basement of Emil Fischer's Institute for Chemistry, because Fischer didn't yet believe that radiochemicals were worth studying.[2]

The pivotal moment in Otto Hahn's career came in September 1907 when he met an Austrian physicist at a seminar. Lise Meitner had trained in Vienna and was the second woman to be awarded a doctorate by that university; having published extensively on radioactivity, she had moved to Germany to expand her horizons. Meitner was petite, shy and a good match for the 'frank and informal' Hahn, a few months her junior; the pair began their thirty-year collaboration in late 1907 (Plate 1). Women were barred from laboratories in German universities, but Fischer accepted her in Hahn's basement workshop and a nearby café let her use its toilet. Two years later, the university relaxed its ban on women and Meitner was allowed upstairs into Professor Hahn's lab (and the new women's toilet, installed at Fischer's command). In 1911, Hahn was appointed Research Member at the new KWI for Chemistry in Dahlem, and a year later Hahn and Meitner moved into their smart new laboratory in the Department of Radioactivity – Hahn as director, and Meitner as 'guest'.[3]

By spring 1914, they were Germany's leading experts on radioactivity, publishing prolifically and closing in on an exciting new radioactive element, the 'mother-substance' of the recently discovered actinium (element number 89).[4] Neither gave any thought to two matters that were still irrelevant to science: Meitner was Austrian, not German, and although baptised Protestant, she had been born Jewish. Elsewhere in Germany, however, such things were becoming important.

With everything primed and ready to explode, it only took a couple of bullets from an assassin's pistol in Sarajevo on 28 June 1914 to trigger the catastrophic reactions that tipped Europe into the Great

War. Across the continent, scientists were pushed into new roles for their country's war efforts. Some found this disturbing; others slipped in as comfortably as a foot into a well-worn shoe.

In England, Sir Ernest Rutherford (knighted on New Year's Day 1914) was quietly tapped on the shoulder to lead a top-secret research programme against the German U-boat fleet, which had been built up to fearsome strength. The Admiralty had already tried out sea lions (too distractable, sent back to their circus) and had considered issuing sailors with sacks and hammers to neutralise any periscopes that popped up within reach. During the war, Rutherford divided his time between a large tank in the basement of his lab in Manchester and the 'tonic' of an Admiralty base on the Scottish coast, developing underwater directional microphones which could pick up U-boats and determine their range, bearing and speed.[5]

Over in Paris, the war scuppered the grand opening of Marie Curie's magnificent Radium Institute on the Rue Pierre Curie, named after her late husband. She'd been alone when she accepted her second Nobel Prize – Chemistry (1911) – for isolating the elusive radium. In April 1906, Pierre had fallen under a two-horse carriage during a heavy rainstorm; Marie took off her widow's veil after she burned the clothes – stiff with mud, blood and brain – in which he'd died, but she never stopped grieving. When war was declared, Marie immediately abandoned her research for even greater things. She donated her 1911 Nobel Prize money (still languishing in a Stockholm bank) to the French war effort; they refused to take her Nobel and other gold medals, a decision that she ridiculed as 'absurd fetishism'.[6] Once a week, she 'milked' her institute's radium for its radioactive emanation, enabling cancer patients to continue their *Curiethérapie* while Paris was under bombardment. And she set up an emergency X-ray service with a fleet of 'radiological cars' (soon nicknamed '*Petites Curies*'), each fitted with a generator, X-ray tube and fluorescent screen in a tiny curtained-off darkroom. Her own *Petite Curie*, an ugly grey Renault *voiture familiale*, took her to field hospitals near the front lines in Flanders. Casualties were delivered, broken bones and bullets were marked up on sketches of the fluorescent image, and surgeons did their best, sometimes operating on the X-ray table. Curie had an apprentice – her daughter Irène, just seventeen and still at school when hostilities began but already attending physics courses at the Sorbonne. Irène quickly learned enough in the back of Marie's grey Renault to take her own *Petite Curie* to the Flanders battlefields. By the end of the war, Curie's network of 200 hospital X-ray units and twenty *Petites Curies* had received a million casualties.[7] But while she was saving lives, her research died.

In Berlin, the KWI for Physical Chemistry in Dahlem was developing a new research theme in common with the British, French and Americans, and in contravention of the Hague Convention of 1907 which had banned chemical weapons. The Germans opted for chlorine – a dense green gas that could kill in minutes – which required a top-class brain to crack the technical problems of deployment in battle. The first field experiment was encouraging (6,000 French troops killed by Operation Disinfection near Ypres, April 1915), as was the follow-up study on the Russian front (over 1,000 dead). Chlorine did the job brutally but not always efficiently; many victims were finished off by a mercy bullet from a German soldier yelling at them to lie down for a better death.[8] One of those who volunteered for the gas-attack Pioneer Regiment 53 was Otto Hahn, the Director of the Hahn–Meitner Radioactivity Laboratory at the nearby KWI for Chemistry. Hahn had already been awarded the Iron Cross for bravery during the brutal offensive which crushed Belgium in autumn 1914. He joined the Pioneer Regiment as a strategist, selecting the best sites for gas attacks. On the Russian front, he was shocked to witness the death throes of chlorine victims and tried unsuccessfully to resuscitate some with an experimental oxygen mask. This horror didn't prevent him from planning the 'very successful' gas attack which killed 600 Italians at Isonzo in October 1917.[9]

Spoils of War

The fifty-one months of the Great War snuffed out 20 million lives, injured another 21 million people and left 7.5 million displaced and homeless. Many of the wounds it inflicted took years or decades to heal, but science recovered remarkably quickly. Some laboratories were up and running within weeks of the armistice being signed in November 1918.

In Manchester, Ernest Rutherford abandoned his underwater microphones and returned to demystifying the atom. In Paris, Marie Curie picked up her heavy research and teaching loads once more. Her daughter Irène, now twenty-one years old and with a military medal for radiographic gallantry in battle, joined the Radium Institute as assistant to Mme Curie and began dissecting the alpha-rays emitted by polonium, the first radioactive element discovered by her parents.[10]

In Berlin, Otto Hahn and Lise Meitner were reunited at the KWI for Chemistry. Meitner's war had echoes of Marie Curie's – she'd served as a nurse-radiographer in the Austro-Hungarian Army but returned whenever possible to continue hunting the mother-substance of actinium. She'd worked alone, as military service had cleared all the men out of the lab. Her efforts were rewarded by a landmark paper describing the new element 'protactinium', published in June 1918, five months before the armistice. Dr L. Meitner was the sole author, but she began by explaining that protactinium was the product of a long-standing collaboration with Prof. Dr O. Hahn. Protactinium was duly slotted into penultimate position (number 91) in the periodic table, immediately before uranium.[11]

At the University of Heidelberg, a resentful fifty-six-year-old man who was disgusted by Germany's humiliation donated a solid gold medal to be melted down to provide food and clothes for German war widows and their families.[12] He was Philipp Lenard, judged 'one of the greatest experimental physicists'; the medal wasn't his Nobel Prize in Physics (1905), for inventing a clever gadget to capture and study the enigmatic cathode rays generated inside X-ray tubes, but the prestigious Rumsford medal awarded by the Royal Society of London. Unfortunately, Lenard's practical genius wasn't matched by his grasp of theoretical physics. His explanation of cathode rays as disturbances in the (non-existent) 'ether' was laughable and quickly forgotten when the rays turned out to be streams of electrons, recently identified by J.J. Thomson in Cambridge. Lenard was pathologically desperate for recognition, thin-skinned, vengeful and so paranoid about the theft of his ideas that he locked his colleagues and students out of his lab. In early 1914, he published a vitriolic pamphlet accusing Thomson and English physicists in general of plagiarising his results and stealing his glory.

During the war, Lenard's paranoia and hatred refocused on enemies closer to home. He had done pioneering experiments on the photoelectric effect (the emission of electrons by metals when bombarded by ultraviolet light) but again tied himself in knots trying to explain his findings. The photoelectric effect was clarified in 1905, in an elegant paper written by an amateur theoretical physicist who worked as a clerk in the patent office in Bern and had never done an experiment in his life. Albert Einstein had used Lenard's own results to solve the very problem that had defeated Lenard. Initially, Lenard was flattered and praised Einstein as 'a deep and far-reaching thinker'; then, as Einstein went from strength to strength, bitterness set in.[13] That same year, Einstein published two landmark papers on the 'theory of relativity', which boiled down to a simple equation, impossible to test experimentally, which Lenard found absurd. This linked energy (E) with mass (m) through a constant which was the speed of light (c) multiplied by itself: $E=mc^2$. Theoretical physicists hailed Einstein's papers as revolutionary, but Lenard simply couldn't understand them.

When Lenard realised that Einstein and the pro-relativity theoreticians were Jewish, he became obsessed by what he saw as a Jewish conspiracy to undermine the pure experimental physics – *Deutsche Physik* ('German Physics') – to which he and other true Germans had devoted their lives.[14] By definition, *Deutsche Physik* could only be done by pure-blood Germans. Everything else, especially 'the physics of the Negro or the Jew', was a travesty, even if undertaken in Germany. Jews could never do *Deutsche Physik*; and *Deutsche Physik* could only thrive in an unsullied, Jew-free German environment. By now, Einstein had shown himself to be a traitor, as one of only four signatories on an anti-war 'Appeal to Europeans'.[15] Lenard spent the rest of the war drawing up plans for victorious Germany to purge itself of Einstein, theoretical physicists and all Jewish scientists.[16]

During the war years, Germany had carried off three non-contentious Nobel Prizes in Physics – Max von Laue (1914), Max Planck (1918) and Johannes Stark (1919). By contrast, the 1919 prizewinner in Chemistry provoked international outrage. He was Fritz Haber, whose creation of 'bread from air' clearly met Alfred Nobel's criterion of recognising 'the greatest benefits to mankind'. But now Haber was branded a war criminal because he'd masterminded Germany's poison-gas programme, designing all the equipment and even the uniform of his Pioneer Regiment and leading his men into battle. He never expressed remorse – 'In peace, for the world; in war, for the Fatherland' – and had set off for the Russian front in May 1915, the morning after his wife Clara, disgusted by his 'abomination of science', shot herself in the chest and died in the arms of their thirteen-year-old son.[17]

Faced with a tricky dilemma, the Nobel Prize committee diplomatically sanitised Haber's reputation ('He was appointed a consultant to the German War Office and organised gas attacks and the defences against them'), and he got his prize.[18]

The war had accelerated the evolution of weaponry. The cavalry charge was extinct, reduced to mincemeat by the machine gun. Paris had been shelled by a super-gun 75 miles away. Bombs had been dropped from flying machines. And over 90,000 front-line soldiers were killed by chemical poisons.

The World Set Free came out to red-hot reviews in the spring of 1914, but H.G. Wells's atomic bomb was quickly forgotten when the real-life horror story began a few months later. Wells set his 'atomic war' in the early 1950s, when mankind had harnessed the energy locked away in the atom to create atomic-powered cars and planes and, inevitably, bombs of unimaginable power. Nowadays, some details seem quaint: the beachball-sized bombs were armed by the pilot pulling out a celluloid trigger with his teeth and dropped over the side of an aircraft. But Wells's vision of the bombs trailing 'serpents' tails of fire' as they ignited in flight and incinerated the world below was as terrifying as anything he'd ever written.[19]

Before the war, few scientists had thought seriously about unleashing the power inside matter. In his public lectures, Frederick Soddy had only considered the peaceful uses of boundless atomic energy. Pierre Curie had acknowledged the risk of atomic power falling into the hands of the 'great criminals who lead the peoples towards war',[20] but the atomic bomb remained pure fantasy. By now, though, a card had been tossed in that could be either a trump or a joker. The speed of light was huge, and multiplying it by itself produced a number so vast that it surpassed comprehension. This meant that if Einstein's equation $E=mc^2$ could be believed, it wouldn't take much 'm' – perhaps just a few kilos – of the right kind of stuff to release enough 'E' to wipe out a city.

3

PURE PHYSICS
1918–1928

Two key members of Sir Ernest Rutherford's team in Manchester spent the war interned as enemy aliens. The first was Otto Baumbach, a German-born master glassblower who created the intricate apparatus for Rutherford's experiments on alpha-particles. Baumbach was sent to Strangeways Prison in late 1914 for bragging that the Kaiser's army would win the war.[1] The second internee took months to make his way home because he'd been imprisoned deep inside Germany.

He was James Chadwick, one of Rutherford's brightest young stars (Plate 2). As a painfully shy grammar-school boy from rural Cheshire, Chadwick won a scholarship to Manchester University. Intending to study mathematics, he found he'd been put in the interview queue for physics but was too timid to protest. He stuck with physics and fell under the spell of Rutherford, who supervised his third-year research project. This turned out to be undoable; Rutherford was disappointed that such a bright student had missed the fatal flaw, only to discover that Chadwick had spotted it immediately but didn't dare to point it out.[2]

Chadwick was awarded a First, followed by a prestigious PhD scholarship that included two years with Hans Geiger, Rutherford's former student whose experiments with alpha-particles and gold foil had shown his chief what the atom really looked like. Geiger had returned to Berlin to run the State Physical-Technical Institute, which is where twenty-three-year-old Chadwick was when Germany invaded Belgium. He spent four years interned in the stables at a racecourse near Spandau. Conditions were harsh, especially when winter brought sub-zero temperatures and 'the agony endured daily as his feet thawed out at about 11 a.m.'. Welcome relief was provided by the talks which the internees gave to entertain each other, and the money sent by the lady in London who administered Chadwick's scholarship. As proof that you can never hold a good physicist down, Chadwick bribed the guards to bring him odds and ends – including radioactive toothpaste containing thorium – to continue working on his PhD. When word filtered out to Walther Nernst, Professor of Chemistry in Berlin, Chadwick was temporarily released into his custody to talk science. After the armistice, it took Chadwick several weeks to pick his way back to England through the wreckage of Germany and the Netherlands.[3]

In 1919, Rutherford left Manchester to succeed his old boss J.J. Thomson as head of the Cavendish Laboratory in Cambridge. He took several tons of equipment, his stock of radium (a pre-war gift from Vienna, which had contaminated the whole building with its radioactive emanation), and James Chadwick as his right-hand man.[4] Under Rutherford, the Cavendish quickly became the Mecca for aspiring young physicists who flooded in from Britain, Europe and beyond. An early wunderkind was Niels Bohr, who had first visited from Copenhagen in 1911 and then spent two eventful years from 1914 to 1916 as Rutherford's lecturer. Bohr brought his wife Margrethe and two young sons and quickly settled into life as an exile in England, complete with a devoted governess, Mrs Ray, who taught the boys English.[5] He updated Rutherford's model of the atom, showing that the electrons didn't circle randomly around the nucleus but were confined to specific orbits or 'energy levels' and could jump between these under particular circumstances (Figure 2). Bohr returned to Copenhagen later in the war, and in 1920 was appointed Director of his own Institute of Theoretical Physics. Two years later, he won the Nobel Prize in Physics for elucidating 'the structure of atoms and the radiation emanating from them'.[6] This was a good time for Rutherford's old boys; the previous year, the Nobel Prize in Chemistry had gone to Frederick Soddy for discovering isotopes.

Meanwhile, Rutherford and Chadwick had been exploring a phenomenon that was almost as bizarre as alpha-particles bouncing back off gold foil. In between working on his U-boat detector, Rutherford had followed up reports that traces of hydrogen appeared when alpha-particles were fired into thin air. He confirmed the observation and went on to blast pure nitrogen with alpha-particles. Hydrogen was again generated, in proportion to the dose of alpha-particles, and Rutherford concluded that the barrage of alpha-particles somehow chipped a hydrogen atom off the nitrogen atom. He and Chadwick widened the search, firing alpha-particles at the nineteen lightest elements (up to potassium in the periodic table) and always finding hydrogen. Rutherford suggested that the hydrogen nucleus – atomic weight 1 and a single positive charge (H^+) – was a basic component of the nucleus of all elements, and named it the 'proton'.[7]

Backed up with precise measurements of atomic weight by the newly invented mass spectrograph, the proton quickly established itself as a fundamental building block of the nucleus and therefore of all matter. The number of protons in an element's nucleus – its atomic number – determined its chemical properties and was perfectly matched by the number of orbiting electrons. But there had to be more to the nucleus than protons, because their total weight fell short of each element's atomic weight. Copper, for example, had an atomic *number* of 29 and therefore 29 protons, but an atomic *weight* of 64. And what about uranium, with 92 protons, an atomic weight of 238 and a deficit of 146?

In discussion with Chadwick and others, Rutherford floated the notion of nuclear particles, as yet undiscovered and unnamed, that would weigh as much as a proton (atomic weight 1) but were electrically neutral. These would be packed inside the nucleus with the protons, in the correct number to top up the element's atomic weight. Like protons, they would be a crucial component of the nucleus.[8] For now, the neutral nuclear particles remained hypothetical – and even if they existed, their properties were pure conjecture.

La Vie Parisienne

After four years of suspended animation, the Radium Institute on the Rue Pierre Curie was at last operating again. Irène Curie, now aged twenty-one and 'calm and marvellously balanced', was outstandingly effective as her mother's 'working companion'.[9] The institute remained chronically short of money and materials, which is why Marie did something that Pierre would never have countenanced. A visiting American benefactress, profoundly moved by the 'gentle woman in a black cotton dress and with the saddest face I have ever seen', asked Curie what she needed most. Radium, Marie said, to continue my research. How much? One gramme. The cost – $100,000 – was soon met by the Marie Curie Radium Fund which sprang up across America, but Mme Curie had to sing for her supper. In early May 1921, she crossed the Atlantic in first class on the *Olympic* with her daughters Irène and Ève. They were met at New York by 'an enormous mob', and set off on a gruelling two-month tour of the United States with public appearances, lectures, awards of honorary degrees and – handed over by President Harding himself – radium, the 'priceless treasure from American friends'. The 'timid little woman' conquered America with her towering genius and heart-rending frailty. She whipped up near-religious fervour, with supplicants crowding to touch her radium-ravaged hands and even her clothes. The travelling circus ('just like Barnum', Ève later wrote) got as far as the Midwest before its star succumbed to exhaustion, leaving her daughters to stand in for her at ceremonial events. 'Too much hospitality', lamented the newspapers. And even while she was off-stage, the mythology of the saintly genius with human vulnerabilities continued to grow.[10]

Back in Paris, the American gift rekindled the Radium Institute's research programme. Irène had slipped into Pierre's shoes as Marie's research partner and took on responsibility for training up Frédéric Joliot, twenty-two years old and three years her junior. Joliot was handsome, charismatic and 'the most brilliant and high-spirited of the workers at the Institute' – and history began to repeat itself.[11] Around them, the team expanded, with Marie running the show 'like those chess champions who can follow 30 or 40 games at once without looking up'.[12] In 1922, she joined Einstein on the League of Nations' International Committee on Intellectual Cooperation. In April the following year, an over-capacity audience at the Sorbonne listened spellbound to her account of the twenty-five years that had passed since P. Curie and Mme P. Curie first reported 'a new and strongly radioactive substance' in pitchblende.[13]

The next big landmark was laid down in 1926 by Irène, not Marie. The daughter had already proved that she wasn't tied to her mother's apron-strings, by building up her own research theme with Frédéric Joliot. Now, she announced their engagement. Marie was initially dismayed at losing her closest collaborator, but 'content to see her daughter's visible happiness'. And Marie's sixtieth birthday on 7 November 1927 was a twelve-hour working day like any other.[14]

Student Unrest

Europe squeezed in two decades of peace between the end of the First World War and the beginning of the next. From the start, there were signs that many of the toxic ingredients which had made everything unravel in 1914 were still present and thriving. In Germany – demoralised by defeat, hyperinflation and mass unemployment – a new and virulent strain of nationalism was taking hold. The appeal of the *Nationalsozialistische Deutsche Arbeiterpartei* (National Socialist German Workers' Party) – 'Nazis' for short – cut across German society from manual workers to professionals. Even universities, those supposed bastions of free thought and expression, were seduced.

Albert Einstein's position should have been impregnable. In 1917, Max Planck appointed him the first Director of the Kaiser Wilhelm Institute (KWI) for Physics. Two years later, an extraordinary experiment proved that Einstein's 'untestable' theory of relativity was correct. At issue was the extent to which the sun's gravity would bend a beam of light: a tiny deflection of 0.83 arc-seconds (the width of a thumb held up 2 miles away) according to classical Newtonian physics, versus Einstein's prediction of at least 1.8 arc-seconds. The experimental apparatus was the total eclipse of the sun on 29 May 1919, and the single measurement on which Einstein's destiny hung was the angle through which a star 'close to' the edge of the Sun's obscured disc appeared to be pulled in from its true position. The moment was captured by two astronomical telescopes 3,000 miles apart, and it took five months to calculate the deflection. The result: 1.9 arc-seconds.[15] Overnight, Einstein was everywhere in banner headlines – 'REVOLUTION IN SCIENCE', 'NEWTONIAN IDEAS OVERTHROWN', 'NEW THEORY OF THE UNIVERSE' – except in Germany, where the reception was muted. Only the non-political picture magazine *Berliner Illustrirte Zeitung* gave the story prominence, with a full front-page photograph of a pensive Einstein ('A new giant of world history').[16] Academic accolades followed thick and fast: the Copley Medal and Foreign Membership of the Royal Society, reserved for the greatest and the best outside Britain, and the 1921 Nobel Prize in Physics – not for relativity, but for cracking the mystery of the photoelectric effect.[17]

Predictably, Einstein's successes riled his enemies in Germany and the anti-relativity, anti-Einstein mood music orchestrated by Philipp Lenard began its crescendo. In early 1920, nationalist students started to disrupt his lectures. In August that year, a trumped-up organisation which named itself the 'Working Society of German Scientists for the Preservation of Pure Science' hired the 1,600-seat Berlin Philharmonic Concert Hall for staging public lectures to tear apart the 'relativity scandal'. Einstein went along out of curiosity; he was bemused by the incoherent ranting of a nobody nicknamed '*Einsteintöter*' ('Einstein-slayer'), and deeply disturbed by the swastika lapel badges, Nazi literature and anti-relativity propaganda by Lenard that were on sale in the foyer.[18] To clear the air, Einstein challenged Lenard to a head-to-head discussion about relativity before the Society of German Scientists and Physicians. The 'Einstein debate', in Bad Nauheim on the evening of 23 September 1920, attracted a full house and heavy press coverage, but was delayed until nearly midnight and only lasted fifteen minutes. Einstein easily fended off Lenard, but now saw the growing energy of the campaign against himself. Afterwards, he wandered the grounds in distress with a friend, pouring out his fears for the future.[19]

The harassment intensified, and in late 1922 the Weimar Republic's foreign minister, Einstein's personal friend, was shot dead by two right-wing army officers because he was Jewish. When Einstein's wife, Elsa, suffered a nervous breakdown, he began to talk seriously about leaving Germany for good. But he stayed, encouraged by his close friends Max Planck and Max von Laue.[20] For now, those two great men were beyond the reach of *Deutsche Physik*, but its momentum was growing. Lenard had acquired a comrade-in-arms: Johannes Stark, winner of the 1917 Nobel Prize in Physics for discovering how an electric field affects the light emitted when electrons jump between orbits around the nucleus. Thereafter, Stark's research had flagged, and he failed to get jobs. This was because others were better and he was difficult to work with, but Stark smelled a conspiracy of Jews hell-bent on wrecking his career. He built his own lab with his Nobel Prize money and pursued his experiments without peer interference. Stark had previously been friendly with Einstein and, while editing a physics journal, had even commissioned him to write a review on relativity. Now, like Lenard, he hated him.[21]

Deutsche Physik secured support in high places because it reinforced the Nazi credo that ethnically pure Germans were racially superior. The Nazis claimed descent from a tall, blond, blue-eyed super-race (the 'Aryans') who had spread from northern Europe to seed civilisation and creativity across the planet. This belief was as deluded as the theory, popular among Nazi 'scientists', that the universe was created when a comet made of ice crashed into the sun. But no matter; the Aryan race became the nucleus around which Nazi thinking and policy crystallised. *Deutsche Physik* chimed with Nazism.[22] At first, Lenard and Stark made a dream team for the Nazis. In 1924, while Hitler was in prison for his failed putsch (and dictating volume 1 of *Mein Kampf*), they wrote an article entitled 'Hitler's Spirit and Science' in which they praised the 'Führer of the sincere' and promised that 'we shall follow him'. After *Mein Kampf* was published in 1926, Hitler and Rudolf Hess visited Stark at home to congratulate him. Nazism and *Deutsche Physik* began to feed off each other like symbiotic organisms, starving their common enemy of essential nutrients.[23]

To the chagrin of Lenard and Stark, relativity didn't crawl away to die in a corner, and their top target remained unassailable. Einstein had been offered chairs abroad but stayed in Berlin, lecturing and publishing, admired by foreigners and supported by a small but dedicated core of allies within Germany. The big two – Planck and von Laue – hadn't budged in their support for him. Planck was the most revered and powerful of Germany's physicists and still untouchable. Also difficult to dislodge would be von Laue, popular and widely respected as the suave, witty winner of the 1913 Nobel Prize in Physics for discovering X-ray diffraction as a tool to determine molecular structure. He had already taken flak for Einstein, standing in for him to give a keynote lecture about relativity at a big conference in Leipzig; he had won the day even though everyone there had been given a venomous anti-Jewish, anti-Einstein pamphlet written by Stark. As a result, von Laue was branded an 'honorary Jew', in the firing line alongside Einstein.[24] And Einstein was still in Germany.

Safe Havens

Despite the best efforts of *Deutsche Physik*, theoretical physics continued to advance in leaps and bounds, because researchers were connected through an extraordinary neural network that transcended national boundaries, propagated ideas, brought like-minded souls together, created opportunities and made things happen.

In 1927, the triennial Solvay Physics Conference in Brussels was devoted to quantum mechanics, with a line-up that included Einstein, Niels Bohr and the pioneers of quantum physics, Wolfgang Pauli and Werner Heisenberg. Bohr and Heisenberg knew each other well. As Bohr's assistant lecturer in Copenhagen, the twenty-five-year-old Heisenberg had recently formulated his 'uncertainty principle' while laying the mathematical foundations of quantum mechanics, work that would eventually earn him a Nobel Prize.[25] And Bohr was much more than Heisenberg's supervisor: mentor, sounding-board, skiing companion and friend (Plate 3).

Through the theoretical-physics network, Heisenberg hosted twenty-one-year-old Rudolf Peierls at his summer school in Leipzig in 1928, and sent him to break new ground at the fringes of quantum theory. Peierls went to study with Pauli in Zurich, where he met Enrico Fermi, a short and energetic Italian theoretician who was diversifying into experimental physics, and a gaunt-faced American who wore a pork-pie hat (even when out sailing on the Zürichsee) called Robert Oppenheimer.[26]

The network was centred on Niels Bohr's Institute of Theoretical Physics in Copenhagen, now based in a mansion – the 'House of Honour' – which an heir to the Carlsberg brewing fortune had bequeathed for the use of the Dane who had made the greatest contribution to the nation's arts, literature or science. Bohr's Institute quickly became an alternative shrine to Rutherford's Cavendish, a magnet for brilliant young scientists and a breeding-ground for Nobel Prize winners. The highlight each year was the summer symposium for the top forty or fifty experts in the topic of the moment, with the 'Great Dane' presiding over conversation and thought interspersed with walking, football and table tennis.[27] And it was an unspoken condition that politics, ideology, religion or anything else that could get in the way of science was excess baggage to be left at home.

4

YEARS OF WONDER
1932–1935

By the beginning of 1932, Sir Ernest Rutherford had surrounded himself with a constellation of rising stars from across the globe. The brightest was James Chadwick, now carving his own path as the Assistant Director of the Cavendish Laboratory. Rutherford also held high esteem for Mark Oliphant (Plate 4), an Australian who was finishing a PhD on how ions interact with metal surfaces. Inspired by hearing Rutherford lecturing in Adelaide, Oliphant had come to Cambridge in 1929 on a prestigious 1851 Exhibition scholarship. He was bright, outspoken and versatile; his bench skills included making a wedding ring for his wife, Rosa, out of a nugget from an Australian goldfield.[1]

On arrival, the Oliphants had been invited to the traditional welcome tea party by Sir Ernest and Lady Rutherford to meet the other members of the group: another two Australians, a Swiss, a Russian and several home-grown prodigies including Chadwick, Patrick Blackett, John Cockcroft and Ernest Walton. This seemingly trivial event would cast a disproportionately long shadow. Four of these names will crop up at key moments during the thirteen-year countdown to August 1945. Without two of them, the atomic bombs that destroyed Hiroshima and Nagasaki could not have been built in time – and the American-led Manhattan Project to make the bombs would never have got off the ground.

The year 1932 started with no inkling it would be a 'year of wonder' for physics, or that two step-change discoveries at the Cavendish Laboratory would transform the understanding of the nucleus. The excitement began with James Chadwick. He had been investigating a mysterious new radiation which Frédéric and Irène Joliot-Curie had created by firing high-energy alpha-particles from polonium at the light metal beryllium. It was powerful enough to knock protons out of legendarily inert paraffin, and the Joliot-Curies believed that it consisted of high-energy gamma rays.[2] Two weeks of experiments persuaded Chadwick that the rays were electrically neutral particles, slightly heavier than the proton. He named the new particle in a tentative half-page letter to *Nature* in February 1932: 'The Possible Existence of a Neutron'. In May, uncertainty was dispelled by his follow-up paper, 'The Existence of a Neutron'.[3]

The neutron was everything that Rutherford had predicted, and much more. With its atomic weight 1 and charge 0, it added mass to the nucleus and so could top up atomic weight above the number of protons. The nucleus could now be pictured as a hybrid berry-like structure composed of two equally sized but differently coloured sub-units (Figure 3). At a stroke, the puzzle of isotopes was solved. An element's chemical properties were determined by its complement of

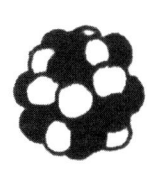

Sodium
11 protons
12 neutrons
Atomic No. 11
Atomic Wt. 23

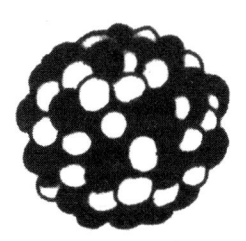

Uranium
92 protons
146 neutrons
Atomic No. 92
Atomic Wt. 238

3. Structure of the nucleus, as envisaged after Chadwick's discovery of the neutron in 1932.

protons (the atomic number), but its weight could be changed without affecting its chemistry by adding neutrons. 'Ordinary' copper had 29 protons and 35 neutrons, giving an atomic weight of 64 (Cu-64); Cu-63 was chemically identical but lighter, with 34 neutrons. In addition to being a fundamental building block of matter, the neutron quickly became an indispensable experimental tool in nuclear physics. Alpha-particles could be used to crack open the nucleus but, being positively charged, were repelled by the protons and skidded away from the nucleus. Neutrons weighed only one-quarter as much, but their lack of charge meant that they weren't deflected and could score a direct hit on the nucleus – with extraordinary results that would soon revolutionise physics.

Excitement over the neutron was still reverberating when the Cavendish recorded its second seismic discovery of the year – the 'splitting of the atom' by John Cockcroft and Ernest Walton. Cockcroft (Plate 5) was a thirty-five-year-old Yorkshireman who had won a scholarship in mathematics to Manchester University, and was mesmerised when Rutherford replaced a lecturer who couldn't control the class. Then the war supervened. Cockcroft never talked about it afterwards even though he wrote over 400 letters home; he'd been a signaller with the Royal Engineers near Ypres and in July 1917 was the sole survivor of his unit at Passchendaele. After the war, he won a scholarship in mathematics to St John's College, Cambridge, followed by a PhD with Rutherford. He polished this off in just over a year and joined Rutherford's campaign to tear open the nucleus with high-velocity particles.[4]

In late 1929, Cockcroft decided to use extremely high voltages to accelerate protons and hurl them at the nucleus forcefully enough to penetrate the envelope of positive charge around the nucleus. With difficulty, he extracted funding from Rutherford (who hated wasting money on fancy equipment) and built a machine to fire protons faster than ever before. Cockcroft brought in Walton, a twenty-seven-year-old PhD graduate from County Waterford who had come to the Cavendish via the Methodist College in Belfast and Trinity College, Dublin.[5] They took two years to build the prototype Cockcroft–Walton accelerator in an abandoned lecture theatre. Protons were generated by a 'canal-ray tube' devised by Mark Oliphant and then accelerated down a 12-foot vertical tube by a colossal 600,000 electron volts. The device had to be approached on all fours, as anything above waist height was at risk of electrocution. The business end, where the accelerated protons smashed into their target, was plugged into a blacked-out wooden box, just large enough to hold a crouching man and a microscope focused on a fluorescent detector screen. A Frankensteinian touch was added by the sparks that leaped between a pair of massive aluminium spheres when the operating voltage was reached. True to the grand era of 'sealing-wax and string' improvisation in British physics, the vacuum seals were made of plasticine.[6]

In mid-April 1932, Cockcroft and Walton began firing protons at the lightest metal, lithium, hypothesising that the impact of the proton (atomic weight 1) on the lithium nucleus (atomic weight 7) would smash it into two halves, both weighing 4 – the helium nucleus, or alpha-particle. Each alpha-particle would be revealed by a tiny glint of light on hitting the detector screen. On the morning of 14 April, Walton crept into the wooden box, drew the blackout curtain and turned up the voltage until sparks crackled between the aluminium balls – and through the microscope saw a shower of 'scintillations' light up the screen. Cockcroft crawled in to look, followed immediately by Rutherford ('The most beautiful sight I have ever seen').[7] That night, the three met at Rutherford's house to write a letter to *Nature* on 'The Disintegration of Lithium by Swift Protons', which was published on 30 April.[8] The story immediately hit the world's press – 'SCIENCE'S GREATEST DISCOVERY' – and stirred up global excitement that recalled the discoveries of X-rays and radium.

The splitting of the lithium atom was sensational because it revealed for the first time the vast energy locked away inside the nucleus. The lithium atom wasn't split neatly like a diamond cleaved by a master jeweller, but was blown into two halves by a hugely disproportionate explosion that released over 16 million electron volts (MeV) – atom for atom, about a million times the energy produced by a high explosive.[9] The equation $7 + 1 = 2 \times 4$ was an approximation, because all the atomic weights were rounded off. Measured precisely, two helium nuclei weigh slightly less than a lithium nucleus plus a proton. That loss of mass (m), multiplied by the square of the speed of light (c^2) comes to 16 MeV of energy (E) – further experimental proof that Einstein's theory of relativity was more than hypothesis.

Bibliocaust

The early 1930s were also good for *Deutsche Physik*. In December 1931, Einstein was finally dislodged from Germany for a year-long lecture tour in the USA. To discourage him from returning, over 100 scientists and philosophers published an anthology of anti-Einstein vitriol. Before sailing to America, Albert and Elsa visited their country cottage at Caputh, south-west of Berlin, just in case. Their intuition was correct; this would be the last time they stood as citizens on German soil.[10]

In 1932, Einstein was offered a permanent chair at the new Institute of Advanced Studies in Princeton, which he accepted with effect from the following year. The Einsteins had intended to revisit Germany in the meantime, but it soon became apparent that this was no longer an option. First the Gestapo raided the Einsteins' Berlin flat 'for spreading atrocity propaganda' and then a telegram, received in mid-Atlantic on the return voyage, told them that their 'little house' in Caputh had been ransacked. Einstein exiled himself from Germany – hence the exultant headline 'HE IS NOT COMING BACK' in the Nazi rag *Völkischer Beobachter* – and handed in his passport to the German Consulate in Antwerp after docking there on 28 March. In revenge, the Reich confiscated all the Einsteins' possessions. Meanwhile, German universities had begun sacking Jewish academics.[11]

The Einsteins retreated first to a villa on the coast of Flanders, where they received a mass-produced pamphlet headed 'NOT HANGED YET', with photographs of Einstein and the Jewish philosopher Theodor Lessing; this was out of date, because Lessing had already been murdered.[12] A final kick in the guts was a personal letter from Max Planck, now forced into a corner. Planck regarded National Socialism and Jewishness as 'ideologies that cannot coexist', adding that he 'must remain loyal to Germany, no matter who is in control'.[13]

The Einsteins went into hiding in England – Elsa with friends in London, and Albert in a log cabin at Roughton Heath in Norfolk. This belonged to Commander Oliver Locker-Lampson, a prominent anti-Fascist MP and friend, and was guarded night and day by Locker-Lampson's gamekeeper and two glamorous lady secretaries, all carrying shotguns. Einstein emerged occasionally, enjoying a beer in the Roughton Arms, sitting for a sculpture by Jacob Epstein and visiting Winston Churchill. His last stand in Europe was with Rutherford and Locker-Lampson in the jampacked Royal Albert Hall, promoting the Academic Assistance Council which was rescuing displaced Jews from Europe.[14] Four days later, on 7 October 1933, he and Elsa sailed from Southampton to the New World and safety.

The Nazis were now the biggest party in the Reichstag. Hitler had become Chancellor in January 1933 – excellent timing for the publication of Johannes Stark's unctuous book *Adolf Hitler: Aims and Personality*. Against unanimous opposition from the scientific assessors, Stark was appointed to direct the Reich's Physical-Technical Institute. Philipp Lenard was awarded the Freedom of Heidelberg, and his book *Grosse Naturforscher* was translated into English as *Great Men of Science*. Rutherford's book review in *Nature* acknowledged the 'masterly presentation' by this 'scientific investigator of great originality and power', but felt that Lenard's judgement was sometimes 'contrary to accepted opinion' (Roentgen, Marie Curie, Einstein and Rutherford were all omitted).[15]

The books by Lenard and Stark sold well, but other authors had fallen on hard times. On the evening of 17 May 1933, a torchlit procession of university staff and students in academic gowns wound through the streets of Heidelberg – marching alongside black-uniformed SS officers and members of the Hitler Youth, all belting out the Nazis' anthem, the 'Horst Wessel Song'. The procession was greeted in University Square by the medieval spectacle of a funeral pyre, outpourings of hatred, paganistic incantations and 'fire-oaths'. But they weren't burning martyrs, just books.[16]

Heidelberg came late to the party. Most of the thirty-four German and Austrian universities that burned books had done so a week earlier, on 10 May. In Berlin's Opera Square, 40,000 people waited until midnight to hear Josef Goebbels, Minister of Public Enlightenment and Propaganda, congratulate the German Student Union for organising this 'action against the un-German spirit' and hastening the end of 'extreme Jewish intellectualism'.[17] The students had demanded the 'purification' of libraries by eradicating everything written by over 2,500 authors who were Jewish, anti-Aryan, leftist, decadent or pacifist – including Karl Marx, Einstein, H.G. Wells, Thomas Mann (Nobel Prize in Literature, 1929) and even Erich Kästner, author of *Emil and the Detectives*.[18] The 'bibliocaust' of over 10,000 books, accompanied by an SS marching band, was broadcast live on German radio, covered enthusiastically by the press and applauded by most ordinary Germans.[19] And Philipp Lenard was there, savouring the spectacle of Einstein's books going up in smoke.

The juggernaut of *Deutsche Physik* rolled on. In September 1933, Johannes Stark announced at the German Physics Society meeting in Würzburg that he intended to take control of all physics journals in Germany. No more papers about relativity and other non-Aryan topics would be published, and force would be used if necessary against editors who disagreed. Stark's plans were ridiculed, and Max von Laue confronted 'the dictator of physics': trying to ban relativity was as absurd as the seventeenth-century Catholic Church putting Galileo on trial for refusing to deny that the planets circled the Sun. Stark was incensed by von Laue's defiance and by the applause from the 'Jews and their fellow-travellers'; another black mark against von Laue, whom Stark had already threatened with serious consequences for walking around town with a package under each arm so that he couldn't give the Nazi salute.[20]

In spring 1934, soon after Hitler appointed himself Führer, Stark manoeuvred himself into the Presidency of the German Research Foundation, and Lenard crowed that the 'relativity Jews' would get no more research funding and were already leaving German universities and 'even the country'. He was right; eventually, over 15 per cent of academic physicists left Germany.[21] Why? To quote another black-listed writer, the nineteenth-century Jewish German Heinrich Heine: 'Wherever they burn books, they end up burning people.'[22]

Stark's latest article on 'National Socialism and Science' insisted that scientists could only be pure-blood Germans, and must serve the Reich by devoting themselves to military and industrial research. One physicist – the country's youngest Nobel Prize winner – was already in Stark's sights as a troublemaker. He was Werner Heisenberg, now the Professor of Theoretical Physics at Leipzig and, aged just thirty-one, winner of the Nobel Prize in Physics (1932) for conceiving the 'uncertainty principle' that underpinned quantum mechanics. Despite warnings, Heisenberg continued to teach relativity and to mention Einstein by name, and in November 1933 he had refused to join a rally of the National Socialist Teachers League. He looked as Aryan as they came but, like Max von Laue, now stood accused of becoming 'Jewish in spirit'.[23]

By now, 'Heil Hitler!' was the salutation which greeted lecturers in all German universities as the students jumped to their feet, right arms raised in the Nazi salute. Academic robes had been ditched; some staff wore Nazi uniforms, including the black and silver of the SS. The swastika was everywhere: on students' armbands, flags flying on university buildings and the covers of scientific and medical journals. And every article inside those journals had been peer-reviewed – not just for academic rigour, but also to remove all authors with Jewish ancestry.[24]

Changelings

At the Radium Institute in Paris, the Joliot-Curie partnership had come of age. Some had questioned why Frédéric Joliot (nicknamed 'Irène's gigolo') had targeted the institute's 'crown princess': was he chasing a soulmate or a trophy wife?[25] Marie Curie evidently had misgivings, as she drew up a pre-nuptial contract which kept the hard-won gramme of radium the property of the institute, not the Joliot-Curies. But she acknowledged that her son-in-law was 'a skyrocket' when Frédéric turned polonium into the world's most powerful source of alpha-particles.[26] At first, his invention produced two tantalising near-misses that won Nobel Prizes for other people. When the Joliot-Curies fired alpha-particles into beryllium, it sprayed out the penetrating radiation which they labelled gamma-rays but which James Chadwick showed to be neutrons.[27] Next, they reported another novel type of radiation consisting of lightweight, high-velocity particles; this gave the American Carl Anderson his Prize in Physics for proving that the particles were like positively charged electrons, which he called 'positrons'.[28]

In January 1934, Frédéric Joliot told a student, 'We were too late for the neutron and the positron' – but added triumphantly, 'Now we are in time.' They had found that aluminium foil bombarded with high-energy alpha-particles poured out a torrent of positrons – which continued after the alpha-particles were switched off, dying away in the classic radioactive decay curve with a half-life of 3.5 minutes.[29] It was as though part of the aluminium had turned into a radioisotope. Working very fast because the positrons faded so rapidly, they laid bare the extraordinary transformation of aluminium into an unstable phosphorus isotope (P-30) which then decayed by emitting positrons.[30]

This discovery was immediately judged to be 'of capital importance for pure science, and for physiologists, doctors, and the whole of suffering humanity'.[31] Why all the fuss? P-30 doesn't exist in nature; the Joliot-Curies had created a novel, artificial radioelement. When they went on to blast boron and magnesium with alpha-particles, they produced artificial radioisotopes of other elements – nitrogen, silicon and aluminium. This meant that researchers were no longer shackled to the few sources of natural radiation; man-made radio-elements could now be used to advance science and medicine. For example, artificial radioisotopes could reveal what specific elements did in the living body. Chemically, they behaved exactly like their natural, non-radioactive counterparts, but even minute quantities could be tracked by their radioactive signatures – acting as 'tracers', like incendiary bullets interspersed with ordinary rounds to light up the trajectory of gunfire.

The Joliot-Curies' brief papers about artificial radioactivity in February 1934 triggered huge excitement. This quickly snowballed and led to one of the fastest ever awards of a Nobel Prize – Chemistry (1935) jointly to Frédéric Joliot and Irène Joliot-Curie for 'the creation of artificial radioelements'.[32]

Nineteen Thirty-Five

The Nobel Prize Ceremony on 10 December 1935 was a bitter-sweet affair. The Physics Prize went to James Chadwick for discovering the neutron. Chadwick, who hated public speaking, later claimed that he hadn't known he was expected to give a speech at the banquet.[33] His nervous ramblings were eclipsed by a polished and moving duet from the Joliot-Curies. Irène introduced the creation of their 'strange substances with enormous concentrations of energy', the latest chapter in the 'young science' of radioactivity. Frédéric picked up smoothly: over fifty artificial radioisotopes had already been created and they would transform chemistry, physics, physiology and medicine. He foresaw scientists 'building up or shattering elements at will' and perhaps even recreating the explosion of a star into a supernova – 'while taking, we hope, the necessary precautions'.[34]

An invisible presence was there that night, like an empty seat at the top table. When Frédéric Joliot reached the point that everyone was waiting for, he simply said that their discovery had provided 'great satisfaction to our lamented teacher, Marie Curie'. The Joliot-Curies' lucky break had come just in time. In early summer 1934, Marie Curie's worsening exhaustion was belatedly diagnosed as aplastic anaemia (bone marrow failure), possibly caused by radiation. Here was a cruel irony: radium could cure cancer and save life, but aplastic anaemia was a death sentence. Marie Curie died peacefully on 3 July 1934, without knowing that her daughter and son-in-law would follow her to Stockholm the following year. She was fifty-seven years old.

In keeping with a life of frugality, her funeral was attended only by family, close friends and colleagues from the Radium Institute. After a brief eulogy, they lined up to drop a red rose onto her coffin; the whole thing lasted just ten minutes. Marie shared a grave with Pierre and his parents in a quiet corner of the cemetery at Sceaux, a village near Paris where they'd been happy. A red rambling rose beside the grave was in full bloom – the first time, the locals said, that it had flowered in the twenty-eight years since she'd planted it after burying Pierre. Her name was added to their joint headstone, the letters bunched up to squeeze in her farewell tribute to 'the country of origin of one of us': 'MARIE CURIE NÉE SKŁODOWSKA'.[35]

On 13 December 1935, three days after the Nobel Prize ceremony, the new Philipp Lenard Institute in Heidelberg was opened with full Nazi pomp. *Nature* published a translation of the account printed in German newspapers, so that readers could judge for themselves this 'unique ceremony'.[36] The Minister of Education opened by stating that 'the Negro or the Jew will view the same world in a different way from the German investigator'. Johannes Stark laid into 'the followers of Einstein' and the 'notorious' Max Planck, who inexplicably still headed 'a celebrated learned institution'. Next came lectures on 'Nordic Race and Science', 'The Essence of German Blood' and 'The Abstract Mathematical Junk of Jewish Physicists'. In closing, Philipp Lenard 'exhorted all to continue energetically the fight against the Jewish spirit and arrogance', exemplified by Einstein. Proceedings ended with '*Sieg Heil!*' and the 'Horst Wessel Song'.

Nature commented that, apart from Lenard, those present were 'comparatively little known in scientific circles in England'. The explanation was simple, if invisible to *Deutsche Physik*. Jews made up just 1 per cent of the population of Germany, but filled about 15 per cent of academic posts. And they were disproportionately good, generating two-thirds of the German research that was highly regarded internationally.[37] *Deutsche Physik* was skimming off the cream and throwing it away, for others to scoop up.

Also in late 1935, the Norsk Hydro Company of Oslo, best known for supplying ammonia to the fertiliser and explosives industries, began selling 'heavy water' at US $2.50 for a 5-ml glass ampoule. Heavy water is fascinating stuff. Instead of hydrogen (H, atomic weight 1) in ordinary water, heavy water contains deuterium (D), a hydrogen isotope with one neutron that brings its atomic weight up to 2. Deuterium was discovered by Harold Urey in New York in January 1932 and won him the Nobel Prize in Chemistry two years later. Heavy water (D_2O) lives up to its name: it is slightly denser than ordinary water and an ice-cube of frozen D_2O will sink in H_2O. It makes up only 0.02% of everyday water but can be concentrated by electrolysis – passing a powerful electric current through water to split H_2O into hydrogen and oxygen, leaving a residue progressively enriched in D_2O.[38]

The American chemist Gilbert Lewis first isolated heavy water in 1933, and sent a few drops to the Cavendish, where Mark Oliphant used his particle accelerator to fire deuterium nuclei ('deuterons', twice as heavy as protons) at various targets. The results were promising. Paul Harteck, an Austrian-born chemist visiting from the KWI for Physical Chemistry in Dahlem, turned his hand to producing enough heavy water to continue Oliphant's experiments. The outcome, in a landmark paper by Oliphant, Harteck and Rutherford, provided the first evidence of a third, radioactive isotope of hydrogen (tritium, with two neutrons, atomic weight 3), created by the 'fusion' of ordinary hydrogen and deuterium.[39]

The burgeoning interest persuaded Norsk Hydro to sell heavy water commercially. They were already producing it as a waste product of ammonia synthesis: the hydrogen to be fused with nitrogen in Fritz Haber's catalytic process was generated by electrolysis in their ammonia factory at Vemork, high in the Telemark mountains to the west of Oslo. Industrial-scale electrolysis demands vast amounts of electricity – which is why the ammonia factory was built near the world's biggest hydroelectric scheme at Rjukan. In 1934, Norsk Hydro commissioned Leif Tronstadt, Professor of Inorganic Chemistry at Trondheim, and industrial engineer Jomar Brum to design novel electrolytic cells for the world's first heavy-water production plant in the basement of the eight-storey ammonia factory.[40]

Sales were initially sluggish, because this was still an orphan commodity waiting for a business opportunity. Nobody would have predicted that, in just four years' time, heavy water would be something to fight and to die for. Or that ten years hence, barely three months after the destruction of Hiroshima and Nagasaki, Paul Harteck would be back near Cambridge, celebrating a Nobel Prize with Otto Hahn, Werner Heisenberg and Max von Laue.

5

BEYOND NATURE?

February 1934–October 1937

The creation of artificial radioisotopes was revolutionary enough to win a Nobel Prize, but the underlying technology – firing alpha-particles at samples of elements – was so simple that numerous copycat experiments were under way soon after the Joliot-Curies' paper appeared in February 1934. Like a proton in the Cockcroft–Walton accelerator, the tempo of discovery was fired up to crazy speeds. One researcher saw his results published in *Nature* within a week of finishing the experiment.[1] He was a thirty-year-old Austrian refugee working in London with just three months' salary left. After that, the only certainty was that he couldn't return home to Germany.

Otto Robert Frisch (Plate 6) was born in 1904 and grew up in the Vienna of Klimt and Mahler. His family tree was Jewish and rooted in Poland; his mother was a professional pianist and his father a polymath businessman who wrote a mini-encyclopedia called *Catechisms of General Knowledge*. Young Robert[2] was fascinated by science and was judged a mathematical prodigy by the age of ten; he was also a talented pianist. As a teenager, he was inspired by hearing Einstein lecturing and by Olga Neurath, a blind, pipe-smoking 'communal godmother' to the city's young mathematicians. The greatest influence in his life was (and would remain) his spinster aunt, twenty-seven years his senior but a close friend, confidante and scientific mentor. She was Lise Meitner, the co-discoverer of protactinium, whose collaboration with Otto Hahn at the KWI for Chemistry had begun when her nephew was five.[3]

Frisch stayed in Vienna for his PhD, awarded at the age of twenty-two for a three-page explanation of why transparent rock-salt crystals turn blue when irradiated. He then spent three dull years measuring luminosity at the State Physical-Technical Laboratory in Berlin, relieved by supper and piano duets in Aunt Lise's tiny flat near the KWI, and colloquia at the University of Berlin where the front row glittered with Nobel Prize winners including Planck and Einstein. Meitner encouraged Frisch but refused to pull strings; she rejected Otto Hahn's suggestion of bringing Frisch into her lab and said only that her nephew was 'not disagreeable' when asked about his application for a post in Hamburg. Frisch got that job: research assistant to Otto Stern, a patrician Jewish physicist who 'invariably held a cigar, except when it was in his mouth'. Stern was pouring out a stream of papers branded 'UzM', the German abbreviation for 'research on molecular beams'. UzM1 was published in 1926, and UzM20 had just appeared when Frisch joined Stern's lab in 1930.[4]

At first, life in Hamburg was 'ideal': wonderful science, socialising and music with rowdy but chaste sorties into the red-light district. Frisch worked hard into the night, sometimes prompted to go to bed by seeing hallucinations of 'queer animals'. He was oblivious to Hitler until spring 1933, when the anti-Jewish laws roared in; Stern was 'quite shocked' because he'd assumed Frisch to be non-Jewish. The Rockefeller Foundation had just given Frisch a grant to work with the brilliant Enrico Fermi in Rome, but had to cancel it as he couldn't now fulfil the condition of returning to a post in Germany. Frisch wrote up his research as UzM30 and presented the findings at Niels Bohr's summer conference in Copenhagen, which had turned into 'a sort of labour exchange' to find jobs for Jewish physicists expelled from Germany. Stern spent the summer touring Europe, trying to 'sell' Frisch to sympathetic collaborators. Marie Curie didn't buy him, but Patrick Blackett created a place at Birkbeck College in London, and the Academic Assistance Council gave him a year's stipend.[5]

When Frisch returned to Hamburg from Copenhagen, Hitler had seized power. The newspapers denied rumours of Jewish persecution and a 'concentration camp' set up at a place called Dachau, but it was time to go. A friend from Stern's disbanded research team – now a Nazi – found Frisch a cabin on a London-bound freighter. When this 'cockleshell' docked at Greenwich after a rough overnight crossing, Frisch had to pay a crew member half a crown to phone Professor Blackett to vouch for him before he was allowed ashore.[6]

Blackett was Head of Physics at Birkbeck College, where Frisch blended happily into the multiethnic research community nicknamed 'the League of Nations'. When the Joliot-Curies' papers on artificial radioactivity burst onto the scene in February 1934, he 'instantly' began bombarding elements with alpha-particles, rushing fast-decaying products to the radiation detector in half a second along a high-speed track built from a Woolworth's curtain-rail. Frisch created novel isotopes from phosphorus and sodium, wrote up the results overnight and, after Blackett rang the editor, delivered the manuscript to *Nature* the next afternoon. The paper was published seven days later – 'probably a world record'.[7] Just before his money ran out, Frisch was rescued when Niels Bohr visited Blackett's lab. Frisch wrote to his mother, 'God Almighty himself has taken me by my waistcoat button and spoken kindly to me,' when Bohr offered him a job at his Theoretical Physics Institute in Copenhagen.[8]

Frisch left England with only 'a tenuous understanding' of the country. At a Friday Night Discourse at the Royal Institution, he had been thrilled by Rutherford in full flow, but had to keep a low profile behind a pillar because he was wearing lab clothes and everyone else was in evening dress. He had seen little outside London, notably the South Coast ('flattish landscape with some agreeable cliffs') and the Lake District ('the rain never stopped'). In late summer 1934, he was on the ferry from Harwich to Esbjerg, bound for Copenhagen.[9]

Meanwhile, Lise Meitner stayed in Berlin. Frisch's abrupt departure robbed her of her closest friend, piano-duet partner and scientific soulmate, and left her insecure. For now, she was protected against the German racial laws by her Austrian passport and Christian baptism. But it was only a matter of time before the rising tide of anti-Semitism would expose her Jewish roots for all to see.

Out of Bounds

Otto Frisch had good reason to be disappointed when the Rockefeller Foundation axed its grant to send him to work with Enrico Fermi in 1933. Roman born and bred, Fermi was the thirty-two-year-old founding Professor of Theoretical Physics at La Sapienza University. He had taken off younger and initially climbed faster than Einstein in quantum mechanics, before becoming a hands-on experimenter – at which point Rutherford wrote to congratulate him on escaping from 'the sphere of theoretical physics'. Fermi was lively and sporty and surrounded himself with a like-minded team, nicknamed the *Ragazzi* [boys] *di Via Panisperna* from the street outside their lab. They called him *Il Papa* (the Pope) and his lab fizzed with confidence, fun and excitement. During the early 1930s, Fermi's lab could have been set in paradise. The Fascists were in control, but it was still possible to turn a partially sighted eye to Germany and imagine that such ugliness could never infect Italy. *Il Papa* Fermi was in stable equilibrium with *Il Duce* Mussolini, who ordained Fermi's election to the Royal Italian Academy in 1929. Quid pro quo, Fermi joined the Fascists. Anti-Semitism wasn't yet on their agenda, and Fermi never hid the fact that his wife Laura, a vivacious science graduate of La Sapienza, was Jewish.[10]

In early 1934, Fermi was one of the many prompted by the Joliot-Curies' bombshell to create artificial radioisotopes, but he didn't use alpha-particles. Instead, his ammunition was James Chadwick's neutron, the uncharged stealth projectile which could sail straight through the envelope of positive charge and smash into the nucleus of even the heaviest elements. And neutrons did the trick. Fermi's first report, showing that neutrons transformed aluminium and fluorine into novel radioisotopes, was published in late March 1934 – a month before Frisch's 'world record' paper in *Nature*, but in an Italian journal which didn't travel well. His next two papers, in top-rank English journals, stirred up feverish excitement because he had seemingly made neutrons achieve the impossible.

Nearly seventy elements had been run through the Via Panisperna production line. Some, including hydrogen and lead, were resistant to change but over fifty were transformed into radioisotopes. After gobbling up a neutron, most nuclei spat out a proton, alpha-particle or electron and generated a radioisotope of a known element lying one or two places away from the original in the periodic table. However, uranium – the 92nd element at the very end of the periodic table – behaved differently and created a huge problem. One of the radioisotopes produced by firing neutrons at uranium, with a distinctive half-life of 13 minutes, wasn't either protactinium (atomic number 91) or thorium (90), the two elements that immediately preceded uranium. Fermi compared the 13-minute isotope with all the elements down to lead (atomic number 82), but didn't check polonium (84). He didn't find a match and concluded that the 13-minute element must be *heavier* than uranium, or 'transuranic'. There wasn't enough to analyse but he predicted that, chemically, 'element 93' should resemble the metal rhenium. Fermi named it 'ausonium', from the Greek for Italy. He also found what he believed to be another new isotope lying two places beyond uranium, and called this putative 94th element 'hesperium' from another Greek name for his country of birth.[11]

The tsunami of excitement that greeted Fermi's results also swept up the Italian Fascists. Could Fermi re-christen his element 93 'mussolinium'? His response was well calculated: surely *Il Duce* would prefer to be immortalised by something that lived longer than a few hours?[12] There was also much scientific unease, because Fermi had invaded forbidden territory by stretching the periodic table beyond the limit apparently decreed by nature. Previous tantalising hints of naturally occurring elements heavier than uranium had all come to nothing. The latest candidate 'element 93' (so-called 'bohemium') had been detected in a pitchblende sample from St Joachimsthal; it been demolished using X-ray spectroscopy by Ida Noddack, a chemist at the State Physical-Technical Institute in Berlin.[13] However, Fermi was a whizzkid who so far had turned everything he touched into gold, and most experts believed his claim that neutron irradiation had created 'transuranic elements' from uranium.

The sole dissenting voice was Ida Noddack. She and her husband-collaborator Walther had discovered rhenium, the element which Fermi thought would be the chemical template for his new element 93.

Noddack was suspicious of Fermi's findings and proved that his chemical method for excluding matches with the 13-minute radioisotope was invalid, because it picked up polonium which he hadn't bothered to test. Even worse was his 'uncritical' assumption that neutron bombardment had created new elements heavier than uranium. Instead, Noddack suggested that neutrons could break the uranium nucleus into 'large fragments', isotopes of known elements that were *smaller* than uranium. Fermi had blinded himself to this possibility because he hadn't gone far enough below uranium in searching for matches.[14]

Fermi didn't respond when Noddack's paper was published, or when she wrote directly to him in Rome. Nobody else expressed interest in the bizarre notion that neutrons might smash the uranium nucleus into pieces, and Noddack soon moved on to other things.

Fermi's transuranic elements were also a red rag to Lise Meitner in Berlin. She was dubious about his 'element 93' and speculated that neutrons could transform uranium into element 91 – protactinium, which she and Otto Hahn had discovered in 1917. The pair hadn't worked together for over a decade, but she asked Hahn to revive their collaboration. He was keen and recruited Fritz Strassmann, a chemistry graduate born in 1907, the year that Hahn and Meitner had started working together. Strassmann was experienced in handling tiny quantities of elusive elements and had done an excellent PhD, only to scupper his career by resigning from the Nazi-dominated German Chemical Society; he wasn't Jewish but despised Hitler.[15] Hahn could only afford to pay him half a salary, but Strassmann pitched in full-time. The trio began working in three dedicated rooms at the KWI for Chemistry in April 1934 and made a superb team. They were obsessive in avoiding radioactive contamination. Chairs were colour-coded to segregate anyone working on high-activity material; tissue-paper was pinned on all doors so that they could be opened without contaminating door handles; and instead of shaking hands, they touched the tips of their little fingers.[16] They began by analysing the products of neutron-irradiated uranium and quickly proved that Fermi's 'new' elements didn't include protactinium. And then – like

Fermi – they believed that they were creating elements heavier than uranium. Over the next four years, they churned out fifteen papers which claimed to add another three elements, numbered 95–7, beyond Fermi's ausonium and hesperium.

They encountered inconsistencies, but never enough to undermine the burgeoning credibility of the transuranic elements. In late 1936, Strassmann showed Meitner a baffling finding. Samples of neutron-blasted uranium apparently contained traces of the metal barium. Meitner's verdict as a physicist was blunt: neutrons couldn't possibly turn uranium into the much smaller barium, and Strassmann's results 'should be thrown in the rubbish bin'.[17] In so doing, Meitner and Strassmann also threw away their share of a Nobel Prize.

Meitner was similarly dismissive about the 'wild speculation' from the Joliot-Curies, who now claimed to have created another 'element 93' by bombarding uranium with neutrons.[18] She and Hahn thought this was rubbish and privately nicknamed the Parisian element 'curiosum'.[19] It never occurred to them that their own 'transuranic' elements had also been created from artefact and misinterpretation.

Meanwhile, Fermi's *ragazzi* were following up a strange anomaly. *Il Papa* had noticed that neutrons transformed elements more effectively when the experiment was done on a benchtop made of wood rather than marble. Some inspired experiments and lateral thinking persuaded Fermi that the hydrogen atoms in wood were responsible. Even better were hydrogen-rich substances including water and especially paraffin wax, which increased the neutrons' power to transform elements by over a hundredfold. It transpired that neutrons worked better because they were slowed down – or 'moderated' – by hitting hydrogen atoms. Hydrogen was a powerful brake, dragging neutrons down to 1/6,000th of their speed when fired out of beryllium bombarded by alpha-particles. 'Slow' neutrons (which still zipped along at several times the speed of sound) stood a much greater chance of being 'captured' by the target nucleus and triggering transformation.[20] Fermi also investigated why some elements are more easily transformed than others. He discovered that each element has its own 'nuclear cross-section', a measure of its susceptibility to being hit by a neutron. This can be pictured as the size of the bull's eye on an archery target: the bigger the nuclear cross-section, the greater the chance of a randomly fired neutron scoring a direct hit. Fermi calculated that the nuclear cross-section of a readily transformed element such as thorium was over fifty times greater than that of lead, which couldn't be transformed at all.[21]

It had been assumed that uranium – the stepping stone to the transuranic elements – was simply element 92, with 146 neutrons crammed into its nucleus to bring its atomic weight up to 238. The first hint of a twist to the tale had come in 1935, when Arthur Dempster in Chicago discovered a new radioisotope of uranium. This wasn't as rare as deuterium (0.02 per cent of the hydrogen in ordinary water) but was still very scarce, making up only 0.7 per cent of naturally occurring uranium. The new isotope had three neutrons fewer than 'normal' uranium, giving an atomic weight of 235.

So far, uranium-235 was just a faint blip on the read-out of Dempster's mass spectroscope.[22] Chemically, it would be identical with 'normal' uranium-238, but its physical properties – half-life, stability, nuclear cross-section – were completely unknown. There was certainly no reason for anyone to wonder how uranium-235 might react when bombarded with neutrons.

Oh, to Be in England

Ernest Rutherford had been knighted on New Year's Day 1914, and his peerage was announced on the same day in 1931. The new lord took the title 'Rutherford of Nelson' after the city in New Zealand where he'd grown up and almost drowned. His coat of arms also proclaimed his origins, with a kiwi and a Maori warrior alongside the die-away curve of radioactive disintegration and Hermes Trismegistus, the patron saint of both knowledge and alchemy.[23]

Rutherford was enjoying the long high peak of his powers at the Cavendish Laboratory in Cambridge: an imperial presence with a court of brilliant young men ready to colonise the world of physics. His right-hand man was the taciturn James Chadwick, discoverer of the neutron and Nobel laureate – and so absorbed in his work that he didn't notice his wife, Aileen, and their twin daughters when they called into his lab one afternoon (he did, however, say 'Hello, boy' to the family dog who accompanied them).[24]

Rutherford hadn't mellowed and still reacted furiously to irritating requests for money, which opened up a serious rift with Chadwick over a new-fangled American machine. The 'cyclotron', invented by Ernest Lawrence at Berkeley, was like a Cockcroft–Walton accelerator bent into a circle to form a high-powered slingshot that fired particles up to unimaginable speeds and energies. Chadwick desperately wanted a cyclotron but the machines were absurdly expensive to build. After several point-blank refusals by Rutherford, Chadwick finally snapped. In 1935, aged forty-four, he abandoned the Cavendish – 'the Mecca for all aspiring physicists throughout the British Empire' – for the Chair of Physics in his wife's home city of Liverpool. He quickly infiltrated the academic and industrial networks of Merseyside and raised almost enough funds to build a cyclotron with a hefty diameter of 37 inches, covering the shortfall from his Nobel Prize money. In Liverpool, Chadwick was happy and successful. His interests broadened, including (to Rutherford's disgust) the discovery of novel isotopes for medical applications including cancer treatment. Despite his disastrous Nobel speech, Chadwick blossomed as an engaging if oddball teacher: a 'tall bird-like figure', illustrating his 'beautifully prepared' lectures on the blackboard, and breaking into a grin when teaching practicals delivered the right result.[25]

In June 1936, Chadwick was external examiner at Cambridge and stayed with the Rutherfords. This should have been a pleasant reunion but the evening was spoiled by another Rutherfordian rant against the cyclotron; John Cockcroft had visited Liverpool to see Chadwick's machine being built and now wanted one too. Chadwick always regretted that he and his old boss parted on a sour note, because this was the last time that Chadwick saw him. Four months later, he was on the train back to Liverpool after a day of national mourning at Westminster Abbey, writing Rutherford's obituary for *Nature*. The 'short, sharp illness' for which 'all possible care' was afforded was actually one medical cock-up after another: an obvious diagnosis that was missed, and a straightforward operation that was put off until it was dangerous. Rutherford died on 19 October 1937, aged sixty-six, from the complications of surgery to free a strangulated umbilical hernia. Like his life, Rutherford's death made headlines around the world; Niels Bohr, in tears, broke the news to a shocked physics conference in Italy. His ashes were laid to rest in the nave of Westminster Abbey, alongside Newton, Kelvin and Darwin. His epitaph could have been a pithy update of his baronial motto, 'To seek the nature of things'; instead, it simply reads 'LORD RUTHERFORD OF NELSON, 1871–1937'.[26]

Posthumously, numerous Rutherfordisms were exhumed to capture the essence of the man: 'All science is either physics or stamp-collecting' and 'If your result needs a statistician, then you should have designed a better experiment'.[27] And to the end, Rutherford remained sceptical about liberating vast energy from inside the atom. Of course, energy was released when the lithium atom was split by one alpha-particle, but Rutherford argued that vastly more was wasted in firing up the billions of unlucky alpha-particles that failed to hit the magic spot. Harnessing atomic energy was therefore 'moonshine'. Otherwise, he explained, 'We should long ago have had a gigantic explosion in our laboratories, with no one remaining to tell the tale.'

Frozen Out

By now, *Deutsche Physik* had become an irrelevant burden to the Nazis, and Philipp Lenard and Johannes Stark were both running out of time. Predictably, Lenard's institute produced nothing and was dying on its feet; Lenard himself turned against Stark, lambasting him for corresponding with the 'Jewish atrocity journal' *Nature*.[28] Stark's self-nomination to succeed Max Planck as President of the Kaiser Wilhelm Foundation failed because too many people hated him. Next, he was sacked as President of the German Research Foundation for refusing to fund research into cosmic-ice theory (which Heinrich Himmler described as 'the intellectual gift of a genius'). Finally, he was hauled before a Nazi court accused of political intrigue, and narrowly escaped prison.

Isolated and resentful, Stark intensified his vendetta against Werner Heisenberg, who had replaced Einstein as the enemy to eliminate. Heisenberg had written to all seventy-five professors of theoretical physics in Germany, demanding an end to harassment by Stark and Lenard; highlighting Stark's growing impotence, most supported Heisenberg. In July 1937, Stark denounced Heisenberg as a 'white Jew' in an anonymous article for the SS weekly magazine, *Das Schwarze Korps*, but failed to bring down his prey. The SS investigated Stark's allegations and concluded that Heisenberg was an asset whose brilliance and charisma could fire up the new generation of German scientists.[29] Moreover, Himmler was taking a keen interest in Heisenberg, possibly for personal reasons. Even Nazi thugs have mothers; Himmler's was an acquaintance of Heisenberg's, who pleaded her son's case, woman to woman. Whichever his motive, Himmler promised Heisenberg a top-class professorship and whatever he needed to advance nuclear physics in the service of the Reich.[30]

As always, Werner Heisenberg attended Bohr's annual symposium in Copenhagen that summer. The conference photo (Plate 7) is one of those eloquent images that, in retrospect, tells a story that changed the world. Taken in July 1937, it shows the cream of Europe's theoretical physicists – fifty-two men and two women – squeezed onto the benches in a small lecture theatre in Bohr's Institute. In the front row, an awkward-looking Heisenberg, chin on his knuckles and avoiding the camera's gaze, is sandwiched between Bohr, his pipe and matches before him, and Wolfgang Pauli. Beside Pauli, Otto Stern, visiting from America, is sharing a comment with Lise Meitner, who has half-turned to listen to him. In the throng behind them are Rudi Peierls; the maverick Hungarian Leo Szilard, then a refugee in London; Mark Oliphant from the Cavendish; Otto Frisch, now settled in Bohr's Institute; and Carl von Weizsäcker, a friend of Heisenberg's whose interests included astrophysics and, bizarrely, astrology.

A couple of faint smiles are visible, but most of the faces are expressionless and some look grave and preoccupied – as if they somehow knew that this would be the last time they could all leave politics and ideology at home and come together as friends to talk science.

6

TO THE BRINK
February 1938–September 1939

Barely two decades after the armistice, Europe was again stewing in the incendiary ingredients that had ignited the 'war to end all wars' in 1914 (Map 1). Germany was back at the flashpoint, heavily armed and psychopathically nationalistic. Non-Aryan *Untermenschen* – Jews, left-wingers, pacifists, gypsies and the disabled – were being purged from the Fatherland or sequestered in concentration camps. By late February 1938, over 1,800 Jewish scientists had fled Germany.[1] Those staying behind were courageous or foolhardy – and courting disaster.

Exodus

Lise Meitner had shared nearly half her sixty years with Otto Hahn in Berlin, but was now seeking a job outside Germany in case her life-insurance policies – Austrian passport and Christian baptism certificate – were torn up. Niels Bohr offered her a post in Copenhagen but she dithered, even after Hitler's troops flooded into Austria on 12 March 1938 and turned her into a citizen of the Reich. Meitner kept up the facade of normal working until May, when word spread that she was Jewish and her colleagues told Hahn to sack her. She accepted Bohr's offer but was refused a Danish visa because her Austrian passport was now invalid. In mid-June, her world fell apart: the Ministry of the Interior ordered her to be dismissed and prevented from leaving Germany, and she heard whispers that the SS knew about her.[2]

Bohr came to her rescue, scripting a thriller that featured Dirk Coster, Professor of Physics at Groningen in Holland. While working with Bohr during the 1920s, Coster had co-discovered element 72 and christened it 'hafnium' after the Latin name for Copenhagen; more relevant to his current mission, he was unflappable, charming and fluent in German. Coster arrived in Berlin on Monday, 11 July, ostensibly to visit Hahn, and met Meitner the following morning. Only Max von Laue and two others knew that this was her last day at the KWI. Meitner followed her normal routine and left for her flat at 8 p.m. She crammed whatever she could into two small suitcases, then sneaked over to Hahn's house. Early next morning, a close friend drove her to the station, where Coster was waiting in a first-class compartment on the train for Holland. Meitner had already panicked in the car, and ahead lay a twelve-hour journey with the threat that their passports could be checked at any moment. Approaching the Dutch border, Coster had to talk her down from another attack of anxiety. Hahn had given her a diamond ring left to him by his mother, in case she needed to bribe the border guards. Luckily, Coster's smooth talking and the Dutch government's letter inviting Professor Meitner to take up a post at the University of Groningen did the trick. They crossed the border without incident and reached Groningen at 6 p.m.[3]

Meitner escaped with only 10 Reichsmarks and was now stateless, but her career still came first. She was indebted to Coster, but Groningen wasn't suitable. Neither was Bohr's Institute. Instead, she chose Manne Siegbahn's new Nobel Institute for Physics in Stockholm. Meitner busied herself learning Swedish and building up a research group, but was isolated and lonely.[4] She exchanged frequent letters with Hahn and with Otto Frisch in Copenhagen, with whom she wanted to spend her first Swedish Christmas. And she took great care of Hahn's diamond ring, forever a memento of a lucky escape.

That summer, clouds gathered over Italy. On 14 July 1938, 180 academics, doctors and lawyers signed the 'Manifesto of Rome', demanding that Italian Jews be stripped of citizenship and barred from governmental and professional posts. The Manifesto crystallised into the 'Racial Laws in the Kingdom of Italy', which Mussolini waved through 'for political reasons'.[5]

Enrico Fermi had already enquired *sotto voce* about jobs in America and the chance to take Laura and their two children to safety. There was now much to fear in Germany. In September, Hitler threatened to reclaim the ethnic German Sudetenland, which the Treaty of Versailles had given to Czechoslovakia. The Munich Agreement of 30 September 1938 permitted Hitler to do this, on condition that he didn't grab more territory. After the occupation of the Sudetenland, anti-Jewish violence in Germany reached new heights. *Kristallnacht* (the 'night of crystal') on 9–10 November 1938 sounds like a fairytale carnival – but the 'crystal' was shattered glass from the Jewish shops ransacked by Nazi mobs, who also burned down synagogues and rounded up Jews for the concentration camps.

The Fermi family had happier recollections of that day. Early in the morning of 10 November, Laura was woken by an international phone call alerting Professor Fermi to news from Stockholm that evening. He had won the Nobel Prize in Physics (1938), for 'identifying new radioactive elements' through his brilliant work with slow neutrons. Fermi, still a member of the Fascist Party, notified the Italian authorities. Luckily, they were more accommodating than Hitler, who had recently prohibited Germans from accepting Nobel Prizes. The Fermi family was allowed to travel to Stockholm, en route to a six-month sabbatical in New York. On 6 December 1938, the Fermis left Rome for the last time on the train for Sweden. Straight after the Nobel Prize ceremony, they sailed to Southampton and then on to New York, unaware that Italian newspapers had savaged Fermi for not giving the Fascist salute on receiving his prize. On 2 January 1939, they reached the New World and were greeted by a bitter East Coast winter and job offers from five American universities. Fermi accepted a chair at Columbia University in New York, and resumed experimenting within days.[6]

He couldn't have predicted that he'd be back at the docks just two weeks later and would hear astonishing news which would rewrite nuclear physics – and would demolish the 'new radioactive elements' which had won him his Nobel Prize.

Revelations

In mid-December 1938, Otto Hahn wrote to Lise Meitner with a conundrum. It was the same anomaly that she'd told Fritz Strassmann to throw in the bin two years earlier: neutron-irradiated uranium contained something chemically identical to barium. Meitner again insisted that neutrons couldn't possibly create barium from uranium, but Hahn wrote again on 21 December. The barium refused to go away. Could she provide 'some fantastic explanation'?[7]

Without waiting for her reply, Hahn and Strassmann submitted their paper to *Naturwissenschaften*, the leading German science journal. Hahn was still uneasy about the 'peculiar results' of the barium which shouldn't have been there, but the editor of *Naturwissenschaften* was told by his boss at the Springer publishing house in Berlin to fast-track the controversial paper.[8] When the proofs arrived after Christmas, Hahn had dreamed up his own 'drastic' way of creating barium from uranium – summarised mathematically as '138 + 101 = 239!', with the '!' to emphasise his surprise. The notion 'contradicts all previous experience in nuclear physics', but could uranium (atomic weight 238, increased to 239 by adding a neutron) somehow have split into barium (atomic weight 138) and another element of atomic weight 101?[9] One of the papers scheduled for publication in *Naturwissenschaften* on 6 January 1939 was held back to make space for Hahn and Strassmann. It was a clever move, because the suggestion that neutrons could smash the uranium nucleus into two fragments immediately caused huge excitement and controversy. By then, though, Lise Meitner and her nephew Otto Frisch had already reached the same conclusion, christened the phenomenon – and calculated that it released a gigantic blaze of energy.

Otto Frisch had been quickly assimilated into Bohr's Institute in Copenhagen. He picked up enough Danish ('a thick soup of vowels interrupted by the occasional glottal stop') to give his first seminar *på dansk* within a month of arriving, although he couldn't yet answer questions. The institute was bursting with intellectual energy that flowed from the top. Visually, Niels Bohr (Plate 3) reminded Frisch more of a peasant than a deity. 'The Great Dane' was heavily built, with large head, hairy hands and bushy eyebrows above grave eyes that 'could hold you with all the power of the mind behind them'. He was fluent in German and English, but his soft voice and strong Danish accent sometimes created uncertainty about which language he was speaking. He'd given up competitive football but still skied, sailed and chopped down trees, and could beat men half his age at table tennis and sprinting up staircases.[10] Bohr's fiftieth birthday on 7 October 1935 was celebrated with a day of world-class physics, a banquet in the House of Honour, and half a gramme of radium presented by the Danish people. Frisch helped to crush beryllium – a brittle, toxic metal – into a powder that would blast out neutrons when bombarded with alpha-particles from the radium's emanation. His own research, creating novel radioisotopes from rare-earth elements, was good but hardly world-shattering. Then, just after Christmas 1938, Frisch took the ferry from Copenhagen to Malmö in Sweden for 'the most momentous visit of my life'.[11]

Every year since 1927, Frisch had joined his aunt Lise for Christmas in her cramped apartment in Dahlem. This time, the two exiles met in a hotel at Kungälv, near Gothenburg. On 28 December 1938, Frisch came down to breakfast and found Meitner poring over a letter from Otto Hahn. Did she think that 239 (uranium plus a neutron) could possibly 'burst' into 138 (barium) and 101? Frisch was sceptical, but the two bounced ideas off each other over breakfast and then while following a snowy trail through the woods. They would have made an odd couple, with thirty-four-year-old Frisch on cross-country skis and his sixty-year-old aunt in stout walking shoes, sitting side by side on a log in the snow with pencils and scraps of paper. Frisch later described it as a gradual, mutual dawning rather than a bolt of inspiration. Sketches took shape: a sphere that stretched out into a cylinder, then pinched in the middle like a peanut and finally tore apart into two smaller spheres.[12]

The enormity of what they'd conceived hit them immediately. Until now, elements had been transformed by chipping fragments – single protons or neutrons – off nuclei. This was fundamentally different: somehow, a single neutron had to smash the massive uranium-238 nucleus into two unequal pieces. They predicted that barium's other half was an isotope of element 36 (the inert gas, krypton), representing the difference in atomic number between uranium (92) and barium (56). The scenario was 'very hard to understand' but could tie in with Bohr's recent theory that the nuclei of the heaviest elements behaved like a drop of water poised to fall, with the nuclear counterpart of surface tension holding in the neutrons and the self-repelling protons. The impact of a single neutron could wreck this delicate equilibrium and, like an overfilled drop breaking away, the uranium nucleus would split in two (Figure 4). It looked feasible on paper, but everything had to add up. Sitting on their log, they calculated that a huge blast of energy would be released when the two fragments of uranium shot apart – about 200 MeV, twelve times that liberated by splitting the lithium nucleus. Moreover, the mass (m) of barium plus krypton was slightly less than uranium plus a neutron; and when this shortfall – just one-fifth of the mass of a proton – was plugged into $E=mc^2$, 'E' came out at 200 MeV.[13]

Meitner and Frisch spent the rest of their Christmas holiday trying and failing to poke holes in their reasoning. They drafted a paper for *Nature*, finishing it expensively by phone after Frisch returned to Copenhagen on New Year's Day 1939. Bohr was preparing to leave for a three-month sabbatical at Princeton with Einstein, but Frisch cornered him briefly on 3 January. Bohr listened in silence, then smacked his forehead and exclaimed, 'What idiots we all have been! This is wonderful, just as it must be.' Two days later, Frisch cycled to the House of Honour to discuss the draft paper. Next morning, he reached the station just in time to hand Bohr the first two pages of the manuscript, all that he'd been able to type up overnight. Greatly excited, Bohr promised not to tell anyone about this 'wonderful' discovery until Frisch and Meitner had published their paper.[14] He was nearly as good as his word.

Meitner and Frisch needed a snappy name for the extraordinary splitting of the uranium nucleus. It reminded them of cells dividing,

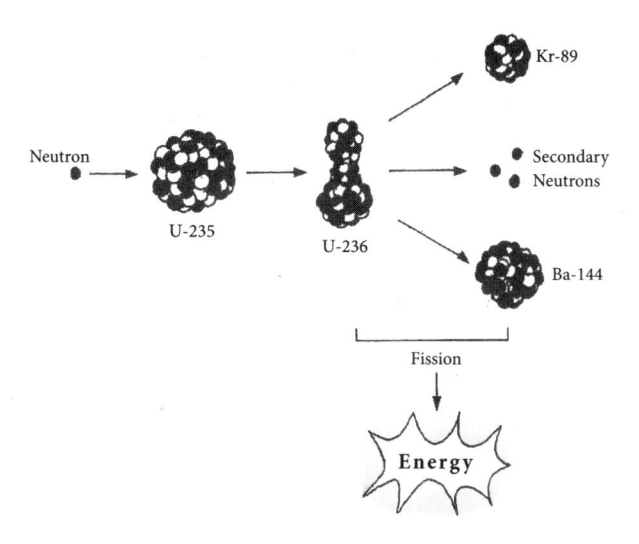

4. Nuclear fission in uranium (U-235).

which a visiting American biologist told Frisch was called 'fission'; that evening, Frisch typed the word for the first time in a paper on physics. He decided that they needed hard evidence of fission and set up a quick experiment to detect the massive energy burst when the uranium nucleus exploded. He found it on the lucky afternoon of Friday, 13 January, spent three days confirming the result and finished a second paper late on 16 January. By 5 a.m. the next morning, both manuscripts were in the airmail box and bound for London – but as Frisch later said, without easy access to Blackett to badger the editor, *Nature* would take its usual month or so to publish.[15]

Niels Bohr had intended to spend three months talking quantum mechanics with Einstein, but underwent a remarkable sea-change between Gothenburg and New York. He spent much of the rough nine-day crossing feeling sick and pacing the deck with his two travelling companions – his engineer son Erik and Léon Rosenfeld, Professor of Physics at Liège. They talked only about the astounding ability of a single neutron to crack the massive uranium nucleus in two. Unfortunately, Bohr was so preoccupied that he forgot to say he'd promised not to share the secret until after the Meitner–Frisch paper was published.[16]

Bohr and his companions were met off the boat on 16 January by John Wheeler, Professor of Physics at Princeton, and Enrico and Laura Fermi. The Fermis drove the Bohrs into New York to stay with friends for a couple of days, while Rosenfeld and Wheeler took the train to Princeton. During the hour-long journey, Rosenfeld talked non-stop about his conversations with Bohr, and the mesmerised Wheeler joined the secret elite – barely a dozen worldwide – who had heard of nuclear fission. Wheeler invited Rosenfeld to the departmental journal club which he was chairing that evening, where he innocently asked Rosenfeld if he had anything interesting to share. Oblivious of Bohr's vow of silence, Rosenfeld delivered the first ever presentation about nuclear fission. Those listening were bowled over and poured out of the room desperate to spread the word that Fermi's so-called 'transuranic elements' were actually fragments of the uranium nucleus.[17]

When Bohr reached Princeton a couple of days later, he was distressed to hear people talking about the secret which Frisch had entrusted to him. Frisch couldn't be contacted, so Bohr decided to tell the world formally in an extraordinary (in every sense) opening session grafted on to the forthcoming Conference on Theoretical Physics in Washington. 'The bombshell burst' at 2 p.m. on 26 January 1939 – twenty days after Hahn and Strassmann's paper was published – in Room 105, Building C of George Washington University. Fission fever instantly swept through physicists across America. One heard the news while having his hair cut; he jumped out of the barber's chair, sprinted to his lab and replicated Frisch's experiments that afternoon. He was too slow. Several of those in Room 105 phoned colleagues immediately after the session and told them to drop everything and look for fission. Even they lagged behind, because researchers at the nearby Carnegie Institution were at their benches before Bohr finished speaking.[18]

'Disintegration of Uranium by Neutrons: A New Type of Nuclear Reaction' by Meitner and Frisch was published in *Nature* on 11 February 1939, and Frisch's experimental paper followed a week later.[19] By this time, labs in America and Paris had confirmed Frisch's demonstration of the energy released by fission. And in Princeton,

Bohr turned his back on Einstein and spent his sabbatical with John Wheeler, writing a massive review article on everything that was known and could be deduced about nuclear fission.

Chain Gang

On 15 March 1939, Hitler broke the Munich Agreement of six months earlier and sent German troops into the Czechoslovak heartland of Moravia and Bohemia. The Nazis labelled the occupied territory a 'protectorate'; native Czechs called it an invasion. The fragile spectre of peace shivered, but in everybody's interests, it clung on for now.

In Copenhagen, Otto Frisch saw no immediate danger. Like all nuclear physicists, he was frantically busy. Nuclear fission, which he'd christened barely three months earlier, now had life of its own and the focus of excitement had shifted to 'the most important point... which we completely missed', as he later said.[20] Uranium contained more neutrons than the sum of its split products, barium and krypton – so what happened to the excess? Perhaps one or more neutrons were spat out when the uranium nucleus was ripped in two? If so, could these 'secondary' neutrons go on to crack another uranium nucleus, generating more nucleus-cleaving neutrons? If fission begat neutrons and therefore more fission, it could set off a self-propagating 'nuclear chain reaction' that would tear through a mass of uranium until it had all been consumed (Figure 5).

While Frisch was thinking about neutrons 'multiplying in uranium like rabbits in a meadow', others were already looking for secondary neutrons and evidence that these could trigger a nuclear chain reaction. One researcher had a prepared mind because, five years earlier, he'd envisaged a nuclear chain reaction driven by neutrons and had even patented a nuclear-power generator. Leo Szilard (Plate 32), a forty-year-old perpetually wandering Jewish physicist, could pinpoint the moment when the notion hit him: standing at traffic lights on Russell Square on a dull London morning in October 1933.[21] He'd had to leave Germany because he was a red rag to *Deutsche Physik*: Hungarian, Jewish, friend and collaborator of Max von Laue and Einstein. Szilard was brilliant, imaginative, quirky and often unplugged

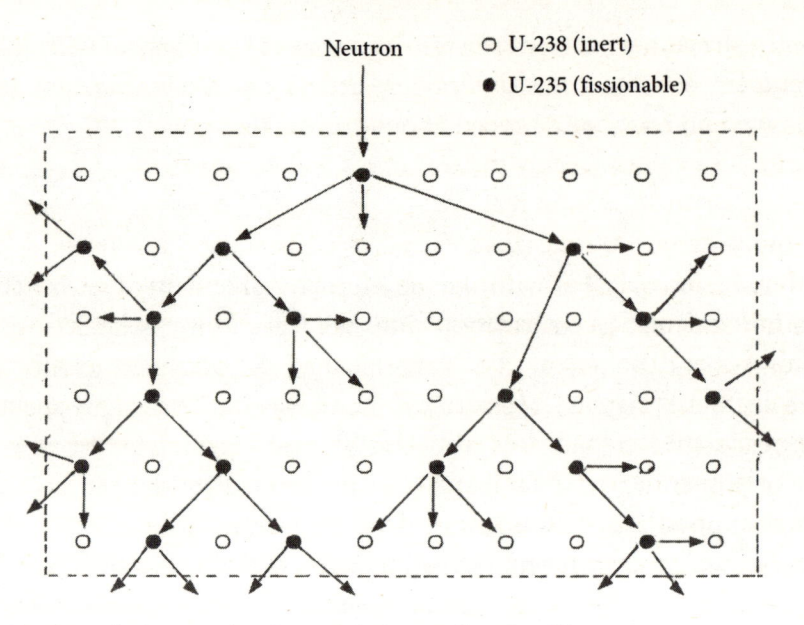

5. Nuclear chain reaction in uranium, confined to U-235.

from reality. His revolutionary refrigerator with no moving parts (co-conceived with Einstein) was ahead of its time but went nowhere; he patented an electron microscope and a cyclotron, but others more practical won Nobel Prizes for making them happen; and his nuclear-power unit, driven by a chain reaction in beryllium and indium, required neutrons to perform tricks forbidden by the laws of physics.[22] Szilard had helped Rutherford to set up the Academic Assistance Council for refugee scientists; unfortunately, Rutherford's reaction to Szilard's chain-reaction idea was to throw him out of his office.[23]

When Szilard attended Bohr's summer conference in July 1937, he was already eyeing up America because he feared another European war. The Academic Assistance Council couldn't find him a job in Britain, so he sailed for the New World on 2 January 1938. For months, he wandered from hotel to hotel in university cities across the USA, living out of the two suitcases that contained all his worldly possessions. In November 1938, he settled into the King's Crown Hotel at Morningside Heights, New York City, a few minutes' walk from Columbia University. Szilard had no hint of a job offer there (or anywhere else), but wangled invitations to give seminars and get his face known.[24]

In late January 1939, he went to Princeton to visit another Hungarian refugee physicist, his childhood friend Eugene Wigner – and was pitched into the furore over the neutron-induced fission of uranium. Szilard saw immediately that uranium, unlike beryllium, really could sustain a nuclear chain reaction. Initial proof of concept was simple – simply demonstrating that excess neutrons were produced when uranium was blasted with neutrons. Enrico Fermi, the Nobel Prize-winning neutron bombarder now installed at Columbia, could verify this faster than anyone but was too busy to help. Fortunately, the Chairman of Physics at Columbia, George Pegram, took a gamble and gave Szilard lab space free for three months. Szilard teamed up with his compatriot Walter Zinn and acquired some beryllium, radium and uranium, and an oscilloscope which would reveal neutrons as flashes of light. One evening in March 1939, Szilard and Zinn set up their experiment and (as soon as they remembered to turn on the oscilloscope) saw the screen light up with the tell-tale flashes. They watched the spectacle for some time, then switched off and went home. Szilard didn't sleep, because he'd followed the nuclear chain reaction through to an obvious conclusion. He said later, 'That night, there was very little doubt in my mind that the world was headed for grief.'[25]

Szilard's paper with Zinn[26] came out in mid-April 1939, and they joined the growing crowd of sharp-elbowed researchers digging into an extraordinarily rich seam that everyone had walked past until three months earlier. Hahn and Strassmann had confirmed that fission generated krypton, barium's predicted other half.[27] Dozens of researchers in America, France, Britain, Russia and Japan were hard at work. Twenty papers had already been published and another eighty would appear before the end of the year.

Frédéric Joliot led the pack, publishing experimental proof of the 'explosive rupture of uranium' on 30 January before Meitner and Frisch's *Nature* paper.[28] The weekly meetings of the Academy of Sciences in Paris often featured something new and exciting from Joliot's team, notably Hans Halban, a suave Austrian Jew who had worked with Irène Curie, and Lew Kowarski, a monolithic Russian-born chemist who had joined Joliot in 1934 (Plate 8). In mid-March, Halban calculated that each fission released on average 3.5 secondary neutrons, which travelled faster than the slow neutrons that had triggered fission. Next, they fired neutrons at a large mass (14 kg) of uranium oxide and found rapid increases in neutron production which suggested 'the possibility of a chain reaction'. After further experiments with even more uranium oxide inside a half-metre copper sphere, their colleague Francis Perrin estimated that a uranium sphere with a radius of 1.3 metres (corresponding to a 'critical mass' of about 40 tonnes) would enable the chain reaction to become self-perpetuating rather than fizzling out when the triggering neutrons were switched off.[29]

Joliot seemed ready to fulfil the prophecy in his Nobel speech: 'scientists, breaking and making atoms at will, one day will succeed in bringing about explosive nuclear chain reactions ... which will liberate enormous amounts of energy.'[30] On 1 May 1939, the publication date of Perrin's paper on the critical mass of uranium, Joliot, Halban and Kowarski patented 'a device for energy production' in which heat produced by the chain reaction would turn water into steam and drive a turbine. The next day, they patented a 'method for stabilising' such a device. Their third patent – No. 445686, registered in secret on 4 May – was to 'perfect an explosive' that would blast out all the energy inside a large mass of uranium in a tiny fraction of a second.[31] An atomic bomb.

A Letter to the President

Everyone struggled to comprehend the energy unleashed by fission. Otto Frisch estimated that a 200 MeV energy burst could flick a grain of sand[32] – trivial, but this resulted from splitting a single uranium atom and a gramme of the element contains 2,500 million trillion atoms. Siegfried Flügge at the KWI for Physical Chemistry calculated that the fission of a cubic metre of uranium could lift a cubic kilometre of water 27 km into the sky; liberating all that energy in a fraction of a millisecond could produce 'an exceedingly violent explosion'.[33] A thousand times more devastating than the same weight of TNT? A million times?

Haunted by the nightmare of a German atomic bomb, Leo Szilard pictured a massive uranium bomb – too big to deliver by an aircraft but easily transportable by boat and yielding 'destructive power beyond imagination'.[34] His fears deepened as Europe began disintegrating during spring 1939 – while detailed papers about fission continued to flood into journals that were still posted to researchers in Germany. *Deutsche Physik* had swept out many excellent nuclear physicists, but a few remained who could turn nuclear fission into a bomb. Otto Hahn and Fritz Strassmann were both anti-Nazi and wouldn't willingly work on a German atomic bomb – but what if they were forced, possibly with a gun to the head, to choose between conscience and survival? Of greater concern was Carl von Weizsäcker at the KWI for Chemistry, whose father was a hardcore Nazi high in the Reich's hierarchy. And what about the brilliant young Nobel Prize winner, Werner Heisenberg? To survive being branded a 'White Jew' in the SS's own magazine, he had to be in favour at the very top.

Szilard decided that dangerous intelligence about atomic research must not reach Germany. But when he asked American colleagues not to publish about nuclear fission, they weren't interested; Fermi, currently perfecting his American slang, simply said 'Nuts!'[35] Joliot's group in Paris also ignored Szilard's plea, sent in 'the longest telegramme anyone had ever seen', which was delivered to Joliot in his bath. Joliot responded tersely – 'PROPOSITION VERY REASONABLE BUT COMES TOO LATE' – and his group continued pushing out their papers and patents.[36] Evidently, the scientist's survival motto – 'publish or perish' – still ruled during a world crisis.

This left Szilard with one morally repugnant option: the only way to outflank Germany was for America to build an atomic bomb first. This was an exceedingly long shot, as America was at peace with Germany, and President Roosevelt was determined not to get embroiled in a European conflict. Moreover, nuclear fission was a million miles from a weapon to excite a military strategist. Szilard drafted a document explaining the risk of Germany exploiting fission to make an atomic bomb. He was helped by Eugene Wigner and Edward Teller, another Hungarian refugee physicist in New York, but needed much more muscle because he was pitching so high.

On 16 July 1939, Teller drove Szilard to Peconic on Long Island, to find a secret address disclosed by an obliging administrator at the

Institute of Advanced Studies in Princeton. Deep in the woods, a lad on a bike pointed out the house belonging to the 'kindly old gent with a long grey mane'. Einstein came out to meet them in dressing gown and slippers; this was his refuge from the 'puny demigods on stilts' in the 'quaint and ceremonious village of Princeton'. He heard them out, helped them to revise the document, and agreed that it could be sent as a personal letter from him.[37] On 2 August, Einstein signed the letter, typed by Szilard. It began:

> Sir: Some recent work by E. Fermi and L. Szilard, which has been communicated to me in manuscript, leads me to expect that the element uranium may be turned into a new and important source of energy in the immediate future. Certain aspects of the situation which has arisen seem to call for watchfulness and, if necessary, quick action on the part of the Administration.[38]

Einstein explained that a nuclear chain reaction in a large mass of uranium would generate 'vast amounts of power' and could conceivably be turned into an 'extremely powerful bomb of a new type'. Carried by boat and exploded in a port, a single such bomb 'might very well destroy the whole port together with some of the surrounding territory'. The US Administration must work closely with American nuclear researchers to secure uranium supplies and 'speed up the experimental work'. Einstein made no direct plea for America to build its own atomic bomb but ended with two chilling omens. Germany had recently stopped exporting uranium; and 'the son of the German Under-Secretary of State, von Weizsäcker', was a researcher at the Kaiser Wilhelm Institute in Berlin, where 'American work on uranium is now being repeated'.

Einstein's letter was addressed to President Franklin D. Roosevelt at the White House. To seize the President's full attention, Szilard entrusted the letter to a friend, Alexander Sachs, a White House adviser who had Roosevelt's ear.[39] And while Europe was being levered to the edge of the abyss, Szilard waited anxiously for the President's blessing.

Lockdown

During summer 1939, some otherwise sensible people decided to ignore the gathering storm. In June, Otto Hahn had 'clandestine talks' with émigré German scientists in London.[40] Werner Heisenberg stayed with his old friend Samuel Goudsmidt at the University of Michigan in July; whatever his real motive for the visit, Heisenberg rejected job offers in America.[41] Cutting it even finer, and despite having spent the previous war interned in a German stable, James Chadwick took his family on a fishing holiday to Sweden in mid-August.[42]

On 1 September 1939, the monumental review of fission by Bohr and Wheeler – 'one of the finest collaborations in physics' – was published in *Physical Review*.[43] It built on the 'liquid drop' model of the nucleus and included an eye-catching and counter-intuitive claim concerning uranium-235 (U-235), the lightweight fellow traveller of 'normal' uranium (U-238) that comprised one part in 130 of the natural element. U-235 still existed only as its fingerprint in the mass spectrometer, but Albert Nier had deduced that it was relatively frisky: its half-life was only 700 million years, compared with 4.6 billion years for U-238.[44] For some reason, U-235 had flashed into Bohr's mind while crossing the Princeton campus after breakfast one morning in early February. Before reaching his office, he'd realised that the 143 neutrons of U-235 made it susceptible to being smashed in two by a single neutron – a property that couldn't be shared by the 146 neutrons of U-238. Bohr argued that U-238 was not fissionable at all, and therefore that the scarce U-235 accounted for all the fission in uranium[45] (see Figures 4 and 5). Wheeler's colleague George Placzek was unconvinced and bet him a cent that this was wrong. Wheeler took him up, offering 'a proton for an electron', and would therefore owe Placzek $18.36 if U-238, and not U-235, turned out to be fissionable.[46]

Otto Frisch's abiding memory of Friday, 1 September 1939 had nothing to do with fission. On that day, Hitler invaded Poland and was given an ultimatum by France and Britain to withdraw his troops by 11 a.m.

on Sunday, 3 September. 'We all sat round the radio,' Frisch later wrote.[47] 'There was a great feeling of tense sobriety when we were told that the deadline had passed, and that the war was on.'

Leo Szilard's worst fears were realised that same day. The Germans had started another European war, which could only stoke their desire to build an atomic bomb. Einstein's letter should have reached the White House a month earlier, but Roosevelt hadn't yet responded. Szilard might have been reassured to know that influential scientists in England shared his concerns and had already acted to obstruct a German atomic bomb. Four months earlier, on 26 April, a confidential memo instructed the Foreign Office to prevent large stockpiles of uranium ore in Belgium from reaching Germany.[48] On 3 May – the day before Joliot filed the patent for his explosive device – the British government produced a top-secret review on 'the possibility of producing an atomic bomb'. This concluded that Britain 'must keep in the forefront of the work and give it high priority'; even though the chances of success were remote, 'the issues at stake are too enormous'.[49]

But Szilard would have been devastated to discover that the British had been left behind. Three weeks before the Foreign Office memo, two German nuclear physicists had separately contacted the Reich Research Council and the War Office, pointing out that nuclear fission could produce a bomb of unprecedented power. In late April 1939, Germany halted all uranium exports, notably from St Joachimsthal in occupied Bohemia. And on 1 September 1939, the day when Hitler invaded Poland, the German Army's Ordnance Office began assembling a group of nuclear physicists to include Werner Heisenberg and Carl von Weizsäcker.[50] The group already had a name: the *Uranverein*, or 'Uranium Club'.

7

WARM-UP
July–October 1939

The radio which told Otto Frisch that 'the war was on' was in Birmingham, not Copenhagen. When Hitler annexed Austria, Frisch immediately sought a safer haven because Denmark was an obvious target for its immediate southern neighbour. In early 1939, Mark Oliphant had visited Bohr's Institute and promised to help. Oliphant was now the Professor of Physics in Birmingham, with a hefty 36-inch diameter cyclotron, funded by Lord Nuffield, under construction. He radiated 'confidence and calmness' and invited Frisch to come and discuss options. Unlike his aunt Lise's exodus from Berlin, Frisch's departure was 'not at all dramatic ... just like any tourist'. He packed two small suitcases, leaving behind his violin and a country steeling itself for invasion.[1]

Frisch started at Birmingham University in July 1939, spending the long hot summer vacation lying in the sunshine and, after Hitler marched into Poland, adjusting to the schizophrenic life of the 'Phoney War'. Inside the university's ivory towers in leafy Edgbaston, a new autumn term started; outside lay the real world of air-raid shelters, blackouts and children carrying gas-masks to school. Frisch was 'auxiliary lecturer' to Professor Mark Oliphant FRS, who lectured with great energy but no concessions. Somewhat like a priestess translating the Oracle's utterances at Delphi, Frisch was kept busy clearing up confusion for those who found Oliphant 'heavy going'.[2]

THE IMPOSSIBLE BOMB

The British Home Office helped to keep Frisch focused on nuclear fission. When war was declared, all refugees from Germany were classified as 'aliens' and potential enemies, whether or not the Nazis had chased them out. Frisch was locked out of the physics labs, now housing a mysterious, top-secret research project which mopped up everyone else in Oliphant's team; he soon worked out that it concerned radar. When another clandestine mission in London carried off Philip Moon, Oliphant's right-hand nuclear physicist, Frisch was left as the only person working on fission, which three months earlier had been the hottest topic in physics. He was allowed to tinker in an empty lecture room and decided to try separating the 0.7 per cent of U-235 from the commonplace U-238, using a 'thermal diffusion' technique recently invented by Klaus Clusius in Munich. Clusius had fed chlorine gas, which contains two isotopes (Cl-35 and Cl-37), into a vertical glass tube with an electrically heated wire running up its centre; the heavier Cl-37 slowly gravitated to the bottom of the tube and could be siphoned off. Some uranium compounds turned into gas at relatively low temperatures but there was no guarantee that Clusius's method could separate U-235 and U-238, which differed in atomic weight by barely 1 per cent. Frisch began by trying (and failing) to separate two gaseous elements, nitrogen and oxygen (atomic weights, 14 and 16).[3]

Autumn gave way to winter, with hip-deep snow outside Frisch's room near the university and water freezing in the glass by his bed. He kept warm by writing a review about fission for the *Annual Report* of the Chemical Society, hunched in an armchair in front of the gas fire with a blanket over his shoulders and the typewriter perched on his knees. Frisch rounded off his article with a comforting message for these terrifying times. Bohr and Wheeler believed that the chain reaction in uranium was 'too small, too brief' to release energy explosively. Like them, Frisch saw no risk of turning nuclear fission into the 'super-bomb' foreseen by novelists and scaremongers.[4]

Birds of a Feather

Frisch had already connected intellectually and socially with a like-minded soul whom Oliphant had brought to Birmingham two years earlier. Rudolf (Rudi) Peierls (Plate 9) was three years younger than Frisch and already a professor. He was also Jewish but German, not Austrian, and had left Germany forever in 1933.

Peierls was born in Berlin in 1907 to well-off parents whose friends included Fritz Haber. He'd always been fascinated by theoretical physics and in 1928, during 'the heroic age of quantum mechanics', began a journey of discovery which took him to elite research destinations in Switzerland, Denmark and Italy. His pithy pen-portraits of the great scientists he encountered are like captions in a family photo album. Max Planck (Berlin, 1925): 'the worst lectures I have ever attended'. Werner Heisenberg (Leipzig, 1928): 'did not look like a great scientist until you saw his eyes, whose sparkle expressed enthusiasm and intelligence'. Niels Bohr (Copenhagen, 1930): 'enormous charm'. Enrico Fermi (Rome, 1932): 'a mind of exceptional clarity'.[5] And on a post-conference outing on the Black Sea (Odessa, 1930), he met Genia Nikolaevna Kannegiesser, a lively physics graduate from Leningrad who was 'more cheerful than everybody'. Their relationship blossomed despite geographical separation and led in March 1931 (after just two weeks of contact time) to what some might have judged 'a rash marriage', but one that would last over five decades.[6]

Peierls visited Fermi's lab in 1932, just too early to witness the transuranic smokescreen which obscured the discovery of fission. He was funded by a Rockefeller Scholarship, which he completed in Cambridge. His reservations about England – 'strange in many ways ... deplorable food' – were offset by the regal presence of Rutherford and the aftershocks that followed the discovery of the neutron and the splitting of the atom. He was offered a post in Hamburg, but the flood of scientific refugees pouring out of Germany swept away any inclination to go home. Lawrence Bragg in Manchester conjured up two years' funding, and the Peierls family (now with a baby daughter) moved there in autumn 1933. With rain, fog and lavatorial brown tiling in the physics department, Manchester was 'hardly an attractive city', but Peierls consolidated his reputation in nuclear physics. In 1934, he attended the Solvay Conference and saw Marie Curie in the flesh: an elderly sixty-six year old, but with 'impressive poise and command of physics' and no qualms about contradicting her daughter and son-in-law in public. Next came two years at the Mond Laboratory in Cambridge and close contact with the nuclear fraternity at the Cavendish, notably James Chadwick ('a jaundiced view of life and his fellow men, but with great warmth behind the facade'), John Cockcroft ('few words'), and especially Mark Oliphant. When Oliphant moved to the Chair of Physics in Birmingham in autumn 1937, Peierls went too. Aged thirty, he looked (and was) young to be the freshly minted Professor of Theoretical Physics; he was also 'unassuming and modest, quiet and often shy'. He and Genia bought a car and a house in the 'very attractive' university enclave of Edgbaston, and he focused on the nucleus.[7]

Like everybody else, the Peierls family carried on as usual during spring 1939 'in the hope that there would be some solution'.[8] Also excluded from Oliphant's secret radar project, Peierls began thinking about fission in uranium. Francis Perrin in Paris had calculated that the 'critical mass' of uranium – the minimum needed for a self-sustaining chain reaction – would be huge, about 40 tonnes. Uranium's nuclear cross-section and other vital statistics were unknown, but Peierls derived a series of equations into which those numbers could be inserted when eventually they were determined. His own best guesses gave a critical mass much smaller than Perrin's, just a fraction of a ton. He presented his findings to the Philosophical Society in Cambridge in June 1939 and the paper came out for everyone to see in October, a month after Britain declared war on Germany. At the time, it was only mathematics and physics; much later, Peierls admitted that the information shouldn't have been published openly.[9]

By then, Peierls was getting on famously with the other 'enemy alien' refugee, Otto Frisch. They were both brilliant thinkers with complementary experience and skills, brought together and encouraged to pool their ideas by being barred from radar research. The Frisch–Peierls collaboration lasted only a few months but was a brief encounter with cataclysmic consequences – like two pieces of fissionable material, each fizzing with not quite enough energy, being united to make a whole that exceeded the critical mass for an explosive chain reaction.

Men from Ministries

In autumn 1939, 'Tizard's Committee' was Britain's best group for dealing with secret super-weapons. 'Tizard' was Sir Henry (Plate 10), whose phenomenally quick mind, attention to detail and people skills enabled him to hold down two impossible jobs at the same time. Since 1933, he'd done 'an enormous amount of war work' in preparing the RAF for conflict, probably with Germany. After hours, he moonlighted as Rector (Principal) of Imperial College in London, often waking the college secretary at midnight and working until morning, with breaks to play billiards. Tizard had studied physics in Oxford and Berlin, then invented and test-flew aviation instruments during the Great War. When peace returned, he did valuable research on aircraft engines and fuel. Shortly after his fellowship of the Royal Society came through in 1927, he jumped ship and joined the civil service with the mission of ensuring that the armed forces made full use of British science. Later summed up as 'not a great physicist, although a very good one', he was highly regarded by academics, civil servants and military men.[10]

Tizard's first committee was a three-man outfit – or four, if you counted Harry Wimperis who had run the Air Ministry Laboratory in London and now acted as committee secretary.[11] Tizard was flanked by two other fellows of the Royal Society, one a veteran Nobel Prize winner and the other on that trajectory. Both had seen active service between 1914 and 1918, had invented instruments of war, and were 'wonderfully intelligent, charming and fun to be with'. Archibald (A.V.) Hill, the polymathic Professor of Biology at University College London, had won the Nobel Prize in Physiology (1922) for elucidating the energetics of muscle contraction; he also wrote the *Textbook of Anti-Aircraft Gunnery* (1925). Bilingual and witty in English and German, Hill co-founded the Academic Assistance Council with Rutherford and enjoyed mocking *Deutsche Physik* and Johannes Stark. To show gratitude for 'Hitler's gift' – the influx of Jewish scientists from Germany – a plastic Hitler doll with a rotating right arm for *Sieg Heil!* was stuck to his office wall with plasticine.[12] The third man was Patrick Blackett, the Professor of Physics at Birkbeck College who would soon take Otto Frisch under his wing. Blackett had come late to science. He'd been the top cadet at Dartmouth Naval College, and by his eighteenth birthday had invented a gunnery targeting device and been under fire in the Battle of Jutland. After the war, he visited Cambridge on an Admiralty scheme to rehabilitate young servicemen and wandered into the Cavendish 'to see what a scientific laboratory looks like'. Bowled over, he resigned from the navy and took a First in physics at Cambridge; after working and falling out with Rutherford, he moved to Birkbeck to study the cosmic rays which later won him the Nobel Prize in Physics.[13] When he joined Tizard's Committee, Blackett was collaborating with the Royal Aircraft Establishment in Farnborough to improve the RAF's standard bomb-aiming sight – which Harry Wimperis had invented during the Great War.

This 'Committee for the Scientific Survey of Air Defence' first met in Room 724 of the Air Ministry, at 11 a.m. on 28 January 1935.[14] It was uniquely placed to feed top-quality science into the Air Ministry. Scientists sat on other government committees but were 'kept on tap, not on top' by ministers, civil servants and military men. By contrast, the three-hour meetings of Tizard's Committee must have been a blast: the air sizzling as three brilliant, inquisitive and entertaining men who revelled in each other's company tossed around ideas and jousted among themselves.

Their biggest agenda item stemmed from a crazy question which Wimperis asked the Committee in early 1935: could Germany have perfected a 'death-ray' capable of knocking aircraft out of the sky? Robert Watson-Watt, head of Radio Research at the National Physical Laboratory, said 'No' – but then found himself thinking about rays bouncing off aircraft. His secret memo, 'Detection and Location of Aircraft by Radio Methods', marked the conception of radar. Pushed along by Tizard's Committee and with John Cockcroft and the Cavendish Laboratory acting as 'nursemaids', radar made astonishing progress. Within months, it could detect enemy aircraft hundreds of miles away, and a defensive radar 'wall' was being planned to protect the English coastline against aerial attack.[15]

There was an unhappy and ominous episode in mid-1936 when Winston Churchill – as yet a manipulative and ambitious Member of

Parliament – levered a personal friend into Tizard's Committee. Frederick Lindemann, the Professor of Physics at Oxford, was German in ancestry and accent but otherwise as English as anyone. He and Tizard had first met in pre-war Germany, and their relationship had terminally soured. Lindemann was an 'arrogant, rude, objectionable' boor who wrecked the committee's cheerful productivity with his 'monomaniacal tension'. Hill and Blackett resigned temporarily in order to force Lindemann out, leaving him to nurse a grudge and plot revenge on Tizard.[16] Like a bad penny, Lindemann would keep coming back – and so would Churchill.

Radar had pushed almost everything else off Tizard's table, but one unrelated project still hung by a thread. Hans Halban's paper in *Nature* on 22 April 1939, suggesting that the fission of uranium spewed out enough neutrons to trigger an explosive chain reaction, had seized the attention of two professors of physics – George Thomson at Imperial College and Lawrence Bragg at the Cavendish in Cambridge. Both decided that Germany was unlikely to turn the chain reaction into a bomb, but the potential danger was so huge that preventative action was urgently needed.[17] Germany must be starved of uranium. After invading Bohemia, the Nazis had seized the pitchblende mines at St Joachimsthal, but making bombs would need much more, probably from the world's biggest uranium mines at Shinkolobwe in the Belgian Congo which were owned by the Union Minière Company of Brussels.

Thomson presented the problem to Tizard, who also saw danger – even though he put the odds against 'a successful military application' of fission at 100,000 to 1. He instructed the Treasury and Foreign Office to close possible loopholes that could channel uranium to Germany and invited Edgar Sengier, the President of Union Minière, to meet him in London. Sengier knew all about Frédéric Joliot's group and their patents for the uranium power generator and bomb. Union Minière had recently shipped 100 tons of uranium oxide to the United States, leaving only a few tons in Belgium; Sengier promised to alert Tizard immediately if the Germans requested any.[18]

Although highly sceptical that Germany could develop an atomic bomb, Tizard set up two uranium research projects in July 1939.

George Thomson would study chain reactions in uranium oxide, the material used by Joliot's team, while Mark Oliphant would investigate methods for extracting pure uranium on the assumption that this would be more fissionable. Tizard started them off with 2 tons of uranium oxide, one bought from Union Minière and the other from Brandhurst and Company, England's only suppliers.[19] Ideally, Tizard would also have brought in A.M. Tyndall, Professor of Physics at Bristol, but he wasn't aware that Tyndall had any interest in uranium. In fact, Tyndall had followed up Halban's paper in great detail and concluded that a uranium bomb would be a big brute – a ton or more – and extremely difficult, but not impossible, to build. He'd devised a way to keep the bomb safe until detonation: divide the uranium charge into two halves, each smaller than the critical mass, then smash them together so that the critical mass was exceeded as the chain reaction ignited. In early May 1939, Tyndall put all this into a secret memo on the uranium bomb for the Chemical Defence Committee on which he sat. Like so many contributions from scientific advisers to His Majesty's Government, Tyndall's memo disappeared into the corridors of Whitehall without trace, response or action.[20]

By autumn 1939, Tizard's men had achieved great things. Radar had grown into the Chain Home network of twenty-nine interlinked stations that stretched down the east and south coasts of England and could pass integrated information about incoming aircraft to airborne and ground-based defences.[21] Tizard's group expanded into the Committee for the Scientific Survey of Air Warfare, under the same chairman.

When war was declared, Chain Home was switched on in anger and new military research priorities swept in. George Thomson was diverted to work on magnetic mines, while Mark Oliphant was fully absorbed by the extraordinary fruits of his team's secret radar research. John Randall and Harry Boot had invented the 'cavity magnetron', a fist-sized electronic radar valve which blasted out microwave radiation of unprecedented power and resolution and might eventually be squeezed inside an aeroplane. Oliphant threw everything – staff, labs, workshops – at the magnetron and other radar projects. Research on fission in Birmingham virtually stopped when Philip Moon, Oliphant's second-in-command, was sent to Imperial to help with Thomson's experiments.[22] This left the two aliens, Otto Frisch and Rudi Peierls, out in the cold. It didn't matter what they did, as long as it wasn't radar.

In October 1939, Tizard's new committee consulted James Chadwick about the feasibility of a German atomic bomb. The Chadwick family had recently returned from their Swedish summer holiday via Amsterdam, reaching Liverpool three weeks after the declaration of war. Chadwick thought long and hard about weaponising uranium and didn't write back until December. He believed that a bomb containing between 1 and 40 tons of pure uranium was feasible and could release incalculable amounts of energy, perhaps as much as the gigantic meteorite which hit Siberia in 1908 and flattened 800 square miles of forest. More information was urgently needed, so he and his Polish refugee research assistant, Joseph Rotblat, would continue experimenting with uranium oxide.[23]

Chadwick was renowned for his impassivity and disdain of hyperbole, so his uncharacteristically dramatic warning was alarming. Moreover, Hitler had stirred up paranoia with a triumphal speech in Danzig on 13 October 1939 in which he reportedly boasted about '*eine neue Waffe gegen die keine Verteidigung gibt*' – 'a new weapon against which there is no protection'. Speculation about this 'super-weapon' ranged from a death-ray to crop-destroying insects, but the prime candidate was a uranium 'super-bomb' which, according to rumour, weighed only half a kilo. It soon turned out that the super-bomb had been conceived by sloppy editing. There was no '*neue Waffe*'; Hitler had only threatened to drop five bombs on enemies for every one that fell on Germany.[24] The Nazis' atomic bomb slipped off the government's list of things to worry about, and complacency returned to roam the corridors of power. Towards the end of 1939, Lord Hankey, a minister in the War Cabinet, dismissed the Nazi atomic bomb with 'I gather that we may sleep fairly comfortably in our beds'.[25]

Tizard also took his eye off the non-existent atomic bomb, preoccupied by the threat of the Luftwaffe dropping huge conventional bombs such as the 1,700-kg horror nicknamed 'Satan'. And when

Churchill asked his friend and personal scientific adviser about the chances of Germany making an atomic bomb, Frederick Lindemann told him that it was practically impossible to make such a weapon[26] – and infected Churchill with lasting scepticism that an atomic bomb could ever see the light of day.

As yet, there was nothing to disturb Lord Hankey's sleep. Anyone watching Germany closely would not have detected signs of work on a uranium bomb. The Nazis had stopped all uranium exports, but hadn't ordered any from the Union Minière. While papers and patents streamed out from America, Britain and France, German researchers showed little visible interest in exploiting the chain reaction. Otto Hahn was pursuing unexciting work on uranium and thorium, while Werner Heisenberg and Carl von Weizsäcker hadn't published anything in 1939. But what was really going on? American, French and British scientists had ignored Szilard's plea not to publish about fission, but the Nazis wouldn't allow that to happen. Any work on a German atomic bomb would be invisible to the outside world.

Across the Pond

At Columbia University, Enrico Fermi and Leo Szilard had intensified their efforts to generate electricity from fission. Their working relationship wasn't always smooth. Fermi was physically and mentally athletic, a classic workaholic but also a team player, husband and father. Szilard, 'a brilliant, paradoxical and lonely man' still living out of two suitcases in a hotel room, was also phenomenally productive but favoured three hours of cogitation in the bath. Fermi was often irritated by Szilard's inertia, but in a charitable moment admitted that he had 'an astounding amount of ideas' and the Fermi–Szilard show stayed on the road. (Another colleague suggested keeping Szilard in a deep freeze and defrosting him when a tough problem came along.)[27]

Their main challenge was how to 'moderate' neutrons, slowing them to the relative snail's pace that was best for fracturing the uranium nucleus. Water, which Joliot's team had used, was quickly rejected. Heavy water, in which the twice-heavier hydrogen isotope deuterium (D) replaces ordinary hydrogen (H), looked promising but was unobtainable in bulk: only tiny amounts were produced in the USA, while the world's sole industrial source – the Norsk Hydro electrolysis plant at Vemork – was out of reach and could soon become Nazi property if Germany invaded Norway. Fermi and Szilard decided on graphite – pencil lead – which occurs abundantly in the USA and performed well in pilot experiments.[28]

The prototype nuclear reactor at Columbia was built from interleaved blocks of uranium oxide and graphite. As this would eventually operate like a heap of burning rubbish, Fermi christened it the 'pile'.[29] On paper, the pile looked more convincing than Joliot's 'boiler' – the hollow copper sphere containing wet uranium oxide – and was waiting for uranium, graphite and money to translate it into reality. Szilard had extracted 200 kg of uranium oxide from the Canadian Eldorado Company, which mined uranium (as a source of radium) in addition to gold. Tons more would be needed for their prototype pile but so far their appeals for funding had fallen on deaf ears.[30]

Operating the pile would be like walking a precarious tightrope, with failure on one side and catastrophe on the other. The chain reaction could fizzle out if the secondary neutrons weren't slowed to the right speed to crack new uranium nuclei; or it could rage out of control and turn the pile into a radioactive inferno. The pile therefore had to be moderated by graphite, to slow neutrons to the optimal speed for fission, and simultaneously regulated by an absorber that could instantly mop up neutrons and kill an overactive chain reaction. For the latter, Fermi and Szilard chose the light metal cadmium which stops neutrons dead, like bullets hitting the sandbank behind targets on a rifle range. Szilard's work confronted him with a great personal dilemma. His experiment with Walter Zinn had yielded important and timely information: the average fission produced 2.3 secondary neutrons, theoretically enough to trigger a self-perpetuating chain reaction. This was exactly the kind of potentially dangerous research whose publication he had tried to suppress. Should he now lead by example? If he grappled with his conscience, it wasn't for long. 'Emission of Neutrons by Uranium' by L. Szilard and W.H. Zinn came out in *Physical Review* on 1 October 1939.[31] Days later, the paper would be on the desks of nuclear physicists in

Germany. They weren't blind or stupid. If their counterparts in America, France and Britain could extrapolate these findings into a nuclear pile or an atomic bomb, then why not the Germans?

RSVP

By mid-October – over a month since the war had begun in Europe – Roosevelt still hadn't responded to Einstein's letter. Szilard had dithered for two weeks after Einstein signed on 2 August before mailing the letter to Alexander Sachs for hand delivery to the President. And eleven weeks later, Szilard was becoming agitated.

The problem lay with Sachs, a respected economist who had helped Roosevelt to devise the New Deal which rescued America from the Depression. Unfortunately, Sachs was scientifically illiterate and had turned into a self-important windbag. Einstein's letter was written clearly for the intelligent non-scientist, and Roosevelt – ex-Harvard, with a legendary grasp of complex new information – couldn't have missed the significance and urgency of the plea from one of the world's finest minds. Sachs could simply have handed the letter to the President and said, 'Einstein thinks this is crucially important. Please read it' – but instead cast himself as 'the right person ... to make the relevant elaborate scientific material intelligible to Mr Roosevelt'. Moreover, believing the President to be 'punch drunk with printer's ink', Sachs would feed him the information 'by way of the ear and not as soft mascara on the eye'. Sachs therefore spun Einstein's 500-word letter into a monologue that would require 'a long stretch' to recite to Roosevelt.[32] This took over two weeks, just in time for Hitler to invade Poland.

Roosevelt couldn't see Sachs until late evening on 11 October 1939, and was dead beat. Sachs lost him soon after launching into his spiel but persuaded him to finish the discussion over breakfast. That night, Sachs 'didn't sleep a wink', agonising over what to say and meditating in a nearby park. At breakfast, he trotted out an anecdote: by turning away an American inventor who'd designed a steam-powered warship, Napoleon doomed himself to defeat by the British navy. Roosevelt asked if Einstein wanted 'to stop the Nazis from blowing us all up'. For once succinct, Sachs replied, 'Precisely.' Roosevelt summoned an aide, gave him the letter and told him that 'this requires action', then called up a bottle of Napoleonic-era brandy from the White House cellars for a toast.[33]

Action followed swiftly and looked promising. A new Uranium Advisory Committee to inform the President would meet on 21 October 1939, to hear evidence from Einstein, Leo Szilard, Eugene Wigner and Edward Teller. Curiously, though, the committee would be 'coordinated' by Sachs, who knew nothing about science – and chaired by Lyman Briggs, the sixty-five-year-old Director of the National Bureau of Standards and 'a real nice Southern gentleman' who was well grounded in soil science but completely ignorant about nuclear physics.[34]

8

LIAISONS DANGEREUSES
October 1939 – June 1940

Conspiracy theorists might have wondered whether Alexander Sachs was an enemy agent tasked with killing off the American atomic bomb. He'd almost trashed Einstein's letter to Roosevelt; now he appointed a three-man committee that was entirely incompetent to judge the potential of nuclear fission.

The Uranium Advisory Committee first met on 21 October 1939, nine days after Sachs sipped Napoleonic brandy with Roosevelt. It might have gone better if Einstein had accepted the invitation to attend, or if any of the committee members had known anything about nuclear physics. Lyman Briggs, ex-soil scientist and chairman, might have heard of uranium but didn't understand the project and wasn't excited by it. The other two were senior but scientifically ignorant ordnance officers from the army and navy.[1] The committee considered a briefing document by Leo Szilard and then grilled him, Eugene Wigner and Edward Teller. The military men didn't hide their contempt for Szilard's seemingly pathetic fear of a Nazi atomic bomb and the absurd notion of a super-weapon too big to be dropped from an aircraft. The army man informed them that 'it usually takes two wars to develop a new weapon' and that 'morale, not new arms, brings victory' – and then erupted in fury when Wigner suggested that this was a good argument for cutting the army's budget.[2]

Briggs sent Roosevelt a full report on 1 November. Seven of the committee's eight recommendations concerned a fission-driven power generator, perhaps to propel submarines; Fermi and Szilard should be funded to build a prototype pile. Their eighth recommendation was not to bother with an American atomic bomb because it was too speculative.[3] Having demolished the rationale for Einstein's letter to Roosevelt, the committee tossed Szilard $6,000 to buy graphite and uranium for the pile at Columbia – barely a fifth of what was needed. Despite the presidential demand for 'action', the committee conducted its business 'very informally' and at a tempo that made *mañana* look hasty. It went into hibernation through the winter of 1939–40 and scheduled its second meeting for late April 1940 – five months after its first, seven months into the war in Europe and almost nine months after Einstein signed the letter to Roosevelt.[4]

News Blackout

In February 1940, Szilard tried again to stop American colleagues publishing nuclear intelligence in plain sight of the enemy. He'd written a theoretical paper that predicted a self-perpetuating chain reaction in a close-packed lattice of uranium spheres embedded in graphite and could have been an embryonic nuclear power plant. Szilard asked the editor of *Physical Review* to record the date of submission and then lock the manuscript away until it was safe to be common knowledge. He also persuaded Herbert Anderson, a PhD student with Fermi, to hold back another sensitive paper – a tough decision for the young researcher, as the proofs were already on his desk.[5]

Fermi still refused to agree to Szilard's 'absurd' embargo, as did everyone else who mattered. In a long review about nuclear fission, Louis Turner at Princeton discussed the optimal conditions for releasing unimaginable amounts of energy.[6] Next came a paper which made George Placzek settle his bet with John Wheeler with a Money Telegram for 1 cent and the message 'CONGRATULATIONS!'. The Bohr–Wheeler prediction that only U-235 underwent fission had been confirmed in an experiment that Rutherford would have approved of, because statisticians weren't needed to interpret the results. Albert Nier in Minneapolis had used his mass spectroscope to deposit separate dots of U-235 and U-238 on platinum foil, which he mailed to John Dunning at Columbia. When Dunning's massive cyclotron bombarded each dot individually with slow neutrons, U-238 produced occasional clicks on the loudspeaker that registered the high-energy bursts of fission, whereas U-235 reacted 'like a machine gun'.[7] Precise measurements showed that U-235 was 10,000 times more fissionable than U-238, and that Otto Frisch and Lise Meitner had done their sums correctly while sitting in the snow at Kungälv fifteen months earlier. Smashing a single atom of U-235 released a staggering 200 MeV of energy – several million times more than a molecule of dynamite.

Szilard was shaken to the core by what followed. In late April 1940, a lively debate about U-235 at the American Physics Society meeting in Washington was featured in *Physical Review*, the highest-visibility physics journal in the world. *Science News Letter* reported that 'exploding uranium atoms may set off others in a chain', while the front page of the *New York Times* highlighted 'the probability of some scientist blowing up a sizeable portion of the Earth with a tiny bit of uranium.'[8] Szilard pleaded with Lyman Briggs to block all American publications about fission, only to be told (weeks later) that scientists, not government, should censor scientific publications.[9] Eventually, he persuaded a big name to withhold a red-flag paper. Louis Turner had deduced that firing neutrons at uranium could create a new, heavier element from U-238 and that this hypothetical transuranic element would be fissionable, like U-235. Turner asked Szilard's advice about the draft manuscript. Szilard was scared; if correct, this was dynamite a million times over, because it could be a shortcut to making a fission bomb. To Szilard's relief, Turner did as he suggested. The editor of *Physical Review* recorded receipt of the paper and mothballed it in his safe until the world was a better place.[10]

Within days, however, the same editor waved through an even more dangerous paper. Edwin McMillan and Philip Abelson had created the first genuinely transuranic element, 'element 93', by irradiating uranium with neutrons. The non-fissionable U-238 could occasionally capture a neutron and hang on to it, increasing its atomic weight to 239 and its atomic number to 93. Element 93 had chemical

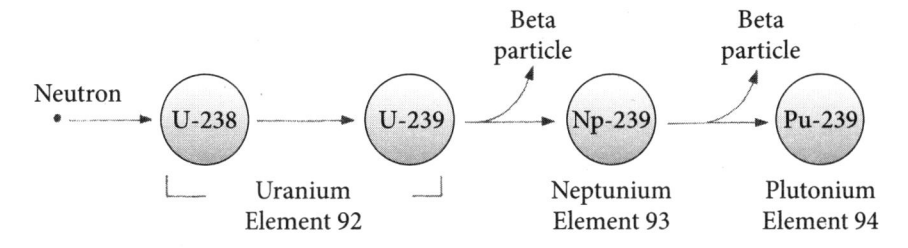

Neutron → U-238 → U-239 →(Beta particle) Np-239 →(Beta particle) Pu-239

Uranium Element 92 — Neptunium Element 93 — Plutonium Element 94

6. Creation of plutonium by irradiating uranium (U-238) with neutrons.

properties resembling uranium and a half-life of 23 days – and decayed into a daughter product with the same atomic weight (239) but an atomic number of 94 (Figure 6). 'Element 94', the second man-made transuranic element, was barely detectable but made McMillan and Abelson speculate about its behaviour.[11] They didn't know that their element 94 was the by-product already predicted by Turner – fissionable and potentially able to make an atomic bomb.

The discovery of the first authentic transuranic elements whipped up a storm of excitement and made some people sense danger. The editor of *Physical Review* received a sharp letter of rebuke from James Chadwick in Liverpool: as the laws of nuclear physics predicted that element 94 would be fissionable, the journal had openly published information that could help an aggressor to develop an atomic bomb. Ernest Lawrence at Berkeley made the same point and warned that 'we are now in many respects on a war footing'. Finally, Szilard's closest critic came on board. Enrico Fermi held back a pivotal paper showing that pure graphite slowed neutrons without absorbing them, and so would support the self-sustaining chain reaction predicted in Szilard's unpublished paper.[12]

The tide now began to turn. Gregory Breit, the Professor of Physics at Wisconsin, knew Szilard well and shared his anxieties. Breit proposed setting up specialised sub-committees to censor material submitted to American science journals. Innocuous papers would be published openly. Risky ones would be held back until it was judged safe to release them, with the original date of submission to ensure that discoveries were properly attributed and 'a suitable acknowledge-ment of the public spirit of the authors'. In the meantime, embargoed papers would be circulated secretly to trusted researchers in the United States. Breit was the antithesis of Szilard – a perfectly balanced, all-American pillar of the scientific community – and his proposals were swiftly adopted and enacted.[13] But was it too late? While Americans had continued churning out papers on fission, these had slowed to a trickle in Europe – as if a publication embargo had been imposed in France and England. And in Germany.

Einstein was again invited but stayed away from the second meeting of Lyman Briggs's Uranium Advisory Committee on 27 April 1940. The Columbia team delivered an update on the pile. Even though Wigner felt that they were 'swimming in syrup' and all were frustrated by the 'snail's pace' of the committee, Briggs reported to Roosevelt that the work was going well and that fission research must be 'pushed more vigorously'. For 'fission research', read 'pile'. The committee didn't even consider an atomic bomb, despite the mounting alarm over U-235 and 'blowing up a sizeable portion of the Earth' – and reports from Berlin that a building at the KWI for Chemistry had been turned over to 'uranium research'.[14]

Roosevelt had decided to replace the Uranium Advisory Committee with a scientifically stronger and demilitarised 'Uranium Committee' which would report to the recently formed National Defense Research Council (NDRC). In the meantime, Briggs was instructed to convene an expert group of six top physicists, including Harold Urey, the discoverer of deuterium, and Gregory Breit. Their meeting on 13 June 1940 was tightly focused and productive, and recommended that the Columbia team should build a troubleshooting pile around one-fifth of the size estimated to sustain a chain reaction. Research must begin urgently on how to concentrate the all-important U-235 from the 99.3 per cent of uranium that wasn't fissionable, because this could be a more efficient energy generator. Total bill: $140,000.

The new Uranium Committee consisted solely of hardcore, American-born scientists. Szilard, Wigner, Teller and Einstein were excluded by a government edict which barred anyone of foreign birth from high-security operations but, by presidential decree, the gentlemanly but ineffective Lyman Briggs remained as chairman.[15] As before, the focus was exclusively on generating energy through fission; the atomic bomb didn't figure at all.[16] Like the military heavyweights in Briggs's original committee, America's top nuclear physicists believed that the bomb was a fanciful waste of time and money.

Paris in the Springtime

In February 1940, the French knew that the *drôle de guerre* – Phoney War – couldn't last much longer. The Germans had swept into Poland five months earlier and must soon swing their attention towards the Low Countries and France. Could the French army, bolstered by the 390,000-strong British Expeditionary Force, beat them back?

At the Radium Institute in Paris, Irène Curie continued investigating the decay products of uranium while answering other calls – two children and a stint as France's Under-Secretary for Scientific Research. Irène had inherited her mother's 'devastating frankness' but not her self-deprecation; unlike Marie's eyes, hers were hard and impatient.[17] Her marriage to 'the skyrocket' Frédéric Joliot had lasted longer than her mother first predicted, although the couple grew apart after he moved across town in 1937 to the new Chair of Physics at the Collège de France. Armed with a big cyclotron and a cloud chamber for visualising subatomic particles, Joliot's team focused on harnessing the energy released by fission. He still put his name on her papers, but she stayed off the stream of publications from the musketeer-like triumvirate of Joliot, Hans Halban and Lew Kowarski. When Frédéric told his wife that the delicate tracery of fission fragments inside the cloud chamber was 'the most beautiful phenomenon on Earth', she replied, 'Apart from childbirth.'[18] Perhaps she just wanted to bring him back into her own orbit.

Joliot, Halban and Kowarski were still telling the world about every step on their journey towards generating electricity from a self-perpetuating chain reaction inside their boiler, a hollow copper sphere filled with a slurry of uranium oxide and water. 'Enriching' uranium oxide to increase the U-235 concentration by 20 per cent, from its usual 0.7 per cent to 0.84 per cent, might do the trick.

So might a better 'moderator', which would slow neutrons without stopping some of them dead. Ordinary water, paraffin wax and graphite were all rejected. In early 1940, they turned to heavy water, knowing that it held promise: several years earlier, while working at Bohr's Institute, Hans Halban and Otto Frisch had found that deuterium braked neutrons without absorbing them.[19] Unfortunately, heavy water was in critically short supply. An expensive dribble could be bought from the USA, but only Norsk Hydro's electrolytic plant at Vemork in Norway could provide enough for an experimental boiler.

Joliot, Halban and Kowarski had finished writing a paper, 'On the Possibility of Producing an Unlimited Chain Reaction in a Mass of Uranium', when they realised that the Nazis must not get hold of their secrets. Belatedly, they followed Szilard's advice and locked their manuscript away in a sealed envelope at the Academy of Sciences until it could be published safely in the Academy's Proceedings.[20] They had decided that heavy water was their best bet – only to discover that the Nazis had got there first.

On 14 February 1940, Joliot was summoned to a meeting with Raoul Dautry, Minister of Armaments. They already knew each other. Dautry had paid for 80 tons of uranium oxide from Union Minière and pulled strings to prevent Joliot's researchers from being called up. Waiting in Dautry's office was a tall, nondescript man in his early forties, balding, bland-faced and heavily bespectacled. He was introduced as Jacques Allier, senior manager at the Banque de Paris and in charge of its Norwegian operation (Plate 11). Norsk Hydro had received 75 per cent of its start-up funding from the bank and was still its biggest customer in Norway. Most of the remaining money had come from the huge German chemical conglomerate IG Farbenindustrie.[21]

Dautry and Allier brought ominous news. Previously, German researchers had ordered a few litres of heavy water each year, but now IG Farben wanted to buy Norsk Hydro's total stock of 185 litres, followed by 100 litres every month. Norway hoped to remain neutral but clearly featured on Germany's hit-list for invasion, and Oslo was already infiltrated by Abwehr military-intelligence agents. It had been rumoured in France that Norsk Hydro was sympathetic to the

Nazis, but its general director, Axel Aubert, proved otherwise. Worried about the 'devilry' that the Germans might work with all that heavy water, Aubert told his staff to stall IG Farben and alerted the French Ministry of Armaments through the Banque de Paris.[22]

Joliot was quizzed about the military implications of fission and heavy water, and it was decided to send Allier to Norway on a secret mission either to buy Norsk Hydro's entire stock of heavy water or to ruin it by contaminating it with neutron-killing pollutants such as boron. Allier travelled to Oslo by train and plane, arriving on 2 March 1940. He took two hefty money orders to buy the existing stock and all the factory's output for the next few years, but they weren't needed. Aubert loathed the Nazis and had already decided to give all his reserves to the French. In return, he asked Allier to quash the rumours of Nazi collaboration and to deliver a personal message to Édouard Daladier, the French Prime Minister. 'If France loses the war,' he said, 'I will be shot for what I have done today. Nevertheless, I am proud to take that risk.'[23]

Allier returned to his hotel and booked himself on a flight to Amsterdam a week later. Meanwhile in Vemork, 90 miles to the west in the mountains of Telemark, the heavy water was decanted into twenty-six canisters made from pure aluminium and boron-free solder, each holding 5–7 litres. The canisters were transported secretly to Oslo, one consignment being driven by night along icy roads by the manager of the ammonia factory in his wood-burning car. Three of Allier's colleagues, who had flown in separately, hid the canisters in a house that the French Embassy had declared 'safe' but was right beside an office, brightly lit day and night, which had been taken over by the Abwehr and was crawling with German agents.[24]

The heavy water was flown out of Oslo airport in two batches. The first thirteen canisters left with Allier and one of his companions on the morning of 10 March 1940. Allier's two-week visit to Oslo while Germany was preparing to invade Norway had aroused suspicion, and Abwehr agents at the airport were watching all arrivals and departures. The Amsterdam flight was halfway to its destination when a grey Luftwaffe fighter materialised alongside and rocked its wings in the signal that means, 'You have been intercepted. Follow me.' The pilot had no choice but to land at Hamburg, where the passengers were ordered off the plane and watched as it was searched from end to end.[25]

At that moment, Jacques Allier, his companion and the thirteen canisters of heavy water were over the North Sea, 400 miles to the north-west and on course for Scotland. Allier had bought seats on a plane that left for Perth around the same time as the Amsterdam flight, and had the canisters transferred across at the last moment. The Germans must have smelt a rat, as the pilot of the Perth flight soon spotted that it was being followed. Allier told him to lose the tail, which he did by climbing fast into cloud and staying so high that Allier passed out because he hadn't been given an oxygen mask. By the time they dropped down towards the Scottish coast, they were alone in the sky.[26]

The next day, the Abwehr agents at Oslo airport were off-guard, and Allier's two remaining colleagues flew to Scotland with the second batch of heavy water. One of them carried a gun, but the journey south was anticlimactic – with twenty-six canisters lined up beside their beds in a hotel in Edinburgh, and then stacked high in luggage racks on the trains to King's Cross and Dover, the ferry to Calais and finally the train to the Gare du Nord in Paris. On 16 March 1940, Frédéric Joliot signed the receipt for the 'numbered drums of deuterium protoxide, received from M. Allier', and 99 per cent of the world's stock of heavy water was locked away in a bomb-proof shelter in the cellars of the Collège de France.[27]

The heavy water had travelled over 1,400 miles since dripping out of the electrolytic cells in the Norsk Hydro ammonia factory. The mission left many questions hanging, notably: 'Who was Jacques Allier?' His colleagues at the Banque de Paris knew him as a clever but reserved law graduate of the elite École des Sciences Politiques in Paris; a devoted husband and father of three; and a deeply religious Lutheran Protestant who modelled his life on his friend Albert Schweitzer. He certainly wasn't Monsieur Freiss, booked on the Amsterdam flight on 10 March, because that passport (using his mother's maiden name) was false. The Germans soon discovered that all the heavy water had disappeared from Vemork, but the Abwehr took much longer to realise that they'd been outwitted by one of their own kind. As soon as war was declared, Jacques Allier had signed up with the Explosives Division in the Ministry of Armaments and with the Deuxième Bureau, the French military intelligence service.[28]

A Nightmare Comes True

In late March 1940, George Thomson began preparing a pessimistic report for Sir Henry Tizard's Air Warfare Committee. Working at Imperial College in London, he and Philip Moon had failed to ignite a chain reaction in uranium oxide. Water blocked too many neutrons to be a good moderator, and so did paraffin oil (they used glass nightlights for children's bedrooms, embedded in the uranium). Thomson and Moon reached the same conclusions as Joliot's group: heavy water or graphite might do the job, and a chain reaction might take off if the U-235 concentration were increased by 20 per cent. But unlike the Parisians, they stopped chasing the elusive self-perpetuating chain reaction.[29]

In Birmingham, the other uranium project launched by Tizard's Committee had also drawn a blank. Mark Oliphant hadn't found time to purify uranium because John Randall's cavity magnetron was now blasting out microwave radiation powerful enough to light a cigarette, switch on an unplugged fluorescent tube, and spot a small boat at sea several miles away.[30]

In Liverpool, James Chadwick was getting into his stride, aided by his newly operational 37-inch cyclotron and his Polish refugee assistant, Joseph Rotblat. Chadwick judged Rotblat 'one of the best I have seen' and saved him from penury when his scholarship ran out by magicking up a lectureship, just as Oliphant had done for Frisch. Rotblat's English was rudimentary on arrival in Liverpool but leapt ahead while he prepared physics lecture notes for the coming autumn term. Having worked on nuclear physics in Warsaw, he covered uranium fission in great detail; Chadwick thought this an acceptable security risk, as leaving it out would look more suspicious.[31]

One day after work, Rotblat cornered Chadwick to share some thoughts about a uranium fission bomb. Fission was most easily triggered by slow neutrons – hence all the effort to find the best moderator to brake the fast neutrons spat out by fission – but Rotblat was convinced that slow neutrons could never cause an *explosive* chain reaction. Their speed of 20 miles per second was agonisingly slow compared with events in the heart of an explosion. A chain reaction driven by slow neutrons would heat and vaporise the surrounding uranium before it could spread through the whole mass. The bomb would blow itself to pieces before it had the chance to 'go nuclear', sacrificing a tiny fraction of its critical mass in a bang that wouldn't be much bigger than with conventional explosives. Everyone was rightly fixated on slow neutrons for slow-burn fission to generate electricity, but would have to refocus on fast neutrons if they wanted to make a bomb.[32]

The young Pole's English was still hesitant. Chadwick listened from behind his legendary mask of inscrutability and responded with a grunt that, even to a native English speaker, could have meant anything. In fact, Rotblat had persuaded Chadwick that fast neutrons had been seriously neglected. Only one person – Merl Tuve at the Carnegie Institute in Washington – had attempted to measure the 'fission cross-section' for fast neutrons in uranium. Tuve's estimate was very small, implying that fast neutrons would find it extremely difficult to trigger fission. Prompted by Rotblat, Chadwick now suspected that Tuve's results were 'too damned low altogether' and might not apply to U-235. He therefore decided to find out for himself how fast neutrons behaved when they hit uranium nuclei – a task perfectly suited to his cyclotron.[33]

Chadwick and Rotblat set to work in January 1940. As the project would indicate whether an atomic bomb could be made from uranium, it should have been top secret. Chadwick preferred 'informal' confidentiality to the straitjacket of the Official Secrets Act, which would have thrown out Rotblat as an alien. Initially, they hid the purpose of the experiments from their co-workers. One was a Quaker and a conscientious objector who refused to do anything that could ultimately harm another human being, but when the project's military implications slipped out over tea one afternoon, he decided to stick with it.[34]

Chadwick's cyclotron proved to be a formidable weapon that demanded respect. To shield the researchers from the lethal hail of neutrons, movable barriers were built from wooden crates containing 100 tons of whale blubber from the Port Sunlight margarine factory across the Mersey; the crates occasionally ruptured, with malodorous results. Also life-saving was the switch which a brilliant but impetuous junior researcher forgot to check before touching a high-voltage terminal. Aged twenty-three, Harold Walke was the first victim of the atomic bomb programme. His death shook the lab and even its normally impassive boss. However, Chadwick – who had complained about other researchers 'dithering' after war was declared – soon had his troops back on course.[35]

In February 1940, Rudi Peierls's application for British citizenship was approved. Being German born, he was still excluded from Oliphant's radar project and occupied himself by theorising about the nucleus and fission.[36] Life had also improved for Otto Frisch, who had abandoned his hypothermic digs for Rudi and Genia Peierls's house in Edgbaston. As a 'Category C' alien, Frisch was subject to a dusk-to-dawn curfew and prohibited from owning a bicycle or large-scale maps or leaving Birmingham without permission. Internment seemed unlikely but, just in case, Genia made him buy some easy-wash shirts.[37]

Frisch had been mulling over thoughts stirred up while writing his review on nuclear fission. Like Chadwick, he now saw holes in the generally accepted picture. Bohr had decreed that the chain reaction in uranium was 'too brief, too small' to produce an explosion. But this was in natural uranium, in which U-238 nuclei outnumbered U-235 by 130 to 1. If, as Bohr had suggested, the rare U-235 nuclei were solely responsible for fission, then could all those non-participating U-238 nuclei interfere with the chain reaction? This made Frisch wonder how a mass of *pure* U-235 would react when bombarded with neutrons. He then went on to estimate how much pure U-235 would be needed for a runaway, explosive chain reaction. He began with Francis Perrin's calculations that had yielded a critical mass of 40 tons of uranium oxide, then blended in Rudi Peierls's

THE IMPOSSIBLE BOMB

modifications and finally his own best guesses for all the variables that he thought looked wrong. 'To my amazement,' he wrote later, 'it was very much smaller than I had expected.'[38]

Frisch's own review on fission for the *Annual Report of Progress in Chemistry* had just been published. In it, he confidently dismissed fears that fission could be exploited to make 'a super-bomb exceeding the action of ordinary bombs by a factor of 1 million or more', because of the 'strong arguments' that such a bomb was impossible.[39] But now Frisch had stumbled on a discovery that changed the game completely. According to his calculations, a super-bomb could be made from pure uranium-235. Its weight? Not 40 tons, or even 1 ton – just a couple of pounds.

9

MEMORANDUM OF UNDERSTANDING
February–June 1940

It wasn't a red-letter day; neither of them recorded the date at the time or could recall it afterwards. 'One day in February or March 1940' Otto Frisch asked Rudi Peierls a question that hadn't occurred to anybody before: what would neutrons do to a mass of pure uranium-235? They tackled the problem together and reached the same shocking conclusion as Frisch had on his own – that a pound or two of U-235 could make a devastatingly powerful fission bomb. They took their calculations to Mark Oliphant, who was immediately convinced and told them to put it all in writing for Sir Henry Tizard, chair of a committee which dealt with ideas that could change the course of the war.[1]

Peierls typed the lot, partly for security but also because all the secretaries were locked away with the radar project. The Frisch–Peierls memorandum 'On the Construction of a "Super Bomb" Based on a Nuclear Chain Reaction in Uranium' was short: a three-page lay summary explaining what the bomb would do and four pages of technical discussion (Appendix I).[2]

How could such a tiny bomb possibly work? In ordinary uranium, the fissionable U-235 was smothered by the vast excess of inert U-238 which would snuff out a chain reaction by stopping neutrons from reaching the next U-235 nucleus – like trying to light a fire using kindling wrapped in asbestos. Slow neutrons, the best at triggering a chain reaction in natural uranium, didn't travel fast enough to cause an explosion. But if Frisch and Peierls were correct, pure U-235 would be easily split by neutrons travelling at any speed, including the fast ones spewed out during fission. A critical mass of pure U-235 would ignite instantly in an explosive chain reaction; they estimated that 5 kg would detonate with the force of several thousand tons of dynamite. The heart of the explosion would be as hot as the inside of the sun. The blast would immediately 'destroy life in a wide area', such as the centre of a city, while the 'very powerful and dangerous radiations' emitted by fission products would kill 'any person entering the area for several days'.[3]

Feasibility? The super-bomb hinged on separating U-235 from the much more abundant and virtually identical U-238. This was currently impossible but 'not insuperable'; they suggested pumping vaporised uranium hexafluoride through a battery of 100,000 thermal diffusion tubes. Cost? Perhaps as much as a battleship but 'certainly not prohibitive'. Design? The U-235 would be in two pieces, each weighing less than the critical mass and therefore 'absolutely safe' if kept a few inches apart until 'the bomb is intended to go off'. Strategic importance? There was no effective protection against the explosion or its radioactive legacy, so the super-bomb would be 'practically irresistible'. Others would have to decide whether to use an indiscriminate weapon that 'would probably kill large numbers of civilians'.[4]

The memorandum was underpinned by a grave warning. The idea of a pure U-235 fission bomb could occur to 'other scientists' and it was 'quite conceivable' that the Germans were already developing their own weapon, perhaps guided by Klaus Clusius, the world's master separator of isotopes. Frisch and Peierls argued that 'the most effective reply would be a counter-threat with a similar bomb'. Therefore, Britain must begin work on its own atomic super-bomb 'as soon and as rapidly as possible' and in total secrecy.[5]

The Frisch–Peierls memorandum had terrifying plausibility. On 19 March 1940, Oliphant handed the document to Tizard, who immediately passed it to George Thomson for comment. Tizard had thought that the chances of developing an atomic bomb were 1 in 100,000, and he remained sceptical. He sought American wisdom via A.V. Hill, recently posted to Washington as the scientific envoy to the British Embassy. Without revealing anything about the Frisch–Peierls memorandum, Hill sounded out his American contacts in nuclear physics. They told him straight: a uranium bomb was a non-starter and 'a sheer waste of time' for hard-pressed researchers in war-locked England. However, George Thomson sided with Mark Oliphant, and their forceful endorsement of the memorandum swayed Tizard. He asked Thomson to assemble a secret sub-committee to consider the super-bomb.[6] But before the new sub-committee took shape, the Phoney War came to an abrupt end and the real one started.

Soft Targets

On 8 April 1940, Germany invaded Denmark and Norway. By 9 a.m. the next morning, Copenhagen was occupied and the Danish government had capitulated. After an initial flurry of bloodshed, the Germans seemed to be on best behaviour, with no further violence and many of the occupying troops speaking Danish.

Bohr's diary illustrated how fast events moved. On 8 April, he'd given a guest lecture in Oslo, followed by dinner with King Haakon and the night train home. En route, the Germans seized the port of Copenhagen, while King Haakon, the Norwegian royal family and members of the Cabinet boarded a train heading north out of Oslo. By lunchtime on 9 April, the streets outside Bohr's Institute were swarming with German troops, and the first were soon inside the building. That afternoon, German officials summoned Bohr to agree that Reich scientists could use his institute's facilities. This was a savage blow for Bohr, godfather to the rising stars of nuclear physics from across Europe and creator of a stateless haven where only science mattered, and where Meitner, Frisch, Peierls and Oliphant talked freely to Hahn, Heisenberg, Harteck and von Weizsäcker. The Nazis could simply have filled Bohr's labs with their own people, but deferred to his status and asked his permission. Bohr allowed German scientists in but refused to let them use the cyclotron. Being Bohr, he had his way – for now.[7]

Elsewhere in Bohr's Institute, a life-saving chemical reaction ran its course. Max von Laue had sent Bohr his solid-gold Nobel medal to prevent the Nazis from seizing it – thereby removing assets from the Reich, a crime which carried an automatic death sentence. The medal, with von Laue's name engraved on it with lethal clarity, was in mint condition when German troops entered Copenhagen. By the time they were combing the institute, only a metallurgist with sharp eyes and a suspicious mind would have looked twice at a glass bottle containing a bright orange liquid. Just in time, George Hevesy had concealed 200 grammes of pure gold by dissolving it in aqua regia, the alchemist's blend of concentrated nitric and hydrochloric acids.[8]

Also hiding in deep cover was Lise Meitner, in Copenhagen to visit Bohr's Institute and her old friends Niels and Margrethe. Protected by her Swedish passport but fearful that the occupation would turn nasty, Meitner quietly finished her business and slipped back to Stockholm at the end of April. The Bohrs had asked her to pass on a message to reassure John Cockcroft and another close English friend, an elderly lady who would otherwise be worried sick. Miss Ray (Maud to the children) had been the Bohrs' nanny twenty years earlier, when Niels worked with Rutherford in Manchester, and had since retired to Kent. Bohr wrote down her address, but something was lost in transmission when Meitner instructed the telegram clerk in Stockholm to send the cable to London.[9]

On the same morning that Denmark was occupied, Nazi paratroopers seized Oslo Airport. Soon after the Norwegian capital fell, a German military convoy headed west into the mountains of Telemark to take control of the Norsk Hydro plant at Rjukan. They met resistance from the Telemark Regiment, spurred on by a telegram from Paris: 'RJUKAN MUST BE DEFENDED BY ALL MEANS', signed 'ALLIER'. Outnumbered, the Norwegians clung on for a few days. When the Germans marched in, they didn't shoot Axel Aubert, general manager of Norsk Hydro, because they needed him. The theft of the heavy water was just a temporary setback. Before long, the electrolytic cells in the ammonia plant at Vemork were pushed harder than ever before, wringing out 100 litres of heavy water each month – the level that had set the alarm bells ringing for Aubert six weeks earlier.[10]

MEMORANDUM OF UNDERSTANDING

What's in a Name?

Sir George Thomson's sub-committee met for the first time on 10 April 1940 – two days after the German invasion of Denmark – in the Royal Society's Committee Room in Burlington House, Piccadilly. The group had no name, terms of reference or administrative support. Thomson (Plate 12) doubled as chair and secretary, the other members being John Cockcroft, Mark Oliphant and Philip Moon. They decided that the Frisch–Peierls super-bomb made sense scientifically and must be a top-secret national imperative. Work should begin immediately on purifying U-235. The group must be consolidated and given a name that wouldn't excite the enemy's curiosity. Perhaps surprisingly, they also agreed that the Secret Service would be asked to 'stop any leakage to Germany which might be escaping from Dr Frisch'.[11]

That last action was prompted by a tall, quiet Frenchman wearing heavy spectacles and an expression that gave nothing away. Jacques Allier had flown in from Paris with intelligence made all the more timely by the Nazi invasion of Denmark and Norway. He explained (with some poetic licence) that Joliot and his team were still ahead in the game; they had nearly perfected the fission boiler and were 'working on the production of an atomic bomb'. However, the Germans were stepping up their work on fission, as witnessed by their sudden thirst for heavy water and the urgent refitting of the KWI for Chemistry for research into uranium fission. Allier had rescued the heavy water from Vemork but predicted that the electrolytic plant which produced it would soon be overrun by the Nazis. Joliot had drawn up a list of German nuclear physicists who might help to develop a Nazi atomic bomb, notably Werner Heisenberg, Carl von Weizsäcker, Otto Hahn and Paul Harteck. French military intelligence was trying to keep them under surveillance; the British should do likewise and both Secret Services should share their intelligence. Joliot was concerned about Otto Frisch's close relationship with Lise Meitner, still in regular contact with Otto Hahn at the KWI for Chemistry. To prevent Frisch from inadvertently leaking secrets, Joliot wanted him kept in the dark and warned off communicating with his aunt.[12]

THE IMPOSSIBLE BOMB

After signing off his minutes – three handwritten sheets of foolscap – Thomson wrote in confidence to tell James Chadwick about the group set up 'to consider the possibility of constructing a uranium bomb by separating U-235'. He added, 'at first sight it seems a bit wild ... but not implausible when you come to look into it' and hoped that 'you will be able to join us'. Chadwick accepted by return.[13]

Thomson's anonymous sub-committee met for the second time on 24 April 1940. Tizard still believed that the super-bomb was 'not in the least likely to be of military importance in this war', but the sub-committee members were 'electrified'. Chadwick and Joseph Rotblat had begun experiments to try to see whether neutrons of any speed could trigger a runaway chain reaction in a small mass of pure U-235. The group considered how to purify U-235, without which the bomb could never exist. Gaseous uranium hexafluoride (hex) seemed the best bet, as various physical methods could – in theory – separate U-235 hex from the slightly heavier U-238 hex. Alternatives to the 100,000 Clusius tubes suggested by Frisch and Peierls should be explored, such as forcing hex across a membrane riddled with tiny pores which might allow the lighter isotope through more easily.[14]

Cockcroft said later that 'most of us hoped that the constants of nature might turn out so that the bomb was impossible', but the super-bomb was already turning into a series of technological challenges for which solutions could be found, and a super-weapon that would ultimately be used. Perhaps the bomb's power could be demonstrated forcibly enough to end the war by exploding one in an uninhabited place rather than over a city?[15]

Thomson's sub-committee came of age during May and June 1940. First, in a petty act of vengeance, it was cut loose from Tizard's Committee. On 10 May 1940 – the day when Germany invaded Belgium and the Netherlands – Winston Churchill replaced Neville Chamberlain as Prime Minister. With Churchill came Tizard's nemesis Frederick Lindemann, now the Prime Minister's personal scientific adviser – which meant that Tizard's fall was as inevitable as that of Belgium. The vindictive Lindemann torpedoed Tizard's Committee for the Scientific Survey of Air Warfare, leaving Thomson's sub-committee to sail on under its own steam.[16]

MEMORANDUM OF UNDERSTANDING

Next, the sub-committee acquired a name. This came from a perplexing telegram which Lise Meitner had sent from Stockholm to the Secretary of the Physical Society in London: 'MET NIELS AND MARGRETHE RECENTLY BOTH WELL BUT UNHAPPY ABOUT EVENTS TELL COCKCROFT AND MAUD RAY KENT'. The telegram was forwarded to John Cockcroft, who wondered whether 'MAUD RAY KENT' contained an encrypted message. Various people, including an amateur codebreaker, speculated about RAY, the letters U and D (the chemical symbols for uranium and deuterium), and the near-anagram RADYUM TAKEN. Nothing convincing emerged – but 'M.A.U.D.' (with full stops inserted to make it look 'more official') seemed a suitably banal name for a group that needed total obscurity.[17]

Sharp-eyed readers will have noticed that two names were missing from the membership of the M.A.U.D. Committee. Otto Frisch and Rudi Peierls were stuck in Birmingham, in the dark and out in the cold, still with no news about the fate of the memorandum which they'd conceived and written four months earlier.[18]

As nobody told him to stop, Otto Frisch continued laying foundations for the British super-bomb. Vaporised hex seemed the best starting material for purifying U-235, but he couldn't get hold of any. The only source in Britain was the huge Imperial Chemical Industries (ICI) – strictly out of bounds to aliens like Frisch. He was luckier with an ingenious idea to find out what neutrons did to U-235, which he left *in situ* in natural uranium and bombarded with neutrons so slow that they wouldn't affect U-238. The neutrons were generated by blasting beryllium with the radiation from radon, and this took him deep underground in the Derbyshire Peak District to a disused fluorspar mine that branched off the legendary Blue John Cavern. Frisch was guided 'over slippery ladders and through narrow, muddy passages' into a chamber containing 'a laboratory table with a lot of glassware on it'. This was the radium store of the Christie Cancer Hospital in Manchester, where the mother radium sample was regularly 'milked' of radon, just as Marie Curie had done to supply cancer hospitals in Paris during the Great War. Afterwards, Frisch lugged a lead suitcase containing a vial of fresh radon on the train back to Birmingham, and set up an experiment with home-made equipment in his disused lecture room. After a thirty-six-hour marathon of recording fission spikes, punctuated by an alarm clock and half-hour naps, his calculations revealed that he'd underestimated the critical mass of U-235, but not by much. The super-bomb wouldn't weigh a pound or two, but perhaps 20 or 30 pounds.[19]

In late April 1940 – after the second meeting of Thomson's sub-committee – an embarrassed but tight-lipped Mark Oliphant finally told them that a committee (unnamed, membership undisclosed) had considered their memorandum. 'The authorities' (unspecified) were 'grateful', but the 'work' (no details) would now be continued by 'others' (anonymous), because 'former or current enemy aliens' were excluded from research concerning national security. Peierls and Frisch were bemused by the 'ridiculous' impasse of keeping a secret from the two men who had originally disclosed it. Oliphant suggested that they write to the chairman (who couldn't be named). Peierls addressed a 'dignified' letter to 'Dear Sir', pointing out that they had thought long and hard about the super-bomb and might conceivably be of some use. While the chairman's reply was delayed by further entanglement with red tape, Oliphant consulted Frisch and Peierls off the record. In late summer, word finally came from on high that they were still barred from the main M.A.U.D. Committee but could join a 'technical sub-committee'. Peierls responded with a commendably British stiff upper lip: 'Very satisfactory.'[20]

Paris in the Fall

During the closing weeks of the Phoney War, the British Department of Scientific and Industrial Research took a gamble in appointing its new Liaison Officer (Paris). The job description was top secret: working with the French authorities to identify French scientific assets which must not be allowed to fall into Nazi hands and, if necessary, removing them to safety in Britain. Depending on viewpoint, the successful candidate was either ideally qualified or totally unsuitable. *Name:* Charles Howard, Jack to his friends. *Age:* 34. *Positive attributes:* first class in pharmacology, Edinburgh; research experience (poisons, explosives) at the Nuffield Institute for Medical Research, Oxford; a speech to the House of Lords on cancer treatment, March 1939; fluent French. *Negatives:* ill-disciplined and irreverent; military career terminated by unspecified misdemeanour; extensive lurid tattooing. *Other information:* the address to the House of Lords was his maiden speech as the 20th Earl of Suffolk and Berkshire.[21]

Jack Howard started work in Paris on 2 March 1940. He brought Beryl Marsden, a lively and unflappable secretary from the Nuffield Institute whose duties included typing and lighting Howard's cigarettes (when stressed, he smoked two at once, mounted in a special double-barrelled holder). Jack and Beryl were close but platonic friends: he was happily married, and her fiancé had recently died. They were offered the Ritz, but instead chose rooms owned by the Ministry of Armaments. After work, they joined the devil-may-care gaiety of a great city that was looking over its shoulder at Poland and waiting for the party to end. Howard didn't keep a low profile. He was loud and inexhaustible, wore a black fedora and enjoyed shooting corks out of champagne bottles. When Nazi sympathisers began following him, he bought two magnum revolvers with a double shoulder-holster and hired a battle-scarred veteran of the French Navy as his personal bodyguard. Howard christened the guns 'Oscar' and 'Genevieve'; anyone who met the bodyguard instantly appreciated why he was known as 'The Gorilla'.[22]

As the bloodstain of Nazi occupation spread towards the Low Countries and France, Howard and his French counterparts drew up the inventory of people and materials to be rescued in the event of a German invasion. By mid-May, with the German advance pushing the French Army and the British Expeditionary Force north towards a showdown at Dunkirk, Howard's shopping-list for urgent export to England took final shape. It included ballistics experts, toxicologists and Frédéric Joliot's three-man team at the Collège de France.[23]

Anticipating the fall of Paris, Raoul Dautry, Minister of Armaments, told Joliot's team to get out of town urgently with the heavy water and all portable equipment. Dautry saw no immediate need for Jack Howard's Doomsday plan to evacuate the group to England, so Joliot decided to move 200 miles south to Clermont-Ferrand. His beloved labs were gutted, unsalvageable documents were burned, and spare uranium oxide was buried at secret locations outside Paris and Toulouse.[24]

On 16 May 1940, Hans Halban headed south with his wife, baby daughter and the back of the family car crammed with their possessions and canisters of heavy water, codenamed 'Product Z'. Lab equipment and the rest of the heavy water followed by lorry. They reached Clermont-Ferrand at 5 a.m. the next morning and deposited the Product Z in the vault of the Banque de France. The manager had been briefed but became so agitated that the heavy water was moved out to a nearby women's prison, and then to the condemned cell in the prison at Riom, just north of Clermont-Ferrand. Meanwhile, Lew Kowarski arrived with a convoy carrying more equipment, having sent his wife and young daughter ahead by train. He and Halban set up an improvised laboratory in the city centre and began planning experiments.[25]

Like the captain of a sinking ship, Joliot was the last man out of Paris. He reached Clermont-Ferrand on 6 June with Irène, their two children and the lab's greatest riches: research records, radiochemicals and the gramme of radium which the Women of America had given Irène's mother in 1921. Their journey had been fraught. The fiasco of Dunkirk had ended favourably for the British but the Luftwaffe started bombing Paris on 4 June, when the Germans were only fifty miles outside the capital. Under a pall of black smoke from burning oil refineries, the exodus of Paris swelled into full flood and jammed all the roads. The Joliot-Curies had problems of their own: Irène had recently been diagnosed with tuberculosis and was waiting to go into a sanatorium at Clairvivre, near Bordeaux.[26]

On 10 June 1940, with the Germans just 20 miles from Paris, the French government fled to Tours and then on to Bordeaux. Having shed 2 million of its inhabitants, the capital was nearly deserted when the Nazis marched in on 14 June. In Clermont-Ferrand, Joliot's team carried on with their experiments until lunchtime on Sunday 16 June, when a car pulled up outside with a guardian angel at the wheel: Jacques Allier, bearing the urgent news from Minister Dautry that France was now a lost cause. They had to transport themselves and their personal and scientific possessions without delay to Bordeaux, where a boat would take them to safety – England initially and then, because it too would soon be invaded, on to America.[27]

Next morning, while the lorries were being loaded, Allier and Halban took a truck to collect Product Z from the prison at Riom. Carrying out the canisters provided fresh air and exercise for residents of the life-sentence block – watched sullenly by the prison governor, who had refused to cooperate until Allier drew his revolver. The Halban and Kowarski families left with the lorries on the 250-mile journey to Bordeaux. Arriving at midnight on 17 June, they found chaos at the docks, with thousands of people trying to leave amid bombs falling. Luckily, strings had been pulled and they were given a permit to board a British ship, scribbled on a page torn out of a discarded school jotter. And there they met the man who had pulled the strings: Jack Howard, Earl of Suffolk and Berkshire.[28]

Howard's last days in Paris would have been eventful even without the Germans closing in. He rounded up over thirty scientists, directed them to the docks at Bordeaux and promised safe passage to England on the destroyer which he'd requested through the British Embassy. While Paris emptied around them, he and The Gorilla scoured laboratories and factories, gathering up equipment, machine tools and blueprints. En route, they cleaned out the city's diamond merchants of several million pounds' worth of gems and industrial diamonds. Donors were given a receipt signed 'Suffolk'; those reluctant to hand over their goods were referred to Oscar and Genevieve and/or The Gorilla. On 12 June 1940, with just a few hours to spare, the human flood pouring out of the doomed capital was joined by a truck loaded with diamonds, industrial and scientific booty, Jack Howard, Beryl and The Gorilla. They reached Tours on the night of 14 June – in time to catch an air raid and the news that Paris had fallen – and finally Bordeaux, now crammed with half a million refugees, at 2 a.m. on Sunday, 16 June.[29]

There was no sign of any British rescue ship. Travel-stained and 'festooned with guns', Howard brazened his way into the headquarters of the refugee French government but failed to persuade Marshal Pétain (who would sign the armistice with Germany three days later) to let him have a French warship. Disguised as a French sailor, he bribed a local skipper to undertake the two-day voyage to England; soon after, that boat hit a mine and sank. A British destroyer sailed in, but only to evacuate the British Embassy; fortunately, the SS *Broompark*, a scruffy Glaswegian coal boat, had also been sent for Howard's mission. While air raids continued, Howard loaded his passengers and cargo, which now included aircraft engines, a lorry, anti-aircraft guns and crates of champagne donated by drunken sailors. After a neighbouring ship sank during an air raid, the *Broompark* was moved a couple of miles upstream and then, early in the morning of 19 June, steamed out into the Gironde River and headed for the Atlantic. On board were 110 souls, including thirty-two scientists and engineers with their families, a Belgian diamond executive, the Norwegian skipper and his crew, and Jack Howard and Beryl Marsden. The Gorilla had melted into the crowd at the docks to seek his next assignment, and the Joliot-Curies missed the boat completely.[30]

The crossing to England was hot and mostly calm, with seasickness (treated with champagne) and scares. Howard decided to lash the diamonds and heavy water to a makeshift wooden raft on the foredeck, so that they could be salvaged if the ship sank; the idea gained credibility when a nearby cargo boat took a direct hit and was blown out of the water. The raft could also have saved Howard, Beryl and several others; even counting the inner tubes stripped out of all the tyres on board, there weren't enough lifebelts to go around. The scares included Morse-code signals flashed at night which looked like U-boats preparing to attack and, on the second day, a low-flying aircraft off the Brittany coast that forced all the passengers into the sweltering, pitch-black death trap of the coal hold.[31]

The longest day, Friday, 21 June, dawned with the *Broompark* at anchor in Falmouth Bay, surrounded by the masts of sunken boats sticking out of the water. After inspection by the Falmouth Contraband Service, Howard went ashore to send a terse telegram (signed SUFFOLK) to the Ministry of Supply, followed by a long phone call detailing Beryl Marsden's list of passengers and cargo. The ministry sent a special train with an armed guard, which left Falmouth for Paddington at 11 p.m. that night. At 5.30 the next morning, the night porter at the ministry was woken up by an extraordinary vision: bearded, dead beat but forceful, clothes thick with coal dust, jacket open and sleeves rolled up to reveal profuse tattoos and two huge revolvers. The vision was debriefed by the staff member on call (a Parliamentary Under-Secretary named Harold Macmillan, who was destined for greater things) and went home to bed.[32]

While the Kowarskis and Halbans waited in London for orders from France to travel on to America, the heavy water enjoyed a heartier welcome. Some canisters were sent to the Cavendish Laboratory in Cambridge; the rest did time in the MI6 headquarters at Wormwood Scrubs Prison in West London before being conveyed to a vault under the library of Windsor Castle which also sheltered the Crown Jewels.[33]

Meanwhile, Frédéric Joliot was in Clairvivre with his family. He'd driven Irène to the sanatorium from Clermont-Ferrand to begin her tuberculosis treatment. His fellow musketeers – both Jewish, and desperate to leave – took a message to Bordeaux, explaining that Joliot was staying in France with his sick wife. After leaving her at the sanatorium, he was racked with remorse because he hadn't said goodbye to his two friends and scientific soulmates, whom he might never see again. He drove straight to Bordeaux, only to find an empty berth where the *Broompark* should have been. He was told that she'd been moved after a bomb destroyed the ship alongside, but nobody knew where. Joliot wandered fruitlessly around the docks for several hours. Finally, exhausted and sick at heart, he drove back to Clairvivre to await orders from his German masters.[34]

Same Coin, Two Sides

On 20 June 1940, while the *Broompark* was tracking round Brittany towards Falmouth, the M.A.U.D. Committee met for the first time at its new base in the Ministry of Aircraft Production. Sir George Thomson was in the chair, with a proper secretary in attendance. It was agreed that James Chadwick would take charge of experimental work on uranium. His contacts at the ICI works at Widnes would prepare metallic uranium and crystalline hex, while the Liverpool cyclotron would be used to measure uranium's 'constants of nature' and to try separating U-235.

During July, Patrick Blackett and Norman Haworth, Professor of Chemistry at Birmingham and Nobel Prize winner (1937), were brought on board. The M.A.U.D. Committee was now on a full war footing, with five Nobel laureates – Blackett, Chadwick, Cockcroft, Haworth and Thomson – and the beginnings of a research plan to build the U-235 super-bomb outlined in the Frisch–Peierls memorandum.[35]

Meanwhile, across the Atlantic, Lyman Briggs's Uranium Committee remained stuck in the doldrums. At Columbia University, Enrico Fermi and Leo Szilard were scaling up their experimental piles of graphite blocks and tins of uranium oxide with the aim of generating electricity for powering submarines and other applications. And the atomic bomb? A waste of time and effort, and nobody was working on it.

10

SEPARATION ANXIETIES
May–August 1940

In spring 1940, while waiting for bombs to start falling on Birmingham, Genia Peierls enrolled as a nursing assistant and Rudi as an auxiliary fireman, complete with uniform, helmet and axe. They became 'very friendly' with their lodger, Otto Frisch, and at weekends escaped with him by train to explore the countryside. Their last outing ended in a village where, although they'd booked, there was no room at the inn or even in the cells at the police station. Someone found them a room and they spent the night fully clothed on a single mattress covered in brown paper, with Rudi trying to sleep in the middle while Genia and Otto 'carried on an intellectual conversation over my head'. Next day, they listened to Churchill on a pub's radio, promising the British 'blood, sweat and tears'. It was 13 May 1940.[1]

After Dunkirk, Britain was swept by vengeful anger against foreigners, and 8,000 aliens were deported to detention camps in Newfoundland and Australia. The SS *Ettrick* sailed from Liverpool to Quebec on 3 July 1940, carrying 1,200 internees lumped together with Nazi prisoners of war as 'scum of the earth'. The internees included Max Perutz, a future Nobel Prize winner who was hauled out of the Cavendish Laboratory in Cambridge, and Klaus Fuchs, a brilliant young theoretical physicist whose admirers included Rudi Peierls.[2] They'd been dealt a lousy hand but were still in the game – unlike the 650 deportees who drowned when a U-boat torpedoed the *Arandora Star* off the Irish coast, twelve hours before the *Ettrick* left Liverpool.[3] Rudi and Genia Peierls spent three nail-biting days while the ship taking their two children to safety in Toronto waited at anchor in Liverpool Bay; seven-year-old Gaby had told them that they 'would be very sad if the radio reports that the ship evacuating little children to Canada has been torpedoed'.[4]

The blitz of Birmingham began on 9 August 1940. Apart from a 'sinking heart' while driving towards fires and bomb flashes, Peierls found firefighting 'undramatic'; a rare perk was hot chocolate served by two steel-helmeted girls in a Cadbury's van.[5] By day, Peierls continued exploring theoretical methods for extracting the 0.7 per cent of U-235 from native uranium. He worked alone because Otto Frisch had recently moved to Liverpool. Peierls had taken him up on the train to see if James Chadwick could provide hex to run through the Clusius separation tubes. Always inscrutable, Chadwick surveyed his visitors in silence for a good half-minute, 'turning his head from side to side like a bird', then abruptly asked Frisch how much hex he wanted.[6] Chadwick's lab instantly became the ideal place for Frisch. Unfortunately, enemy aliens were excluded from Liverpool because its 10 miles of docks were the gateway to the Atlantic and housed the Royal Navy's underground Atlantic Command Centre. Chadwick pulled strings, and the Chief Constable of Merseyside (a good friend of Chadwick's wife, Aileen) granted Frisch exceptional clearance to stay in the lab after the 10.30 p.m. curfew, and to own a bike so that he could cycle home in the blackout.[7]

Frisch's first months in Liverpool were punctuated by the Luftwaffe; the city's bombardment was second only to the London Blitz. He became used to hearing air-raid sirens wailing for real, sheltering in the basement and finding familiar buildings reduced to 'empty shells' overnight. In quiet moments, he practised the piano ('worse than the guns', according to a fellow-lodger). After one terrifying bomb that whistled for so long he thought his name must be on it, the landlady ran away. So did he, picking his way by the light of burning buildings to a friend's house on the outskirts of the city. Another near miss was the one-ton parachute bomb which blew out all the windows in the physics department but magically left standing the much-loved, red-brick Victoria Tower to watch over the university campus.[8]

Hex was the very devil to work with, and Frisch failed to make the Clusius tubes separate U-235. He got to know Joseph Rotblat, Chadwick's thirty-one-year-old assistant from Warsaw. Rotblat was 'kind and outgoing' and not at all bitter, although he epitomised the agony and resilience of European Jews more poignantly than Frisch. When Germany invaded Poland, Rotblat managed to escape to take up a scholarship with Chadwick but had to leave behind the girl he'd married as a student because she had acute appendicitis and was too ill to travel.[9] A year later, he was still waiting for news of her fate; it was only after the war that he learned that Tola Rotblat had been murdered in the Majdanek concentration camp in 1941. Through Rotblat, Frisch sampled Liverpool life outside the lab, once playing Chopin's *Grande Polonaise* 'badly' but passionately on a carious piano for a roomful of emotional Polish soldiers.[10]

Frisch's frustration in the lab peaked after he invented an ingenious instrument to pick out alpha-particles of specific energies. This proved conclusively that Clusius's thermal diffusion method – on which he and Peierls had pinned the feasibility of their super-bomb – achieved 'no measurable separation' of U-235.[11]

Holey Grail

Luckily, the ill-fated Clusius tubes weren't the only irons in the fire. Down in Birmingham, Rudi Peierls had assembled a team of whizz-kid mathematics students equipped with mechanical calculating machines (a rare luxury donated by His Majesty's Stationery Office). Each student was only told enough to help him crack an apparently isolated problem, but, being whizz-kids, some undoubtedly worked out that they were trying to solve the make-or-break conundrum of how to purify U-235. Each potential method was pulled apart to see if it could do the job. If so, how many enrichment steps would it take? Now imagine an industrial-scale separation plant based on that method. How big would it be? How expensive to build and maintain? How much power and how many people would be needed to run it? And how long would it take to make 50 lb of U-235?[12]

The most promising candidates were ultracentrifugation – spinning gaseous uranium so fast that the marginally heavier U-238 would gravitate outwards – and membrane diffusion, in which gaseous uranium was forced through a thin barrier perforated with millions of tiny pores through which the lighter U-235 would pass more easily. The need for gaseous uranium saddled them with hex, which is solid at room temperature but boils into a dense gas when heated to 60°C. The hex molecule (chemical formula UF_6) is an ugly brute, armed with six atoms of highly reactive fluorine, and gaseous hex is vile stuff. It chews through rubber and metals; reacts explosively with water; blocks pipes by solidifying on cooling; cauterises the skin, eyes and lungs; and is lethal if inhaled at low concentrations.[13] Tacking six fluorine atoms on to uranium also made it even harder to pull out U-235: elemental U-235 is 1.3 per cent lighter than U-238, while [U-238]F_6 and [U-235]F_6 differ by only 0.9 per cent.

Peierls concluded that membrane diffusion was the best bet but needed confirmation from someone who did experiments. Frisch was fully occupied in Liverpool, so Peierls consulted another Jewish refugee physicist who was unemployed because his research had been axed. Civil servants advising the M.A.U.D. Committee were determined to exclude him, so Peierls went to Liverpool to obtain Chadwick's blessing. Franz Simon (Plate 13) was close to Chadwick's age, almost fifteen years older than Peierls. Born in Berlin, he studied physics there and served his country with distinction on the Western Front. Simon never talked about the war (he was gassed, wounded and hospitalised for a year) or the act of heroism which won him the coveted Iron Cross (First Class). Post-war, he focused on ultra-low-temperature physics and built up a top-class research group in Breslau, which brought him to the attention of Frederick Lindemann, then Professor of Physics at Oxford and another pioneer in low-temperature physics. During the early years of fascism, Simon was protected against anti-Jewish persecution by his war-veteran status and Iron Cross, but he foresaw a grim future when Hitler became chancellor. In 1933, he handed in his passport and Iron Cross and left

Germany with his wife, Charlotte. Their escape was organised by the Academic Assistance Council and Lindemann, who wangled him an ICI scholarship in low-temperature physics at the Clarendon Laboratory in Oxford.[14]

When their British citizenship was approved in late 1938, Franz (now known as Francis) and Charlotte Simon were fully integrated into life in Oxford, and his lab was among the best in its field. He still had a faint German accent and styled himself the 'Vice-President of the Broken-English Speaking Union' because he felt his command of the language wasn't quite perfect. His 'impish' sense of humour, rather than ignorance of Englishness, probably explains why he once asked a keen huntsman how many foxes he'd shot that day. Strangely for a man whose research flourished at 250°C below zero, Simon hated the cold; his reactions to falling temperatures in the lab were affectionately sent up by his team as '70°F: hat. 68°F: muffler. 65°F: home'. Then the war came along. Low-temperature physics wasn't a wartime priority and Simon had been born in Germany, so his lab was shut down and he was marooned with nothing to do – and no family, as he had sent his wife and children to safety in Canada.[15]

When Peierls visited Simon in early summer 1940, he found a frustrated man, desperate to unleash his pent-up brainpower on an important problem that others couldn't solve. The purification of U-235 fitted the bill perfectly, and Peierls returned optimistic to Birmingham. Simon agreed that membrane diffusion seemed the best option. This was another German invention, developed during the 1920s by Gustav Hertz, who separated the noble gas neon (Ne) into two isotopes, Ne-22 and Ne-20, by forcing it through porous pipeclay. The process had to be repeated hundreds of times to isolate the pure isotopes, which differed in atomic weight by 10 per cent. This augured badly for the extraction of U-235, barely 1 per cent lighter than U-238.

Simon's team began by trying to separate carbon dioxide from water vapour. According to legend, the experiment was inspired by soda water and Charlotte's kitchen strainer: pumping a mixture of the gases through 'Dutch cloth', a metal gauze woven from hair-thin copper threads. The finest Dutch cloth can filter out bacteria, and its holes could be narrowed by hammering the gauze.[16] Even these tiny pores were astronomically big: scaling the hex molecule up to the size of a marble, the pore would be as wide as the Albert Hall. But it was a start.

During summer 1940, Rudi Peierls was challenged to stress-test the assertion that a small mass of pure U-235 could make a devastating super-bomb. An American theoretician had hypothesised that most of the energy of a uranium fission bomb would be wasted as heat. George Thomson consulted an English physicist outside the M.A.U.D. Committee who sided with the American and estimated that the explosion would only be half as powerful as Frisch and Peierls had stated. Peierls and his whizz-kids dismembered the equations, ran everything through their calculating machines, and confirmed that 50 lb of pure U-235 would be equivalent to several thousand tons of TNT; by contrast, over 28 tons of unseparated uranium would be needed, even if it could be made to explode. Helped by Frisch chipping in from Liverpool, Peierls prepared ten detailed documents for the September meeting of the M.A.U.D. Technical Committee. These covered the yield of the super-bomb and possible methods for U-235 separation, notably gaseous membrane diffusion. This was important enough for Oliphant to provide secretarial help; the redoubtable Miss Hytch typed up Peierls's words, engraved on a rotating wax cylinder inside an archaic Dictaphone.[17]

To ensure that the documents had a soft landing, Peierls took Francis Simon's advice and went to see a man who, single-handedly, could kill the super-bomb. Frederick Lindemann, now king of the castle as Churchill's personal scientific guru, had insisted that there was 'no danger' of an operational uranium fission bomb for several years, but continued to gatecrash M.A.U.D. meetings and demanded copies of all its papers.

Simon engineered an audience for Peierls with Lindemann in the Cabinet Office rooms in Whitehall in early June 1940. Peierls brought advance copies of the documents for the forthcoming M.A.U.D. meeting. He also made a personal appeal to Lindemann that would have moved any normal man, let alone a founding member of the Academic Assistance Council which had rescued so many refugee scientists from Hitler. The Nazis must never get the super-bomb, Peierls said. He, Frisch, Simon and many other Jews wouldn't survive a German invasion of Britain. In that event, Peierls was putting all his trust in Lindemann to send everything that Britain knew about the super-bomb to American scientists for them to continue the crusade against Nazism.

Lindemann was a man of few words and indecipherable body language. Peierls later confessed that 'I did not know him sufficiently well to interpret his grunts correctly'. However, Lindemann didn't seem 'particularly impressed' and left Peierls hoping desperately that 'the whole thing would be taken seriously'.[18]

Home from Home

After a bumpy start, Hans Halban and Lew Kowarski restarted their boiler research in England during summer 1940. Britain's welcome hadn't met their expectations as internationally renowned scientists: a perfunctory 'That's all' from the Earl of Suffolk as he handed over the coal-soiled *Broompark* evacuees to a harassed civil servant.[19] The Kowarskis went to stay with friends and the Halbans to a posh hotel in Mayfair. Halban and Kowarski met John Cockcroft and other senior British scientists informally that evening, and more formally the following day. Cockcroft arranged for Halban, whose fluent English made him the obvious spokesman, to visit the fission research groups in Liverpool, Birmingham, Cambridge and London. Halban and Kowarski were both invited to the next meeting of a secret committee called 'M.A.U.D.', at the Royal Society in early July. In preparation, the committee wanted a detailed report about the Parisians' nuclear research.

Halban and Kowarski gave Cockcroft their report – 'BR94', written in the bedroom of Halban's hotel – which claimed that the boiler was close to fruition and explained that their fission bomb had been abandoned because it was 'too difficult'. Problems then became apparent. Halban's tour of England's fission research centres revealed that nobody was working on – or even faintly interested in – harnessing fission to generate electricity.[20] Shortly before the M.A.U.D. Committee meeting, Cockcroft told Halban and Kowarski to come to the Royal Society as planned but to wait outside the meeting room until they were called in. On the day, they sat in the corridor like the last candidates at a job interview. Finally, the doors opened and the committee members dispersed in silence, leaving Cockcroft to explain that French and British interests in fission had diverged critically. Paris was fixated on the boiler, whereas Britain was throwing everything at the super-bomb. Also, Joliot's patents could jeopardise collaboration: if the boiler was perfected through research done in England, how would Britain claim its rightful share of intellectual property and royalties? But to encourage collaboration, temporary lab space was waiting for them at the Cavendish Laboratory in Cambridge, together with Marie Curie's radium, their heavy water and a housewarming gift of uranium oxide. But because Halban and Kowarski had no interest in a fission bomb, they couldn't join the M.A.U.D. Committee.[21]

Always volatile, Halban was incensed by what he saw as a patronising brush-off and threatened to go straight to America. Kowarski believed that the Cavendish was the best option in what was left of free Europe, and persuaded him to give it a try. In mid-July, Halban and Kowarski borrowed an elderly car and went to Cambridge ahead of their families. This was 'something of an adventure' because all signposts and place names had been removed to disorientate invaders – and, being aliens, the Frenchmen would have been imprisoned if caught with a road map. So Halban drove, and Kowarski navigated, having memorised the sequence of pubs and other landmarks along the route.[22]

They were given a warm welcome, lab space, equipment, consumables and salaries, but found only two like-minded souls. Norman Feather and Egon Bretscher had followed Rutherford's star to the Cavendish, from Yorkshire via Trinity College, Cambridge (Feather), and Zurich via Edinburgh (Bretscher). They began collaborating in 1936, when Bretscher became suspicious about the transuranic elements which Meitner and Hahn had supposedly discovered. He nagged Feather to use his advanced X-ray 'fingerprinting' method to measure the atomic weights of the reaction products. Feather found that the 'transuranic' elements were actually isotopes of known elements lighter than uranium – but too late, because Hahn, Strassmann, Frisch and Meitner had already published. Feather and Bretscher remained interested in fission, and Feather had started investigating what neutrons travelling at different speeds did to uranium. They were left high and dry when the war swept all their colleagues away into radar. As a Swiss alien, Bretscher had to report to the police every morning, while Feather was saddled with all the undergraduate teaching. They worked in a different building from Halban and Kowarski, but spoke the same dialect of physics and provided intellectual stimulation. Feather greatly admired the boiler, having prophesied that nuclear power could abolish the 'crime' of wasting energy by burning coal.[23]

Cambridge was good to Halban and Kowarski and their families. The city was largely spared while London, Liverpool and Birmingham were being blitzed; the university sustained 'no significant damage', although a young researcher at the Cavendish was 'completely blown up' when his digs took a direct hit. Kowarski became fluent in idiosyncratically elegant English 'with amazing speed' and both Frenchmen were quickly absorbed into university life; hearing dons pontificating about the future of Europe at the high tables of posh colleges must have made an intriguing counterpoint to their memories of Bordeaux and SS *Broompark*. Rudi Peierls and Otto Frisch came to visit and all became friendly. Halban, 'a skilful organiser and a man of the world', saw himself as the alpha male, but both Frisch and Peierls warmed more to the other half of the double act. The Russian 'bear of a man' struck a personal chord with Frisch because he too could have been a concert pianist – until a growth spurt made his fingers too big for the keys and forced him into science.[24]

During the autumn, Halban and Kowarski reconstructed what they could of their lab and began building their most ambitious boiler yet: a constantly rotating aluminium globe containing heavy water and a one-metre sphere of uranium oxide blocks. This should confirm that the chain reaction could begin to spread and would indicate how much uranium and heavy water were needed to trigger a self-sustaining chain reaction for generating nuclear energy.[25]

During the summer of 1940, the M.A.U.D. Committee stuck two fingers up to the 'innate prejudice of civil servants' against foreigners and must have sent tremors through the Ministry of Aircraft Production.[26] M.A.U.D. would split into a top-level Policy Committee to advise the minister, and a Technical Committee that dealt only with hard science. The Policy Committee comprised Thomson, Chadwick, Cockcroft, Appleton and Francis Simon (in charge of separating U-235), with Lindemann in attendance. The Technical Committee included Mark Oliphant, Philip Moon and Norman Feather, outnumbered by aliens: Frisch, Peierls, Bretscher, Halban and Kowarski. M.A.U.D.'s remit would comprise two separate strands of fission research: 'slow neutrons' to designate the continuous, controlled burn inside the electricity-generating boiler; and 'fast neutrons' for the U-235 super-bomb.

Why the change in heart over the Parisians, whose research was worthy but irrelevant to the bomb? Thomson had been lobbied by Frisch, Peierls and Chadwick to bring in the Frenchmen, who had impressed everyone in Cambridge. There was also the 'vague intimation' of an easier route to a super-bomb without having to isolate U-235. Halban and Kowarski speculated that fast neutrons spewed out by the fission of U-235 could occasionally induce a magical transformation of U-238, the dross that Simon was trying so hard to get rid of. In contrast to U-235, incoming neutrons wouldn't smash U-238 in two but might sometimes be captured to create artificial elements heavier than uranium. These man-made products of U-238 were predicted to be unstable, and one in particular – atomic number 94, atomic weight 239 – could be fissionable and potentially even more explosive than U-235 (see Figure 6).[27]

At the time, nobody in England knew that this had been foreseen by Louis Turner in his paper entitled 'Atomic Energy from Uranium-235', locked away unpublished by the editor of *Physical Review*. Then, in mid-July 1940, a delayed, month-old issue of *Physical Review* finally reached Britain. It contained the paper by Ed McMillan and Phil Abelson, reporting that they'd created 'element 93' by blasting uranium with neutrons. Their speculation about how its daughter product element 94 might behave was what provoked Chadwick's furious confidential letter to the editor, accusing him of printing an open invitation for the Germans to start making an atomic bomb.[28]

And now that the Nazis had swept through much of Europe and seized the Bohemian uranium mines, the Norwegian heavy-water plant, Bohr's Institute in Copenhagen and Joliot's lab at the Collège de France, that possibility might already have become reality.

UXB

Few of those on board *Broompark* might have guessed that Jack Howard, Earl of Suffolk, had a wife (Mimi, a former film star) and a baby due in August 1940. While Howard was bonding with his new son, he was diagnosed with progressive arthritis and discharged from active service. Disgusted and hungry for excitement and danger, Howard signed up for bomb disposal.[29]

The unexploded bomb (UXB) was a terror from hell. It could be a dud or biding its time until woken by a delay fuse up to three days after impact. The science of bomb disposal was invented on the hoof by trial and error (too often fatal), by people of extraordinary bravery. You should be 'strong, unmarried, able to run fast, of strong character and prepared for the afterlife'. Your bible – *A Manual of Bomb Disposal* – covered everything from stick incendiaries to the two-ton 'Satan', fifty different types of fuse, and devilishly clever anti-handling devices invented specifically to kill you. The tools of your trade included drills, high-pressure hoses to flush out high explosive, and a stethoscope to pick up the faintest buzz of internal clockwork which (unless you ran like hell) would be the soundtrack to the last fifteen seconds of your life. With average luck and skill, you might expect to outwit between twenty and forty bombs.[30]

Jack Howard hand-picked his own bomb-disposal team: Beryl Marsden as record-keeper and Fred Hard, who worked for a house-removal company, as driver. Fred's van was purchased and painted black, with 'UNEXPLODED BOMBS' replacing 'Pickford's'. Racing through the blackout with a police escort, or parked outside the Ritz for a team dinner after a hard day's work, the van looked the part. Jack Howard didn't, with his plus-fours, white duffel coat, black fedora, cigarettes and his trusty revolvers, Oscar and Genevieve. He swore a lot but seemed disciplined and methodical, treating each bomb like an experiment to be analysed and learned from. With Fred assisting like a trainee surgeon, Howard would dictate his findings to Beryl, standing nearby and around a corner if available. Jack, Beryl and Fred – nicknamed 'The Holy Trinity' – remained lucky, and clocked up their twentieth bomb in late 1940.[31]

11

MEN WITH MISSIONS
June 1940–January 1941

Frédéric Joliot returned to Clairvivre on the morning of 19 June 1940, despondent after his wasted journey to Bordeaux. Irène had settled into the sanatorium and their children were safe with friends in Brittany. Stuck in limbo, Joliot watched events unfolding and waited to be called to Paris.[1]

On 22 June, Marshal Pétain signed his armistice with Germany and the puppet French government in Vichy immediately danced to the Nazis' tune. The next day, Hitler toured Paris like a conquering emperor and declared the capital a 'free city'. It was anything but. Joliot learned later that a squad of German scientists swept into his lab soon after the Nazis entered Paris, looking for uranium, heavy water and the cyclotron, currently awaiting repair. Most of the interlopers were unknowns from the German Army's Ordnance Department, headed by a Kolonel Schumann who spoke no French. Two names were familiar: Walther Bothe, Professor of Physics at Heidelberg, and Wolfgang Gentner, a physicist drafted in as Schumann's interpreter.[2]

Gentner brought a ray of hope. He'd come to the Radium Institute in 1932 – the first German to work there since the Great War – and followed Joliot to the Collège de France. The two were standing side by side at the eureka moment which led to the Joliot-Curies' Nobel Prize. They'd fired neutrons at an aluminium target, and Joliot was baffled when the Geiger counter continued to click furiously after the neutron beam was switched off. Gentner had built the counter and insisted that it wasn't malfunctioning – which made Joliot realise that they'd created a new radioactive product. From Paris, Gentner moved to Heidelberg with Bothe, who was desperate for the new-fangled cyclotron and sent him to Berkeley to learn about the machine from its inventor, Ernest Lawrence. When the war started, Gentner was back in Heidelberg, building Bothe's cyclotron.[3] Now he'd reappeared in Joliot's lab with Nazi scientists who wanted a cyclotron. Five years ago, Gentner had been a close colleague and friend, but could Joliot trust him now?

In mid-July, the Germans closed the Collège de France to all French staff and summoned Joliot with an ominously cordial invitation to discuss research collaboration. The day began in a lecture theatre packed with Joliot's staff and German Army scientists, and was stage-managed by Kolonel Professor Erich Schumann, flanked by Walther Bothe and Kurt Diebner, the military administrator of the KWI for Physics. Schumann projected himself as a cultured man; a remote descendant of the composer, he wrote 'indifferent' military marches and had been a Professor of Musical Acoustics in Berlin. Through Gentner, Schumann heaped praise on the Collège and Joliot, whose team would now work in harmony with German researchers.[4]

Then Joliot was taken into his own office by Schumann, accompanied by Gentner, Diebner and Bothe. The door closed and the gloves came off. It was a classic good guy–bad guy routine, but with just one guy. The *kultiviert* Schumann was a career Nazi who led weapons research in the Army's Ordnance Department. He laid into Joliot, with Gentner translating.[5] Where was the uranium? No idea, replied Joliot; the French Ministry of Armaments had taken it all away. The Abwehr knew that a French agent had stolen the heavy water from Rjukan. Who was he? Joliot didn't know. Where was the heavy water? On an English boat that sank during an air raid; Joliot had seen the empty berth and wreckage. And the laboratory records of the fission experiments? Also lost when the boat sank.

The interrogation left Schumann flustered and none the wiser. Gentner waited until the others had gone, then told Joliot that they must talk in secret. They met that evening in the back room of a café on the Boulevard Saint-Michel and Gentner (who had told Schumann he was going to sample Parisian nightlife) poured out information that could have cost him his life.[6] He loathed the Nazis and wanted them defeated. Schumann and Diebner led the so-called *Uranverein*, a network of nuclear physics researchers centred on the KWI for Physics and coordinated by the Army's Ordnance Department. The *Uranverein* included Heisenberg, von Weizsäcker, Bothe and Paul Harteck, and existed to exploit nuclear fission for the military. Bothe needed a cyclotron to measure the nuclear constants of uranium, essential knowledge for building nuclear energy generators and atomic bombs. The Heidelberg cyclotron was nowhere near ready, so they would use Joliot's. Schumann wanted Gentner and Joliot to run the cyclotron together and supervise fission research at the Collège de France.

Joliot and Gentner agreed to accept Schumann's offer, showing that the two old colleagues were working together again; Joliot would also help to repair the cyclotron on the condition that it wasn't used for military purposes. But unknown to Schumann, Joliot and Gentner would collaborate on a different agenda that could drag them both in front of a German firing squad.

On a Plate

During autumn 1940, the uranium super-bomb had the chance to enter a beauty contest with twenty other technological wonders that could, their promoters claimed, help Britain to win the war. These top-secret military secrets would be shared with the Americans so that the USA could 'reach the highest levels of technical efficiency' before, as seemed increasingly likely, it was dragged into the conflict.[7]

At the start of the war, Anglo-American military and scientific collaboration was virtually non-existent. Although the USA was 'benevolently neutral' to Britain, many Americans preferred Aryans to Anglo-Saxons and wanted Hitler to triumph. On becoming Prime Minister in May 1940, Winston Churchill began making overtures to President Franklin D. Roosevelt. Referring to himself as a 'Former Naval Person' (First Lord of the Admiralty, 1911–15), Churchill asked Roosevelt for 'fifty or sixty old destroyers' to help prepare for a German invasion. The Royal Navy was duly loaned sixty American destroyers, all veterans of the Great War, to be returned or replaced in kind when peace was restored.[8] However, other negotiations foundered. Because of a squabble over patents, the US Air Force refused to hand over its ground-breaking Norden M1 bomb-sight, even when the top-secret British Asdic system for tracking submarines underwater was offered in return.[9]

The wily Henry Tizard had already seen the urgent need to make scientists cooperate across the Atlantic, and to that end had seconded A.V. Hill as scientific attaché to the British Embassy in Washington. Hill arrived in late March 1940, armed with his scientific credentials (Nobel Prize, Joint Secretary of the Royal Society), legendary people skills and a hit-list of top US scientists compiled by their British counterparts. While touring American universities and companies, Hill spread goodwill, hoovered up intelligence and identified two hot topics: radar and nuclear fission. The USA believed its radar to be world-leading; in reality, it was 'very good unless intended to be used in war' and far behind the British.[10] American research into fission focused exclusively on generating electricity. Nobody in the USA was working on an atomic bomb because everyone who mattered – scientists, policy-makers, the military – had written it off.[11] The military didn't need it; if America were sucked into the war now devouring Europe, then it would fight and win with its formidable conventional weapons. Physicists saw no hard evidence that the bomb was anything other than science fiction. Even if it worked, the atomic bomb would play no role until the next war, or even the one after that.

Gaping holes in war research on both sides of the Atlantic convinced Tizard that the two countries must collaborate. He soon discovered that various important people wanted him to take the idea to the Americans. Roosevelt signalled his approval in mid-July 1940, and on 1 August Churchill told Tizard that he wanted him to lead this 'important' initiative.[12]

Tizard immediately began organising the secret 'British Technical and Scientific Mission to the United States'.[13] As radar had top priority, Tizard's deputy was John Cockcroft, who was now helping radar supremo Robert Watson-Watt to bring the Chain Home defences into operation. Tizard also recruited Edward Bowen, called 'Taffy' because he was Swansea-born and a graduate of that university. At twenty-nine years old, Bowen was the brightest of the young stars in radar, developing compact, high-powered equipment that could fit inside an aircraft and spot small vessels on the sea. Tizard chose the rest of his mission team with similar canniness – three technical experts, one from each of the services, who could speak man to man with their American peers and had witnessed their gadgets in action in settings such as Dunkirk.[14]

Tizard and Cockcroft drew up the shortlist of British technological treasures, and Cockcroft locked them away in a black lacquered metal document box, the size of a small suitcase, which he bought from the Army & Navy Stores in Victoria. The top-secret booty in the 'Black Box' included a cavity magnetron; 'proximity fuses' that detonated a shell as it passed near an aircraft, converting a miss into a kill; the clever 'predictor' which aimed anti-aircraft guns at where the target would actually be; blueprints for Frank Whittle's revolutionary jet engine; and films of various inventions operating in the field and in battle.[15]

Cockcroft regarded Tizard's mission as 'magnificently organised', but Tizard fought many skirmishes off scene. He believed that trying to trade secrets – Asdic for a bomb-sight – was doomed to failure, and that the mission must 'primarily generate goodwill' by offering British science freely in the hope that the Americans would reciprocate.[16] This raised hackles, with the Admiralty moaning that 'anything we tell the Americans will go straight to Germany'.[17] There was also funny business at the very top. After approving the mission on 1 August, Churchill cancelled it the next day, then chopped and changed for another week and finally gave Tizard the green light on 9 August, three weeks after Roosevelt. Churchill's strings were probably pulled by Frederick Lindemann, now running the Statistics Section that fed Churchill the predigested information on which he made his decisions. Lindemann's passion for grudges defied logic. After being unhorsed from his committees, Tizard wrote to Lindemann, 'I am much more interested in defeating the enemy than in defeating you.' Lindemann didn't bother to reply.[18]

August 1940, when Tizard's mission was launched, was a challenging time to leave England. The Battle of Britain and the London Blitz were in full swing, piling unrelenting pressure on the RAF and Chain Home, and fuelling fears that Germany would invade in September. Tizard and Cockcroft shook hands on a £5 bet about what they would find on returning home.[19]

On 14 August, Tizard went ahead by flying boat to lay the ground: breakfast at Paddington, lunch in Poole, tea beside the River Shannon, and a late breakfast (Bovril, biscuits, fruit) over Newfoundland after a harrowing night flying low over the storm-lashed Atlantic.[20] Tizard's decision to visit Canada first was astute. The country, a fighting ally from the outset, had felt neglected and gave him a warm welcome with particular interest in fission research, one of its own priorities. When Churchill delivered his famous lines, 'Never in the field of human conflict was so much owed by so many to so few,' Tizard was on the train from Ottawa to Washington. He set up base in a hotel, noting that 'A good many people do not know what I am here for'. After meeting Roosevelt ('very nice . . . a most attractive personality'), Tizard began a charm offensive against hard, bright Americans from universities, industry and the forces.[21] Some were openly suspicious of British motives, but Vannevar Bush – Chairman of Roosevelt's NDRC – trusted him from their first meeting over dinner at the upper-crust Cosmos Club. The US Forces had warned Bush not to tell Tizard anything until given clearance, so (in Bush's words) 'We were careful not to be seen together . . . to avoid suspicions of some sort of conspiracy.'[22] While the Luftwaffe hammered London and the American military stared him down, Tizard remained courteous and patient, waiting for the Black Box of tricks to win the day.

The box arrived with Cockcroft, Bowen, the other mission members and armed guards to prevent it from falling into enemy hands.[23] On 28 August, Bowen spent a rough night in a hotel at Marble Arch with the box under his bed. Next morning, it was lashed to the roof of a taxi by a stroppy cabbie. At Euston, it disappeared into the crowds with an over-zealous porter and was deposited in the first-class compartment on the Liverpool boat train reserved for Bowen and a silent, pipe-smoking man carrying *The Times*. At Liverpool's Lime Street Station, the box containing the 'most precious cargo' to reach America during the war was marched away by the armed guards to the Gladstone

Dock.[24] The *Duchess of Richmond* cast off early on the evening of 29 August shortly before the Luftwaffe arrived, and entered the Irish Sea against the backdrop of Liverpool burning.

The Atlantic crossing took nine days, two days longer than usual because of zigzagging to thwart U-boats. Also on board were Royal Navy crews sent to bring back the sixty elderly destroyers loaned by Roosevelt. On discovering that 'the man who split the atom' was there, they asked Cockcroft to give a lecture. He told them about nuclear energy and invented a new unit of measurement for the occasion: the 'battleship-foot' was the amount of energy needed to lift a 50,000-ton ship that distance out of the sea, and it was contained in a cup of water.[25] His talk went down well. The Black Box would have gone down even better. On disembarking at Halifax, Nova Scotia, Bowen noticed that rows of holes had been drilled in each end; in mid-Atlantic, Cockcroft had realised that the box would float and took decisive action to consign it to the depths in case of trouble.[26]

The Black Box's treasures thrilled the Americans and seemed to break down barriers obstructing the two-way flow of ideas and inventions. Cockcroft wrote, 'There is a burning desire to give us all the help we need,' while noting that the mission had made the Americans 'realise how far they are behind'. Tizard held back the best until the end: Bowen's latest radar detector, which sat in the nose of an aircraft and could pick up the conning-tower of a surfaced U-boat several miles away. The cavity magnetron at its heart, produced by Cockcroft with a magician's flourish, was 1,000 times more powerful than America's nearest equivalent. The Americans were stunned, but not for long. The Bell Telephone Company took the magnetron away for dissection – the prelude to the mass production of magnetrons in the USA and the filing of patents by Americans who claimed to have invented it.[27]

His mission accomplished, Sir Henry Tizard returned to Blitz-ravaged London on 20 October 1940. He and Cockcroft had achieved great things – frictionless exchange of science and technology between America and Britain, to be facilitated by a London bureau of the NDRC and a British Central Scientific Office in Washington, both to be established in early 1941. Tizard was too wise to expect thanks from on high. Churchill didn't find time to acknowledge receipt of his report on the mission.[28]

John Cockcroft remained in America, £5 richer because Britain hadn't fallen to the Nazis; Tizard had bet against Chain Home and everything else he'd masterminded since 1935.[29] Cockcroft took another six weeks to sweep up, and his brief now included nuclear physics. They hadn't made room in the Black Box for the Frisch–Peierls memorandum and Peierls's documents for the M.A.U.D. Committee; unlike Tizard's other treasures – tried, tested and mostly battle-ready – the super-bomb didn't even exist as a rough sketch.

Cockcroft found that, even though America was not at war, the research priorities of Vannevar Bush's NDRC were distorting the scientific landscape. His old friend Merl Tuve had abandoned nuclear fission to design proximity fuses for anti-aircraft shells, while Ed McMillan had dropped elements 93 and 94 and devoted himself to radar. At Columbia, Enrico Fermi and Leo Szilard had built an 11-foot pile of graphite and uranium, hopefully inching closer to a chain reaction which could generate electricity. Others were exploring how to concentrate U-235 using diffusion, centrifugation and Ernest Lawrence's cyclotron, aiming to produce better fuel for nuclear energy generation.[30]

In America, the atomic bomb was conspicuous by its absence. Fermi and everyone else of note dismissed the notion as a waste of time, effort and money.[31] Two weeks before he returned to England, Cockcroft went to Washington to tell Lyman Briggs and his Uranium Committee everything about the nuclear research being conducted in England. He later noted, 'atomic energy was less urgent than radar'. Unlike the contents of the Black Box, neither the Frisch–Peierls super-bomb nor the Halban–Kowarski boiler excited any interest or offers 'to give all the help we need'.[32]

Year End

In Oxford, Francis Simon's plans for separating U-235 galloped ahead when the Minister of Aircraft Production finally coughed up £5,000 to support his 'irregular' efforts. In mid-December 1940, Simon produced a 'striking memorandum' entitled 'Estimate of the Size of an Actual Separation Plant'.[33] Simon shaped his separation plant around the area of diffusion membrane calculated to produce 1 kg per day of 99 per cent pure U-235: a staggering 70,000 square metres (17 acres), divided between 18,000 production units. U-235 would be progressively enriched through twenty stages, driven by 70,000 tons of machinery covering 40 acres and drawing enough electricity for 6,000 houses. Cost: £4 million (roughly a battleship) up front, then £1.5 million per year to run. Timeframe: eighteen months to build, then twelve days to begin producing U-235. Feasibility? Simon's conclusion began simply, 'We are confident . . .' and all those who read the memorandum shared that confidence. The map of Britain was scoured for sites to put a top-secret, 40-acre plant that was ravenous for electricity and processed dangerous materials whose identity could never be revealed.

Lew Kowarski and Hans Halban were now well established in Cambridge. The boiler had evolved considerably during the eighteen months since they picked up the whisper of excess neutrons that hinted at 'une réaction en chaîne'. To keep the heavy water circulating around the central globe of uranium oxide, the spherical aluminium container was continually rotated, like an oversized glitter-ball that someone had forgotten to switch off after the last dance. Neutrons fired into its heart ignited a tantalising flare of secondary neutrons sprayed out by fission. This fizzled out on the brink of taking hold but suggested that size mattered. Kowarski and Halban estimated that 5 tons of uranium oxide, moderated by 600 litres of heavy water, would 'go critical'.[34] A working boiler would have to wait, because England at war couldn't provide enough of either ingredient – but this was the tipping-point between a scientific pipe-dream and a manageable technological challenge. In normal times, the breakthrough would have triggered chain reactions of excitement in scientific journals and newspapers. Kowarski and Halban contented themselves with writing a report for the M.A.U.D. Policy Committee, and the thought that Frédéric Joliot, the absent musketeer, would have been thrilled.

Next door in the Cavendish Laboratory, Egon Bretscher and Norman Feather continued their two-pronged assault on fission.

Feather had tried to measure uranium cross-sections for neutrons of various energies, while Bretscher followed up Halban and Kowarski's 'vague intimation' that the boiler could generate fissionable elements heavier than uranium. By now, McMillan and Abelson had published their creation of 'element 93' in neutron-irradiated uranium, and Bretscher was intrigued by its decay product, element 94. From first principles, he predicted some of its properties: 94 would be a hard, dense metal chemically distinct from its grandparent, uranium. Then he applied the Bohr–Wheeler formulae which had prophesied that U-235 would be fissionable – and was as shocked as Otto Frisch had been on calculating what neutrons would do to pure U-235.[35] Element 94 would undergo fission, probably more easily and explosively than U-235. Bretscher and Feather wrote up their findings in a top-secret feasibility paper on a fission bomb using element 94, which Bretscher would present at the next M.A.U.D. Policy Committee meeting in January 1941.

Early in the New Year, the two anonymous transuranic elements underwent a rite of passage. Nicholas Kemmer, a young Russian physicist lodging with the Bretschers, had been let into the secret. Over dinner, Kemmer suggested that '93' and '94' deserved proper names, inspired by the heavens. Uranium had been named after the farthest-flung planet known in 1789; the two new elements should therefore take their names from the two outermost planets discovered subsequently. He christened element 93 'neptunium', and its daughter, number 94, 'plutonium' (see Figure 6).[36]

Neptunium was a transient curiosity that melted away in two weeks, but plutonium was screaming for attention. If predictions were correct, it could be created as a by-product of fission inside the Halban–Kowarski boiler and, being chemically distinct, should be easily separated from uranium. Plutonium was the ideal explosive for an atomic super-bomb.

At the Collège de France, Joliot and Gentner kept a careful eye on the experiments planned for when the cyclotron began operating again. Most of the German scientists respected the gentlemen's agreement not to use it for military purposes, but it was rumoured that Walther

Bothe, Gentner's former chief in Heidelberg and a leading light in the *Uranverein*, intended to slip in some cross-section measurements on uranium.[37] Despite the devoted attention of a team of young German engineers, the cyclotron was still bedevilled by technical problems with the copper piping through which cold water rushed to carry away the explosive surge of heat when its massive magnet was switched on.

During October 1940, staff and students at several French universities took to the streets to demonstrate against the German occupation. They were urged on by leading academics including Paul Langevin and his son-in-law Jacques Salomon, both revered as world-class scientists and anti-fascist intellectuals. Langevin, a wise old man and formerly Professor of Physics in Paris, had been Joliot's mentor and very close to the Curies. Salomon was a brilliant theoretician, the French counterpart of Werner Heisenberg and, aged thirty-two, already tipped for a Nobel Prize.

The Nazis tolerated the unrest until the protests turned violent in November. Ten students and lecturers were shot dead and hundreds arrested. The sixty-eight-year-old Langevin was thrown into solitary confinement without reading or writing materials, which incensed Joliot and fortified his hatred of the Nazis. Then, unexpectedly, Langevin was transferred from prison to house arrest in Troyes, 100 miles south-east of Paris. A German expert on the French *esprit* had warned the Nazi High Command in Paris that they were playing with fire by making Langevin a martyr. Langevin's saviour? Kolonel Erich Schumann's trusted aide and interpreter, Wolfgang Gentner.[38]

Meanwhile, the Abwehr had failed to identify the French secret agent who stole the heavy water from Rjukan. In December 1940, an Abwehr general interviewed a potential source of clues: a trusted liaison officer with the Vichy government who had looked after Norwegian accounts at the Banque de Paris. The general took the ex-banker into his confidence. The only leads were hotel and flight bookings in the false name of Freiss. Any idea who he might be? Impassive as always, Jacques Allier shook his head.[39]

12

IN THE DARK

March 1939 (reprise)–May 1941

In early spring 1941, the two new portals for exchanging science across the Atlantic – the British Central Scientific Office in Washington and the NDRC Liaison Office in London – opened their doors. In January, three months after John Cockcroft briefed Roosevelt's Uranium Committee about British fission research, the Americans received every last fact about the U-235 super-bomb and the Halban–Kowarski boiler. George Thomson sent the Frisch–Peierls memorandum, all the M.A.U.D. Committee papers, Rudi Peierls's portfolio of updates and Francis Simon's plans for 'an actual separation plant' to Lyman Briggs and Vannevar Bush for distribution to the Uranium Committee and the NDRC. In addition, American nuclear physicists were invited to attend future meetings of the M.A.U.D. Committee.[1]

This was the first trial of the special two-way relationship which Tizard hoped would enable American and British researchers to whisper every new and exciting secret into each other's ears. It was also an act of faith by a beleaguered country in a time of desperate need. The Battle of Britain had been won – narrowly – but the Blitz continued, and Nazi-occupied Europe began just 22 miles across the English Channel. Britain was fighting for survival or extinction; for most Americans, the war remained a distant spectator sport.

Frustratingly, weeks went by with no American reaction to the M.A.U.D. Committee's offerings.[2] As Cockcroft had observed at first hand, the Americans were pouring everything into Fermi's pile at Columbia, and an American atomic bomb wasn't even on the drawing board. Perhaps they didn't want to get sucked into someone else's war or tied into an expensive project that could turn out to be science fiction. Their silence marked the first failure of the Anglo-American accord forged by Tizard's mission. The British had handed over all their nuclear secrets, in exchange for nothing

The Americans should have had plenty to share, notably rapid progress with Fermi's pile and burgeoning knowledge about element 94. They hadn't yet christened it 'plutonium', but were far ahead of Egon Bretscher and Norman Feather in every other respect. In December 1940, Glenn Seaborg, a twenty-eight-year-old instructor in physics at Berkeley, followed up Phil Abelson and Ed McMillan's paper reporting the creation of '94'. Seaborg blasted high-energy neutrons at a sheet of paper smeared with uranium oxide paste and detected the distinctive radioactive signature of '94'. Over several weeks, he and colleagues harvested samples too tiny to weigh and by late January 1941 had accumulated enough for chemical analysis. On the stormy night of 23 February, they proved that it was different from all other elements. At the end of May, they bombarded a microscopic speck – half a millionth of a gramme – of '94' with neutrons and found that it was even more readily fissionable than U-235. Seaborg pushed out six papers that defined the chemistry and physics of '94'. All were accepted immediately in secret by *Physical Review* and remained under wraps until the publication embargo was lifted after the war, in 1946.[3]

Within the hallowed walls of Berkeley, secrecy was strict – '94' was referred to as 'copper' (to avoid confusion, copper became 'honest-to-God copper') – and all information about '94' was hidden from the outside world, including the M.A.U.D. Committee.[4] For now, though, the British had other worries closer to home. Their pursuit of the U-235 super-bomb was driven by fear that a Nazi atomic bomb was being developed, and paranoia was being ramped up by a deafening, unfathomable silence from Germany.

Inside Knowledge

Months before the Americans censored publications about fission, most German nuclear scientists had vanished from *Naturwissenschaften* and other journals. Like many in Britain and America, they might simply have stopped publishing because they were diverted onto secret war projects unrelated to fission – but there were worrying straws in the wind that the Nazis had something nuclear to hide. It was like guessing the play to be performed from the props being assembled on stage: extensive rebuilding of the KWI for Physics in Berlin – now under army administration – for 'uranium research'; the Nazis' new hunger for uranium and their raging thirst for heavy water; and their desperation for a cyclotron. Were the Germans working on a nuclear energy generator? Or an atomic bomb?

Who would be involved in a German bomb? The chief suspects were Werner Heisenberg and Carl von Weizsäcker. Heisenberg, the golden boy of German science and indecently young Nobel Prize winner (Plate 3), would naturally lead German efforts to exploit fission in peace, and probably in war. He wasn't overtly pro-Nazi, but they must like him because he'd emerged triumphant from Johannes Stark's attempted savaging. Von Weizsäcker wasn't in the same league intellectually, but his noble family (denoted by 'von') was revered by the Nazis and his father was Secretary of State in the Reich's Foreign Office. Heisenberg and von Weizsäcker headed the lists of potentially dangerous German physicists compiled by both Jacques Allier and Rudi Peierls. Question marks hung over both men, especially Heisenberg. He was still lecturing and teaching – but oddly, wasn't acknowledged in a research paper that he should have supervised.[5] How was the golden boy really spending his time?

By now, the intermittent scraps of intelligence from Otto Hahn, Lise Meitner and other old allies had dried up, because of the personal risks involved. A few clues emerged from analysing when and where the top suspects appeared, what they presented at meetings, and even their undergraduate lecturing timetables. The British Secret Intelligence Service, also known as Military Intelligence

Section 6 (MI6), was put on the case, with variable results. MI6 informed Cockcroft that Heisenberg had been on their files since lecturing in Cambridge in 1935; he must still be in Britain because they hadn't noticed him leave. Peierls commented later, 'If this was a fair example of British intelligence, the outlook seemed grim.'[6] However, MI6 had already proved its worth by arranging for John Cockcroft to talk to a secret source in Berlin who had unparalleled access to the German scientific war machine. The introduction was made by Frank Foley, a short, quiet Englishman with round glasses who tended to pop up in unexpected places. Two years earlier – before the formation of M.A.U.D. and even the declaration of war – the informant had brought Cockcroft intelligence that was pure gold, but was clearly skating on thin ice.[7] Now, when he was needed most, he'd fallen silent.

The man who joined Cockcroft for lunch in the Athenaeum on 10 March 1939 was in his mid-forties, thin-faced, softly spoken and fluent in English. He was exceedingly good company and entirely comfortable dining in posh clubs with the scientific elite. Paul Rosbaud (Plate 14) was the chief scientific adviser to Springer Verlag, the German publishing empire which poured out books, trade magazines and academic journals including *Naturwissenschaften*. Austrian-born, Rosbaud had studied physics in Graz and Berlin, leading to a PhD in metallurgy. Academically, he was very good but not brilliant, and soon discovered that he preferred writing and networking to toiling at the laboratory bench. While editing a metallurgy journal, he impressed Ferdinand Springer, who offered him the dream job of scientific adviser in 1932, shortly before Rosbaud's thirty-sixth birthday. Rosbaud became a roving ambassador, impresario and publicist, indispensable to both Springer and the growing flock of 'his' scientists. He was fun, polymathic, a good listener and a quick learner; scientists in many disciplines found that he knew their work, spoke their dialect and was stimulating and useful. In the cut-throat, pre-war world of the publication rat-race, scientists saw Rosbaud as a peer and someone to be seen with. His close friends included Max von Laue, Otto Hahn and Lise Meitner.[8] By the late 1930s, Rosbaud was influential and renowned for spotting winners. In December 1938, it was he who directed the editor of *Naturwissenschaften* to replace one of the papers slated for the first issue of 1939 with the article by Hahn and Strassmann which heralded the era of nuclear fission.[9]

Rosbaud and his wife, Hilde, were socialites who enjoyed culture with a dash of decadence in Weimar-era Berlin, but were scared by the Nazis because Hilde was Jewish and Paul had something to hide. After the *Anschluss* of Austria, Rosbaud went to the British Embassy on Berlin's Tiergartenstrasse to consult Frank Foley, the Chief Passport Control Officer. Foley spirited up visas for Hilde and their daughter Angela, who were hastily packed off to London, but Rosbaud stayed in Berlin because he still felt safe and had work to do. Foley didn't only issue passports. He was an MI6 operative who ran a secret network of contacts inside Germany, and was Rosbaud's handler.[10]

Over lunch in the Athenaeum, Rosbaud gave Cockcroft an up-to-date picture of fission research across Germany.[11] He understood all the science and knew who was doing what and where, how the work was funded and how it was going. At the KWI for Chemistry, Otto Hahn and Fritz Strassmann had been joined by Josef Mattauch, who slipped into Lise Meitner's job and office and even occupied her flat. Karl Wirtz, an expert in electrolysis and heavy-water production, worked with von Weizsäcker at the sister KWI for Physics, while Hans Geiger was at the Technical University of Berlin. Heisenberg was in Leipzig, forging ahead. So was Paul Harteck, Professor of Physics in Hamburg and another expert on heavy water. In Heidelberg, Walther Bothe was still trying to finish building his cyclotron.

Rosbaud brought unique insights to the enemy: their personalities, motivation, attitude to the Nazis, and weaknesses. He'd already demonstrated where his own loyalties lay. On the evening of 12 July 1938, Lise Meitner wasn't alone in her cramped flat in Dahlem while cramming her worldly goods into two suitcases. Her trusted friend Paul Rosbaud was with her, helping practically and emotionally. He drove her to Otto Hahn's flat for her last night in Germany and to the station next morning; as Meitner's chauffeur-cum-psychotherapist, he talked her down from panic attacks in the car and while walking her to the Amsterdam train where Dirk Coster, her guardian angel, was waiting to spirit her away.[12]

Overall, Rosbaud concluded, there was worrying activity in Germany but no immediate threat. Cockcroft took notes and asked Rosbaud to keep him closely informed. He didn't have to wait long. In early May 1939, just five weeks later, Rosbaud delivered news that Cockcroft had hoped not to hear. Rosbaud regarded his source as totally reliable: Josef Mattauch, Lise Meitner's replacement at the KWI for Chemistry. On 30 April, Mattauch had visited Rosbaud in a state of restless excitement and told him about a meeting held the previous day behind locked doors in the Ministry of Culture on Unter den Linden. Hahn had been invited but had gone unexpectedly to Sweden to lecture, so Mattauch deputised. The meeting was top secret, but because Rosbaud was practically one of the team and a wise man in whom everyone had always confided, Mattauch told him everything.[13]

After Mattauch left, Rosbaud wrote a detailed report to give Cockcroft on his next trip to London a few days later. Cockcroft was away, so Rosbaud entrusted the document to Professor R.S. Hutton from Cambridge, an old metallurgist friend. They met 'in a safe place, on a bench in the Mall'.[14] The report was waiting for Cockcroft on his return, and it made shocking reading. Germany's fission research had crystallised into a clear, urgent threat. The meeting which Mattauch attended had created a research network that would make Germany the first country to build an atomic bomb.

Uranverein

The Nazis' atomic bomb wasn't conceived by Hitler, his ministers or his military top brass. Instead, German academic physicists had seized the initiative, in two separate approaches to the Reich's War Office.

In mid-April 1939, Professor Georg Joos in Göttingen wrote to Bernhard Rust, the loutish Minister of Education who had Nazified schools and universities. Joos explained that the energy blasted out by fission could be turned into either electricity or a bomb of staggering power that could guarantee military supremacy. Rust bounced the letter on to Abraham Esau, radiophysicist and a true Party animal who directed the Reich Research Council and the State Physical-Technical Institute. Esau set up an exploratory meeting on 29 April and invited Otto Hahn, Hans Geiger, Walther Bothe, Joos and a couple of others – but not Werner Heisenberg, presumably to prepare the ground before trying to net the big one. Josef Mattauch filled in when Hahn absented himself.[15] The meeting focused on a practical plan of attack: who could turn theoretical physics into an atomic bomb and what they would need to would make it happen. Selected researchers in Berlin, Göttingen, Leipzig and Heidelberg would participate. Exports of uranium ore from St Joachimsthal would immediately be blocked and other sources investigated. Preliminary experiments would begin on uranium oxide, to which Esau gave the cloak-and-dagger codename 'Substance 38' (after its chemical formula, U_3O_8). The enterprise christened itself *Uranverein*.[16]

Before he could do anything, Esau was pushed aside by the second prong of the scientists' attack, launched in mutual ignorance of the first, which had impaled someone more important. On 24 April 1939, Paul Harteck in Hamburg had written to Kolonel Professor Erich Schumann, chief of Weapons Research in the army's Ordnance Division, suggesting that nuclear fission could 'produce an explosive many orders of magnitude more powerful than conventional ones'. He added, 'The country which first makes use of this will have an unassailable advantage.'[17] Harteck's letter instantly hit the target because Schumann had recently created the hypothetical post of 'Head of Atomic Explosives'.[18]

Harteck was on an intellectual tier below Heisenberg but had clout and credibility. Born in Vienna and not yet forty, he'd trained as a physical chemist with Fritz Haber and won a Rockefeller scholarship in 1933 to work with Rutherford and Oliphant at the Cavendish – where he became an expert on heavy water and helped to create tritium, the third isotope of hydrogen. Harteck had already advised the Army Ordnance Division about explosives and knew Schumann as a dilettante who sometimes prioritised music over physics and weaponry.[19] Schumann's untried 'Head of Atomic Explosives' was thirty-four-year-old Kurt Diebner, the new Nazi Director of the KWI for Physics, who had badgered Schumann into inventing the 'Atomic Explosives' post to beef up his own standing. Diebner and Schumann now began planning to launch the *Uranverein* and build the bomb which would ensure Germany's victory in the coming war. And Schumann elbowed Esau out and took over the entire operation.[20]

During autumn 1939, the *Uranverein* met at a series of 'Schumann Conferences' on the military uses of uranium, held in the army's Ordnance Department. Otto Hahn hated the whole thing and privately confessed, 'If my work leads to a nuclear bomb, then I shall kill myself' – but with the army enforcing attendance, he couldn't wriggle out to go lecturing in Sweden. The first Schumann Conference was held in Berlin on 16 September 1939, barely a fortnight after Britain and France declared war on Germany. Hahn, Mattauch, Harteck, Bothe and Geiger were there, but not Heisenberg; he was being softened up to overcome his allergy to Diebner, whom he regarded as second-rate. Strangely, for reasons unspecified, Schumann was also absent. Proceedings began with a speculative Abwehr briefing about nuclear research by France, England and America. This profoundly disturbed Geiger and Bothe, who declared that 'It must be done' if there was the opportunity for Germany to develop an atomic bomb first. Harteck was asked to lead the *Uranverein* but declined, leaving the top job for the obvious candidate.[21]

Heisenberg was enticed into the second Schumann Conference, held ten days later on 26 September 1939. The *Uranverein* now sprawled across Germany, from Hamburg in the north to Munich in the south, and from Heidelberg near the French border to Leipzig out east (see Map 3). Its centre of gravity was the KWI for Physics in Berlin, where the Max Planck Haus was being converted into a fission research institute. Nearby, in the grounds of the Institute for Biology and Virus Research, a new building already nicknamed the 'Virus House' had been erected above a brick-lined well intended to accommodate an experimental uranium reactor. Harteck's group in Hamburg took charge of separating U-235 from native uranium. Walther Bothe in Heidelberg would use his still-incomplete cyclotron to measure nuclear constants in uranium, and also test graphite as a possible moderator to slow neutrons for fission. As the best moderator appeared to be heavy water, Karl Wirtz in Berlin would ensure that the Norsk Hydro plant at Vemork, part-owned by the German company IG Farben, would produce enough heavy water for the *Uranverein* and that none was siphoned off into enemy hands. The biggest responsibility was left to Werner Heisenberg, who would work out how much uranium was needed to make Germany's bomb, and in which form: Substance 38? Pure uranium metal? Uranium somehow enriched in U-235? In private, Bothe and Diebner had grave reservations about Heisenberg's role, muttering that a theoretician who hadn't done a single experiment in his life could never galvanise a scattered research network to produce a weapon of war.[22]

On 6 December 1939, Heisenberg presented a comprehensive report on the *Uranverein* and its military ambitions to the Army Ordnance Department. He concluded that uranium fission could be exploited for large-scale energy production and that enriching uranium in U-235 was the key to both efficient energy generation and 'explosives several orders of magnitude greater than the most powerful explosives yet known'. Heisenberg's report was well received, and the *Uranverein* was given a bright green light to proceed.[23] Later that month, IG Farben told Axel Aubert, General Director of Norsk Hydro, that it wished to buy the company's entire stock of heavy water and all its future output. Awkward questions followed about the intended use of such huge amounts of an esoteric by-product; then, in early March 1940, the Oslo branch of the Abwehr picked up unusual activity centred on a visiting Frenchman called Freiss. That name turned out to be false, as did his booking on the morning flight to Amsterdam the following week. The theft of the heavy water was galling, but the Norsk Hydro plant was now safely under German control and pouring out more than ever before.

And so to September 1940, when Schumann and Bothe, with Wolfgang Gentner in tow as interpreter, swept into the Collège de France and secured Frédéric Joliot's agreement for German scientists to use his cyclotron, although not (yet) for military purposes. By October 1940, the Virus House near the KWI for Physics was finished, and three different types of experimental reactor – *Uranbrenner*, or 'uranium burner' – were under construction in Berlin, Hamburg and Leipzig. These contained Substance 38 mixed with either paraffin, ordinary water or dry ice (solid carbon dioxide) as moderators. None of the designs sustained a chain reaction, but these were early days and their best bet – heavy water – was yet to be tried. And as 1940 ended, Bothe in Heidelberg was preparing to test the purest graphite obtainable in Germany, in case a backup moderator was needed.[24]

Timing is everything. The meeting on 1 September 1939, when Hans Geiger and Walther Bothe declared that Germany must make the bomb first, took place several months after Leo Szilard's identical moment of epiphany and a month after Einstein signed the letter to Roosevelt. The Germans quickly made up for lost time. When Lyman Briggs's Uranium Advisory Committee first met on 30 October 1939, the *Uranverein* and the Virus House were up and running. In February 1940, when Fermi and Szilard received start-up funding for their experimental pile, the three 'uranium burner' prototypes were already being built in Germany.

What about the British? The *Uranverein* was launched seven months before the M.A.U.D. Committee first met and was already working on M.A.U.D.'s top priority of purifying U-235. The Nazis had money, people, purpose-built labs and freedom from air raids and the threat of invasion. Logic dictated that they would get their bomb first. By January 1941, British paranoia over the German bomb programme was at fever pitch and stoked by fear of the unknown because no reliable information was coming out of Germany. After sending Cockcroft his report of Esau's exploratory meeting in April 1939, Paul Rosbaud had gone quiet. MI6 had given him an alert message to trigger his extraction from Berlin – '*Das Haus steht auf dem Hügel*' ('The house stands on the hill') – which would be broadcast on the 6 o'clock news of the BBC German Service.[25] But with no communication from Rosbaud or anything else to act on, the wake-up call wasn't sent.

On 31 May 1939, with anxiety running rampant, John Cockcroft told Henry Tizard about rumours that 'German military authorities are taking uranium explosion very seriously and are making arrangements for an immediate trial with 100 kilos of uranium'.[26] The British still believed they held the trump card – the realisation that the critical mass of pure U-235 was only a few kilos – but this was a slender advantage which the Nazis could snatch away at any moment. Rudi Peierls was worried that 'any competent nuclear physicist' could come up with the right answer if he happened to wonder how a chunk of pure U-235 would react when bombarded with neutrons.[27] Perhaps someone like Werner Heisenberg, who had won his Nobel Prize at the age of just thirty-one – younger than James Chadwick, Frédéric Joliot, Marie Curie and even Albert Einstein.

13

MINORITY REPORTS
January–September 1941

At New Year 1941, the M.A.U.D. Committee had spent seven months considering whether uranium could influence the war, now into its second year. Proceedings ran smoothly, with occasional tantrums from Hans Halban, and Frederick Lindemann still attached like a limpet mine that could go off at any moment. George Thomson wisely delegated legwork to James Chadwick and John Cockcroft, both skilled at picking brains, changing minds and banging heads together. Chadwick was direct, terse and inscrutable. Cockcroft usually wore a half-smile, as if remembering a private joke. Like Chadwick, he said little, and 'Yes' sometimes meant 'I heard that' but signified 'No'. Copious notes in tiny handwriting in a black-bound Filofax notebook helped him to deal with several problems simultaneously.[1] One general described Cockcroft as 'the greatest and most helpful of scientists', and Churchill took him to inspect damage and lift spirits after air raids (these trips were hidden from Lindemann, who would have been apoplectic).[2]

Chadwick despised 'bogus security', believing that only trustworthy people would join the enterprise – 'any spy could have walked into the laboratory and asked questions' but hadn't so far. M.A.U.D. documents were classified by one or two rubber-stamped red stars, denoting SECRET or MOST SECRET. MI5 vetted potential recruits to root out Nazi sympathisers and 'enemy alien Communists who might sabotage the programme'. The Ministry of Aircraft Production concealed M.A.U.D. from everyone except the few who needed to know. Inexplicably, they also barred Lord Hankey, Chairman of the all-important Scientific Advisory Committee to the War Cabinet, who only heard about M.A.U.D. after it was dissolved.[3]

There were several potential showstoppers.[4] Uranium ore from the Belgian Congo had dried up and stocks were low; confidential feelers were put out to the Eldorado Mining Company in Canada. Uranium was dangerous. Sintered (powdered) uranium metal, the usual formulation, sometimes caught fire spontaneously, while gaseous hex dissolved vacuum seals and fouled up the turbines which would force the gas across the diffusion membranes. The membranes didn't yet exist; the best 'Dutch cloth' (with an astonishing 160,000 holes per square inch) had been made exclusively in Belgium and was now unobtainable.[5] And although the Manchester-based heavy-engineering firm Metropolitan-Vickers had been commissioned to build a pilot 20-stage U-235 separation plant by the end of the year, the prototype separation unit – just one of the 1,900 in Francis Simon's plans – was still under construction in the Clarendon Laboratory.[6]

U-235 was now the only nuclear explosive on the table in Britain, because M.A.U.D. had recently killed off Egon Bretscher's proposal to develop plutonium as a better option. Bretscher had to miss the Policy Committee meeting where he was due to make his case and was devastated to discover that Rudi Peierls had shot down his research plans behind his back.[7] As Bretscher's wife, Hanni, later said, 'It was a very, very sad meeting for Egon.'[8] Lindemann, also absent, later delivered the coup de grâce: 'Who thinks they can win the war with an element that nobody has ever seen?'[9]

Peierls was spectacularly wrong. In Berkeley, Glenn Seaborg was confirming Bretscher's predictions that element 94 was more fissionable and easier to purify than U-235. Months later, someone in America eventually decided to tell the Brits but Bretscher had already lost the race – and Seaborg was over the horizon when Bretscher and Feather filed two secret patents with Halban and Kowarski to make plutonium in the boiler.[10]

The Americans' silence over plutonium was symptomatic of a deeper malaise. Over twenty American scientists visited Britain after the NDRC opened its London office in February 1941, to facilitate the Anglo-American exchange of military scientific secrets envisaged by Henry Tizard.[11] They attended M.A.U.D. and other top-secret committees and went home laden with 'information of great value' about radar, submarine detection, rockets, explosives and chemical warfare. In return, the Americans gave virtually nothing – 'only a few reports', none of them exciting or useful.[12] London and Washington seemed to be connected by a one-way valve that only permitted westward flow.

A rare concession reached M.A.U.D. in March 1941: Merl Tuve's estimates of the fissionability of U-235, calculated using an indirect method similar to Otto Frisch's. Peierls slotted Tuve's values into his own equations and confirmed Frisch's calculation that a hugely destructive super-bomb could be made with only 8 kg of pure U-235.[13] Chadwick still wanted direct measurements on U-235 and asked Albert Nier in Minnesota to send him a dot of the pure isotope. No reply. Chadwick suggested experiments that Nier could do himself, if he were reluctant to send such precious material across the Atlantic. Still no reply.[14]

In late May 1941, Rudi and Genia Peierls took in a new lodger. Not yet thirty, he had a high forehead accentuated by frontal balding and was impassive behind wire-rimmed spectacles. He was 'aloof, austere and shy' and had an undisclosed illness which periodically necessitated days away 'to see a doctor'. No question of his hiking into the hills with the Peierlses, let alone sharing a bed with them.[15]

The lodger was born near Darmstadt in 1911 and life hadn't been kind to him. His early adulthood was wrecked by Nazi persecution because his family were Quakers. He studied physics at Kiel, where fellow students tried to kill him for belonging to the German Communist Party ('I escaped,' he said simply). Helped by the Academic Assistance Council, he fled to England in 1933, did a PhD in Bristol and then worked with Max Born in Edinburgh. His mother committed suicide in 1933 in a way guaranteed to scar those left behind (she drank acid), and his sister finally evaded the Nazis in 1939 by jumping under a train. In May 1940, despite being a naturalised British citizen, he was arrested and sent to Canada on the *Ettrick* for five miserable months of internment. After petitioning by Born and Peierls – 'brilliant ... one of the two or three best theoretical physicists of his generation' – he was repatriated to Edinburgh in December 1940.[16]

Peierls needed him to help crack the problems of separating U-235, and offered him a post in Birmingham on 'secret war work' in early May 1941. The new recruit was cleared by MI5, signed the Official Secrets Act and started on 29 May with a two-day briefing all about the bomb from one of Chadwick's team in Liverpool.[17] His name was Klaus Fuchs (Plate 15).

Job Done

During spring 1941, the Ministry of Aircraft Production pressurised George Thomson to bring M.A.U.D.'s deliberations to a head – perhaps because its super-bomb was set to bust the government's uranium budget 300 times over. In late June 1941, Thomson circulated a draft report to the M.A.U.D. Technical Committee members, giving them five days to pull it apart. The group's last meeting took place on 2 July at the Royal Society, attended by Lindemann and a visiting American physicist, Charles Lauritsen from Caltech. Chadwick argued that the U-235 super-bomb must be the primary focus and that the Parisians' boiler was a bolt-on project. Thomson gave Chadwick the task of finalising the report, which took him two weeks of twenty-hour working days while 'completely addled' by exhaustion and frustration (Nier's dot of U-235 was still 'awaited from the US'). Chadwick often had to shelter in his cellar because the Liverpool blitz was in full swing.[18]

Thomson signed off the final report on 15 July 1941 and sent it to the minister for onward transmission to Lord Hankey and his Scientific Advisory Committee to the War Cabinet. Copies emblazoned MOST SECRET were delivered to the London office of the NDRC for the urgent attention of its chairman and his deputy, Vannevar Bush and James Conant, and of Lyman Briggs, Chairman of the Uranium Committee in Washington (recently renamed the 'S-1 Committee').[19]

When the M.AU.D. Committee broke up for the last time, everyone around the table would have recognised that M.A.U.D. had served its purpose. Few if any would have predicted the shape of things to come.

The M.A.U.D. Committee's final report was split into two parts: the nineteen-page 'The Use of Uranium in a Bomb' and, much briefer, 'The Use of Uranium to Generate Power'.[20] The committee had begun 'with more scepticism than belief' but was now convinced that uranium could make 'a very powerful weapon of war' which could be decisive in the current conflict.[21]

The design of the bomb? Externally, the device would look like any 1-ton bomb to be dropped from an aircraft; internally, the only certainty was the part that would go bang. It was absurdly small: a 4-inch sphere, the same size as a 10-pound cannonball from the Battle of Trafalgar but weighing two-and-a-half times more because it was pure U-235.[22] You could hold it in your hand, but not for very long. The sphere was bisected and its two halves held apart 'at a safe distance until the explosion is wanted'. High-explosive charges would then blast the halves together much faster than a rifle bullet. Their combined weight would instantly exceed the critical mass for U-235 – so a single stray neutron, generated by the cosmic rays which constantly bombard the Earth or the spontaneous cracking of a U-235 nucleus, would fling open the gates of hell. The 'final avalanche of the chain reaction' would run its course in one hundred-millionth of a second.[23]

Impact? Ten kilos of U-235 would explode with the force of 1,800 tons of TNT, four times the weight of the bombs that had flattened the centre of Coventry. The only man-made precedent was a munitions ship containing 2,000 tons of high explosives which blew up in Halifax, Nova Scotia, in December 1917. The circle of total devastation was a mile and a half across, with very little standing for another mile beyond that. A wide area would be left radioactive and uninhabitable.[24] Its psychological impact would be wider still, with the realisation that this was the work of a single bomb, not a skyful of bombers, and that more could come.

Cost? Eye-watering: £8.5 million, including £5 million (one-tenth of Britain's weekly war expenditure) for a factory producing 1 kilo of U-235 per day. But this would pay for thirty-six bombs, each costing £236,000 – a 40 per cent saving on £392,000 for 1,800 1-ton TNT bombs.[25]

How long? Without 'unforeseen difficulties', two years. 'Messrs ICI', confidential industrial partners and world leaders in complicated chemical factories, predicted that a separation plant would begin construction in August 1942 and start producing U-235 a year later; the first bombs would be ready to drop by the end of 1943.[26]

Risks? The separation of U-235 was 'a matter of great difficulty' and the make-or-break diffusion membranes were still conjecture; a promising American mesh, perforated by chemical etching, hadn't yet been obtained for testing. Metropolitan-Vickers hoped to find 'solutions' for some 'developmental difficulties' in their designs for the separation plant.[27]

Practicalities? The project depended on industrial collaborations with Metropolitan-Vickers and especially ICI, which provided the guy-ropes that kept the edifice standing. ICI personnel, including the chairman Lord Melchett, had visited all the research groups and attended crucial M.A.U.D. meetings. ICI was 'prepared to be in executive charge' of the high-risk separation plant.[28] Most importantly, success hung crucially on American cooperation. Britain alone couldn't make the bomb while at war and under bombardment. America, still at peace, could provide technical know-how, manpower and a safe haven for the separation factory. So far, the Americans had concentrated on generating nuclear energy in Fermi's pile and ignored the British U-235 bomb;[29] their minds would have to be changed.

Patrick Blackett thought that the bomb wouldn't be ready by the end of 1943 but his dissent was outweighed by a dramatic change of heart by a former enemy of the project.[30] Frederick Lindemann, now ennobled by Churchill as Lord Cherwell of Oxford, had finally been persuaded that the bomb could work.[31] Lindemann demonstrated one quality of a good scientist – 'when the facts change, I change my mind' – but his peerage hadn't made him any nicer.

The M.A.U.D. report's short section on using uranium to generate power read as though the committee had washed their hands of it. Even though nuclear energy would undoubtedly be the future and Halban and Kowarski currently led the world, it was not 'worth serious consideration' in war-torn Britain. Halban and Kowarski should therefore continue their work in America.[32] Chadwick had convinced Harold Urey at Columbia that the Parisians had enough promise for the USA to house them, and Urey persuaded the US government to put up $1 million for a heavy-water factory.[33] This plan was also consistent with Minister Dautry's original intention to transfer Joliot's research group to North America.

However, ICI threw in a counterproposal to keep the boiler research in England and under its control. ICI argued that nuclear energy would become a reality surprisingly soon – perhaps propelling ships 'by the end of the War' – and was too important to hand to the Americans. Being 'one of the few companies in the world capable of undertaking this research and commercial development', ICI would own the project – and would appoint Dr Halban to direct the research.[34]

This revelation shocked Kowarski, who regarded the chain reaction as 'the Philosopher's Stone ... far more than a Nobel Prize' and wanted to take the boiler to America.[35] Halban, a Viennese aristocrat and (in Kowarski's words) 'the caricature of a Prussian officer', had always treated Kowarski as his inferior. When they'd first met in Paris seven years earlier, Halban was Joliot's right-hand man whereas Kowarski was a jobbing gas engineer, part-time PhD student and temping as Joliot's *'petite dactylo'* ('little secretary'); he first knew of Cockcroft by typing a letter to him which Joliot had dictated. Kowarski only became a junior lecturer in 1935, when Joliot recognised his aptitude for electronics and lateral thinking.[36] Now, Kowarski was fed up with being the dancing bear while Halban played ringmaster with ICI. And when he discovered that Halban had held clandestine meetings with senior ICI personnel, Kowarski smelled conspiracy: the 'lame duck' M.A.U.D. Committee had been stitched up by 'ICI disguised as the Department of Scientific and Industrial Research'.[37]

Old Friends

While bringing new wizardry to the understanding of gaseous diffusion, Klaus Fuchs also helped Peierls to track *Uranverein* members by sifting through journals, conference proceedings and other low-grade information seeping out of Germany.[38] This was all they had to go on; Paul Rosbaud in Berlin was still silent, after fifteen months. Hints of Nazi malefaction included intense activity in the 'fission research institute' at the KWI for Physics, and purchases of uranium ore from Portugal and heavy-duty fans like those in Simon's separation plant.[39] However, the key players in the *Uranverein* seemed undisturbed – exactly as M.A.U.D. Committee members would have appeared to German eyes.

Werner Heisenberg and Carl von Weizsäcker kept cropping up at Nazi showcase conferences across Europe. Von Weizsäcker visited Copenhagen in March 1941, talking in fluent Danish about the insides of stars. In late September, he was back with Heisenberg to speak at an astrophysics conference, after which Heisenberg returned to Leipzig and von Weizsäcker to Berlin. All seemed to go quiet again but something strange had occurred: a meeting between Heisenberg and Bohr which left unanswered questions and a bad taste in many mouths. We shall never know exactly what happened; neither man kept records at the time and their later accounts diverged widely.[40]

While working in Bohr's Institute in 1926–7, Heisenberg had enjoyed special intellectual and personal closeness with the Great Dane. But when he'd sat beside Bohr for the 1937 Summer Conference photo (Plate 7), the world was changing and Heisenberg had already declared his hand. Reich officials had recently sent him to Edinburgh to persuade Max Born, his former boss and a Jewish refugee, to return to Berlin. When Born refused, Heisenberg taunted him for being a Jew and spat at his feet.[41]

Whatever the real reason behind Heisenberg's meeting with Bohr in September 1941, he laid the ground badly by telling some of Bohr's colleagues that the war was 'a biological necessity'.[42] Between 15 and 21 September, Heisenberg met Bohr three times, twice in the evening at the Bohrs' house together with Margrethe and their son Erik. It was almost like old times. They reminisced, Heisenberg played Mozart on the piano, Bohr read aloud, and after midnight Bohr walked Heisenberg to his tram-stop. One night, probably Thursday 18 September 1941, everything fell apart. It seems likely that Heisenberg asked Bohr if scientists should work on nuclear fission in wartime – and then told him that making an atomic bomb was entirely feasible, even if technically difficult, and could decide the outcome of the war. Bohr had always believed that the bomb was a practical impossibility and was so shocked that, according to Heisenberg, he couldn't continue a coherent conversation. The two men contradicted each other about what came next. Heisenberg later claimed that he'd come to ask Bohr to tell physicists everywhere to stop atomic-bomb research. Bohr insisted that he'd felt threatened, believing that Heisenberg had come to pick his brains about Allied research on the bomb and make him side with the Nazis because they would win the war. And when Heisenberg told him that atomic weapons could decide the outcome of the war, Bohr had felt 'great anxiety'.[43]

The father figure and his prodigal son didn't meet again until after the war. Heisenberg called the encounter 'a disaster'. When Lise Meitner heard about it, she became 'very melancholy'. Heisenberg was no longer a 'decent human being'. He'd broken into the temple, defiled the altar and abused the high priest.[44]

Progress Reports

In autumn 1941, Paul Rosbaud was finally reactivated to investigate a secret German war-research project. He was given a codename – *Greif* ('Griffin') – and a new MI6 handler whom he knew as 'Theodor'.[45] Rosbaud had got on well with Frank Foley, his original MI6 contact who had fixed visas for Rosbaud's wife and daughter in 1938, and introduced Rosbaud to Cockcroft in April 1939. Foley was described as 'an active little man' who worked fourteen-hour days at the British passport offices in Berlin and Oslo, where he was transferred in July 1939. In that role, he arranged visas for some 10,000 German Jews to escape the Nazis. If the Abwehr had searched Foley's house in Berlin, they might have found Jewish fugitives. And if they'd been in the right place at the right time, they could have spotted him helping to load thirteen canisters of heavy water into a plane preparing to leave for Perth at Oslo Airport and later on the gangway of the *Broompark*, supervising the evacuation of French scientists from Bordeaux. Foley was now busy in England, interrogating Rudolf Hess, Hitler's deputy, who had fled Germany in a Messerschmitt fighter and crash-landed in Scotland. Having studied philosophy in Hamburg before the Great War, Foley was fluent in German and Hess couldn't detect any accent.[46]

Foley's replacement as Rosbaud's handler was another short Englishman with round spectacles and a small black notebook like John Cockcroft's. Eric Welsh ('Theodor') was based in London with MI6, but his mind and heart were in Norway where he'd spent half his life. He moved to Bergen in 1919 as a junior research chemist with a paint company, mastered Norwegian, married a local girl and ultimately become the company's technical director. When the Germans invaded in April 1940, Welsh's family fled to England, but he stayed behind to link up with the nascent Norwegian Resistance.[47] Back in England, he was snapped up by MI6 to create a network of agents in Norway. 'Welsh's gang', extracted by Scottish fishing boats for training in Britain and returned by the same route, were fiercely loyal and willing to 'endure hazards and tough conditions'. Their tasks included tracking the *Tirpitz* and other German battleships sheltering in Norwegian fjords, coordinating resistance against the Nazis, and sabotage.[48]

Researching paints had taken Welsh to the barnacle-encrusted hulls of trawlers, the inside of sardine tins and the Norsk Hydro plant at Vemork. This was why he was now handling Rosbaud for MI6. One of Welsh's gang had recently escaped to England with ominous news. Karl Wirtz, of the KWI for Physics and the *Uranverein*, had visited Vemork and ordered Norsk Hydro to increase heavy-water production to 1,500 litres per year, its highest level ever.[49] MI6 immediately expanded Welsh's brief to include the German atomic bomb and decided to cut off heavy water from the Germans. Leif Tronstad, the messenger who brought the bad news from Norway, would lead the sabotage mission. Tronstad was uniquely qualified for the job:

PhD at the KWI for Physical Chemistry in Berlin and now Professor of Inorganic Chemistry in Trondheim; expert on deuterium and heavy water; and co-designer of the electrolytic equipment at Vemork. In autumn 1941, the exiled Norwegian High Command gave Tronstad the rank of Major and responsibility for intelligence, espionage and sabotage. He settled into the secret headquarters of the Special Operations Executive on Baker Street in London, and began learning the dark arts needed to blow up the heavy-water plant at Vemork.[50]

All Change

The summer of 1941 piled frustration on Mark Oliphant. His physics department in Birmingham had produced two potentially war-winning discoveries but both were slipping through his fingers. John Randall was still struggling to patent his cavity magnetron, while money poured into American companies which mass-produced the devices.[51] Even more infuriating was the complete absence of any American response to the final M.A.U.D. report on the U-235 super-bomb, which left the British wondering why the material had 'neither reached the right people nor helped the Americans as much as had been expected'.[52]

Bringing the Americans on board was always destined to be difficult because their nuclear physicists didn't believe that an atomic bomb could be made in the foreseeable future. They had also ignored many prompts to think again. Vannevar Bush and James Conant at the NDRC and Lyman Briggs at the Uranium Committee had received all the M.A.U.D. meetings' minutes, final report and supporting documents.[53] And American scientists visiting Britain had brought back first-hand reports about the U-235 super-bomb.

In March 1941, a month after the NDRC's Liaison Office opened in London, James Conant was sent there on Roosevelt's instruction. After audiences with the King and Winston Churchill, Conant met Frederick Lindemann at his club and was lectured about the super-bomb. At that time, Lindemann still thought the bomb improbable but had shortened the odds against it. This was a revelation for Conant, who hadn't previously heard of 'even the remote possibility of a bomb'; he returned to America better informed but still sceptical.[54] A month later, Kenneth Bainbridge from the Radiation Laboratory at MIT went to London, primarily to discuss microwave radar. In April, Bainbridge joined the meeting of the M.A.U.D. Technical Committee. Surprised by the progress reports presented, Bainbridge informed Bush by telegram that the British were taking the bomb very seriously and would need American cooperation with the separation of U-235. Back home, Bush simply noted that fission research was unlikely to yield 'defense results of great importance'.[55] Finally, on 1 July 1941, Charles Lauritsen attended the last Technical Committee meeting at which the draft M.A.U.D. report was presented and discussed. Lauritsen, originally an astrophysicist and currently working on proximity fuses, had no direct involvement with fission research. On his return, he went to see Bush and stressed the 'new urgency' of the British work on the super-bomb. Bush did nothing then, or soon after when he, Conant and Briggs received copies of the M.A.U.D. meeting minutes and the draft report.[56] Any residual doubts should have been dispelled by the final M.A.U.D. report, which was sent to all three at the end of July. George Thomson later said, 'We were the first to say that the bomb would definitely work,'[57] but if any of this intelligence had an impact – or had even been read – nobody let the British know.

These three men – Bush, Conant and Briggs – were the crux of the problem. A seed as delicate as the U-235 bomb had to fall on fertile ground, and they seemed disinterested in its germination. Vannevar Bush, 'a shrewd, spry Yankee of fifty … plainspoken but with a disarming twinkle in his eye and a boyish grin', was a brilliant electrical engineer who became Dean of Engineering at MIT and then President of the privately funded Carnegie Institution (Plate 16). He was politically adept and enjoyed 'a symbiotic relationship' with Roosevelt. Unfortunately, Bush had a blind spot for nuclear physics; he admitted to being 'puzzled as to what, if anything, should be done in this country' because he couldn't tell if the chain reaction was important.[58] James Conant (Plate 16) was a charismatic and reforming President of Harvard, who trained as a chemist and worked on poison gas during the First World War. He was well qualified to be

Vice-Chairman of the NDRC – but also knew nothing about the physics underpinning the U-235 bomb.[59] Lyman Briggs, Director of the National Bureau of Standards and another political appointment, was charming but 'slow-moving ... he really wasn't able to grasp fundamental concepts' – which included nuclear physics. And although Bush agreed that Briggs was 'operating only at peacetime speed', he left him chairing successive Uranium Committees.[60] Charles Darwin, head of the British Central Scientific Office in Washington, observed the S-1 Uranium Committee at work in July 1941. During the six-hour meeting, he saw 'very little of substance', with only a few minutes discussing the separation of U-235. 'They are very nearly stuck,' he concluded.[61]

Things might have been different if Bush had listened to advice and hadn't been swayed by a personality clash. After attending the M.A.U.D. Technical meeting in April, Kenneth Bainbridge had suggested that Bush should send to Britain 'a representative who is thoroughly familiar with the uranium problem'.[62] An ideal candidate would have been Ernest Lawrence from Berkeley (Plate 16) – Nobel Prize winner, inventor of the cyclotron and the intellectual equal of Thomson, Chadwick and Cockcroft. Unfortunately, the forthright Lawrence had fallen out with the egocentric Bush. Lawrence had been commissioned by the Uranium Committee to investigate the fissionability of element 94 (plutonium). He became frustrated by the committee's inertia and was 'appalled' by the slowness and ignorance of its chairman. When he complained, Bush immediately slapped him down with 'I'm running the show'.[63] And so the three men who knew nothing about nuclear fission maintained their stranglehold over the two groups of scientists – the NDRC and the Uranium Committee – who would tell Roosevelt whether the atomic bomb was fact or fiction.

In late August 1941, Mark Oliphant braved the Atlantic airspace to update the Radiation Laboratory at MIT about British advances in radar. He carried an official letter from George Thomson, authorising him on behalf of the M.A.U.D. Committee to make 'discreet enquiries' into the Americans' apparent apathy over the super-bomb.[64]

Accompanied by Charles Darwin, Oliphant called unexpectedly on the Chairman of the S-1 Committee. One piece of the puzzle immediately snapped into place: the committee had never seen the final M.A.U.D. report. Briggs had received it, glanced at it and couldn't understand it. Instead of circulating the document to the committee members who could have made sense of it, he locked it in his safe where it lay undisturbed until Oliphant strode into his office.[65]

The confrontation with Briggs ('this inarticulate, unimpressive man') left Oliphant 'amazed and distressed'. Oliphant visited members of the S-1 Committee to persuade them to believe in the bomb. He failed. Samuel Allison, one of Enrico Fermi's colleagues at Columbia, was 'surprised' that Briggs had kept them in the dark but did nothing, and Fermi remained convinced that the atomic bomb was a waste of time and effort. Oliphant concluded that American interest in the bomb was 'in the charge of non-nuclear scientists and is therefore being badly mismanaged'.[66]

Finally, in late September, Oliphant went to Berkeley to talk to Ernest Lawrence. They'd been friends since 1937, when Oliphant visited Lawrence's new cyclotron with a view to building one at the Cavendish. He'd stayed for weeks, helping to refine the design. When Lawrence was awarded the 1939 Nobel Prize in Physics for inventing the cyclotron, he received a congratulatory telegram from Oliphant and the ill-fated Harold Walke, just back in England from a spell in Berkeley.

Lawrence gave Oliphant a sympathetic hearing and promised to look into it but hadn't reported back when Oliphant left for England a few days later. There, further disillusionment awaited Oliphant, because the Americans weren't alone in doing nothing about the M.A.U.D. report. It had disappeared into the byzantine labyrinth of Whitehall two months earlier, but nobody was prepared to talk about its fate.[67] Even in peacetime, the delay would have been unacceptable; to Oliphant, for whom 'slow' meant one-tenth of the speed of light, it was outrageous and insane.

14

TUBE ALLOYS

September 1941–January 1942

A week, supposedly a long time in politics, was one-fifth of the time that the M.A.U.D. report took to traverse the constipated bowels of the Ministry of Aircraft Production. It eventually reached Lord Hankey, Chairman of the Defence Services Panel to the War Cabinet, on 27 August 1941. 'Maddening,' said George Thomson.[1]

Luckily, Thomson had slipped Hankey a surreptitious copy as a precaution, and Hankey had also been nobbled over lunch in the Athenaeum by Lords McGowan and Melchett (the Chairman and Vice-Chairman of ICI) and Cherwell (alias Frederick Lindemann).[2] Next came a five-page letter, handwritten for security, from Charles Darwin, head of the British Central Scientific Office in Washington, assuring Hankey that Vannevar Bush and James Conant, the drivers of the US nuclear programme, wanted full collaboration on both the bomb and the boiler as 'a joint project between the two governments'.[3]

Immediately after receiving the report officially, Hankey convened a top-level Defence Services Panel which included three Nobel Prize winners and the President of the Royal Society. During seven frantic days in mid-September 1941, the panel met seven times and grilled Chadwick, Tizard, Lindemann and four other expert witnesses. Its verdict was crisp and urgent. The U-235 bomb merited 'the very highest importance... and all possible steps should be taken to push on with the work'. Without delay, Francis Simon's U-235 separation prototype must be built and tested, while the Royal Arsenal at Woolwich must develop a gun device to fire two U-235 slugs together at incredible speed. ICI couldn't control the boiler, because the future of nuclear energy was too important to 'fall into the hands of private commercial interests'. Success hinged entirely on help from the Americans. The full-size U-235 separation factory could only be built in North America, while the boiler must be a collaboration with the USA or possibly Canada.[4]

Sir John Anderson, Lord President of the Council, received the report on 25 September 1941, just two weeks after Hankey's panel's first met. Anderson didn't look like a magician who could turn science-fiction fantasy into a weapon of war. He was a quiet Scot ('often alarmingly devoid of small talk') with heavy features reminiscent of 'a slightly sad bloodhound', and his pinstriped trousers, wing collars and gold watch-chain recalled 'the war before last'. But his mind was ferociously sharp, his nickname in Whitehall was 'Jehovah', and Churchill counted on his wisdom. He even had scientific credibility, as a brilliant young researcher who had studied the chemistry of uranium in Leipzig before entering the Civil Service.[5] Anderson chaired the top-level Scientific Advisory Committee and was its conduit into the War Cabinet. He was 'greatly excited' by the 'sensational' M.A.U.D. report and the War Cabinet promptly approved all the recommendations of Hankey's panel, including building a U-235 separation factory in the USA.[6]

Revamp

M.A.U.D. had laid the theoretical foundations for the bomb and the boiler; turning them into reality demanded a new organisation with different people and skills. First, Anderson filled the top job. Exceptional credentials were demanded: experience in running complex projects, infectious drive, and charisma to unite scientists, engineers and industrialists around a massive, uncharted venture with no guarantee of success and against a timetable that would be terrifyingly tight even in peacetime. Anderson saw no need for a Nobel Prize or even first-hand knowledge of nuclear physics.

In retrospect, 'no better selection could have been made' but Anderson's choice would later whip up outrage on both sides of the Atlantic. Wallace Akers (Plate 17) was fifty-three years old and possessed 'wide knowledge, unbounded energy, even temperament and absolute integrity', together with 'ingenuity, tact and organising ability of a high order'. Like Anderson, he'd studied physical chemistry (taking a First at Oxford) and had done good academic research, but then defected to industrial chemistry at ICI. As ICI's Research Director, Akers had sat alongside M.A.U.D. scientists while leading the company's bid to take over the U-235 separation factory and boiler.[7]

Akers began by recruiting thirty-five-year-old Michael Perrin as his Assistant Director of the unnamed organisation which didn't yet exist. Perrin, another Oxford-trained chemist from ICI's London headquarters, was renowned for having boosted the reputation of ICI and British industry by turning polythene, a discovery left for dead on an ICI lab bench in Cheshire, into a global money-spinner. Having worked intensively with M.A.U.D., Perrin 'knew the men and the work' better than anyone.[8] He'd begun on the fringes of ICI's war effort with camouflage paints, polythene bullets (no good) and luminous materials that found uses during blackouts ('Why not buy your friend's dog a "Lustre" lead for Christmas?'). Perrin was sucked in as Chadwick's collaborations with ICI spread beyond its factories in Cheshire. On 8 January 1941, Perrin attended his first M.A.U.D. Technical Committee meeting – in ICI's Nobel House on Smith Square in Westminster – to discuss 'Energy-X' (the Halban–Kowarski boiler). In no time, he was brainstorming diffusion membranes and prototype units for U-235 separation, exploring other separation methods, and supplying uranium compounds for making hex and metallic uranium. During 1941, Perrin attended the M.A.U.D. Technical Committee four times, as well as seventeen meetings on isotope separation, twenty-five on uranium and hex, and over fifty concerning the boiler.[9] He was well liked, and his unflappable calm, discretion and skills of reconciliation were invaluable in dealing with Halban's tantrums and deep gloom when experiments didn't work.

On 18 October 1941, Akers and Perrin were released by ICI for the duration and became civil servants in the Department for Scientific and Industrial Research, although Akers continued to advise the company part-time; some noted the 'Gilbertian' irony in having ICI men run the show after Hankey's panel had crushed the company's attempt to take over the boiler. The nascent organisation was given rooms at 16 Old Queen Street – a few minutes' stroll from the Houses of Parliament – and £100,000 to fund its first six months.[10] A cover name was needed which, like Vannevar Bush's 'S-1 Section', would reveal nothing and deflect the curiosity of enemies. Akers and Perrin toyed with 'X Alloys' and took 'Tank Alloys' to Anderson, who suggested 'Tube Alloys'. The name looked 'meaningless and unintelligible ... with a specious probability', and it stuck. The 'Tube Alloys Directorate' embraced the whole enterprise. The Tube Alloys Technical Committee dealt with hard science; chaired by Akers, it comprised Chadwick, Peierls, Simon, Halban and Roland Slade, a senior research chemist from ICI. Contact with the War Cabinet was through the Tube Alloys Consultative Council, with Anderson (chair), Hankey, Lindemann, Sir Edward Appleton (Director of the Department for Scientific and Industrial Research) and Sir Henry Dale, President of the Royal Society.[11]

The Tube Alloys Technical Committee met for the first time in Old Queen Street on 11 November 1941. Wallace Akers's skills in dealing with difficult people had already been tested. Appleton had enraged Chadwick by inviting him to join in a clumsy letter which implied that Akers and Perrin would control the scientific work. Akers immediately smoothed things over; of course, Chadwick was in the driving seat, with purely administrative assistance from Akers and Perrin.[12]

The inaugural agenda was crowded: secrecy, securing uranium and heavy water, Simon's U-235 separation plant, the Halban–Kowarski boiler, patents and how to work with the Americans. Akers was a consummate chairman – wise, witty, impartial – and the meeting ended with actions agreed, new questions clearly identified, and feathers unruffled. Unfortunately, more avoidable unhappiness quickly followed. The only M.A.U.D. veterans invited were Chadwick, Peierls, Simon and Halban; none of the others knew that the meeting was taking place, or that a new organisation had replaced M.A.U.D., or that it was called Tube Alloys. Hankey informed the rest that they weren't needed in 'autocratic and discourteous' letters which he sent out slowly and randomly, causing hurt and resentment. Mark Oliphant, who received his letter in late November, protested that it was ridiculous to have the committee run by 'commercial representatives completely ignorant of essential nuclear physics'. Hankey was even ruder to George Thomson, ex-Chairman of M.A.U.D., and John Cockcroft. He didn't write to them until over a month after the first meeting of the new organisation, whose name was withheld from them. Cockcroft strongly disapproved of the appointment of Akers and Perrin and blamed the debacle on 'ICI and Halban, I suppose'.[13]

But, like everyone else, even Oliphant and Cockcroft were 'won over' when they met Wallace Akers in person: 'thoughtful, donnish... a gentleman... perfectly charming', eloquent about fine wine, art and music, and above all a good listener. The volatile Oliphant quickly succumbed over lunch and admitted that Akers was 'an excellent person to be in charge of the work'.[14]

Tube Alloys was properly immersed in secrecy from the start. M.A.U.D.'s rubber-stamped red stars were superseded by 'MOST SECRET', 'TO BE KEPT UNDER LOCK AND KEY', and 'IMMEDIATE – DECYPHER YOURSELF'. Only those in the Tube Alloys Directorate were party to the entire programme; grass-roots workers saw only their square inch of the big picture, and most asked no questions.[15] (One inquisitive soul, puzzled by an unredacted 'UF$_6$' in a secret document, bought the *Penguin Dictionary of Science* at a railway station and worked out what he was really doing while on the train home.)[16] Science-journal editors were briefed in secret and told to refer any submissions about nuclear fission to the Chief Technical Censor, who had a hotline to the Tube Alloys office. Relevant patent applications were reviewed by the Comptroller General of Patents and filed secretly if necessary. New recruits to the bomb or boiler projects were intensively vetted, with some lapses. Maurice Wilkins, who had just finished a PhD on radar in Oliphant's department, was hired despite being a Communist and having waved provocative banners

('BOOKS NOT BOMBS!' 'SCHOLARSHIPS NOT BATTLESHIPS!') at pacifist rallies before the war. One afternoon, Wilkins walked into Oliphant's office (stubbing his toe on a sack of uranium oxide) and asked to join the fission project. Oliphant soon had him vaporising uranium metal, trying to separate U-235 inside shiny steel columns which ran up the wall of the main undergraduate lecture theatre from Wilkins's lab in the basement.[17]

Uranium remained a perpetual headache. British stocks were dangerously low at just 12 tons of uranium oxide, compared with the 100 tons which Union Minière had recently sold to America and Germany's estimated 155 tons (and more waiting underground in the reopened mines at St Joachimsthal). The Canadian government helped Britain by quietly buying up the Eldorado Mine Company and its uranium mines. The owner of Eldorado was 'entirely reliable and well disposed', according to his close friend the Canadian Minister of Munitions – who happened to be the anonymous trader acquiring large blocks of Eldorado shares as inconspicuously as possible.[18]

Francis Simon's U-235 separation plant was the weak link in the Tube Alloys chain, whose failure could kill off the super-bomb altogether. An ideal site for the top-secret British pilot plant had been found: a former Ministry of Defence poison-gas factory hidden in dense woodland near Mold in North Wales. Rhydymwyn, known as 'Valley' for reasons of security as well as ease of pronunciation, was readily accessible from the Metropolitan-Vickers factory in Manchester and ICI plants in Cheshire. As Valley was already kitted out with barbed wire and gas-masks (essential for working with hex), the cover story of a new poison-gas plant would be easy to embroider.[19]

The all-important gaseous diffusion membranes were now being made at a printing works in Bradford which normally ran off popular magazines like *Stitchcraft*. The process, inspired by someone clever looking closely at a blown-up photograph on an advertisement hoarding, involved microprinting bacteria-sized dots onto a copper sheet which was then etched away with acid. The young women on the production line were trained to hold up the fragile copper sheets, freshly washed from the acid bath, in front of a pinprick of bright light. Those that showed a satisfactory diffraction pattern were dried using a hand-held hair dryer, shipped away to an undisclosed destination (ICI Billingham in Teesside) and would eventually be slotted into the U-235 separation units.[20] These units were still in evolution; Rudi Peierls solved one design logjam with the help of cardboard, paper, screws and origami.[21]

If he hadn't already been included, Francis Simon would now have been added to Hitler's 'Black Book' of German fugitives to be rounded up and executed when the Nazis invaded England.[22] He had a thirty-strong group of scientists and engineers at the Clarendon Laboratory, collaborators at ICI and Metropolitan-Vickers, and a budget of £50,000, half the total for Tube Alloys. But in reality, Simon's team was marching on the spot. They didn't even have any hex, because it was so difficult to synthesise and handle, so a surrogate mixture of vaporised iodine and bromoform had to be used in separation experiments. Making enough hex to produce 10 kg of pure U-235 remained a technological nightmare and perhaps an impossible dream. And they still hadn't built a working two-stage separation prototype, let alone the twenty-stage pilot plants at Rhydymwyn which would lay the ground for the 10,000-stage, 40-acre monstrosity that the Americans would build somewhere in the USA.[23]

Seeing the Light

Everyone who read the M.A.U.D. report understood that the U-235 super-bomb was doomed to failure unless the Americans collaborated. It was therefore unfortunate that Charles Darwin's handwritten letter to Lord Hankey,[24] claiming that the Americans were keen to cooperate, was either wishful thinking or a calculated lie. Darwin had seen for himself the dysfunctional S-1 Section Uranium Committee and America's exclusive focus on Fermi's progress towards generating power with his uranium–graphite pile at Columbia. Nobody there believed that atomic bombs could be produced in the foreseeable future and certainly not in time to help win the war, should – God forbid – the USA be dragged into the conflict in Europe. And the knowledge that had put Britain on a war footing over the U-235 super-bomb – the Frisch–Peierls memorandum and all the M.A.U.D. Committee papers – had no discernible impact when shared with Bush, Conant and Briggs.

In fact, an influential American had been excited by the concerted British efforts to develop the U-235 super-bomb, but that excitement had failed to propagate. In mid-April 1941, the radar specialist Kenneth Bainbridge attended a M.A.U.D. Technical Committee meeting and on returning to the USA persuaded Vannevar Bush to look seriously at uranium fission.[25] Bush ruled supreme as America's 'czar of research', heading the new Office for Scientific Research and Development (OSRD) which Roosevelt – at Bush's suggestion – had recently established. The NDRC, now chaired by James Conant, had become an appendage of the OSRD and oversaw Lyman Briggs's S-1 Section Uranium Committee.[26] Bush now sought an authoritative opinion about the feasibility of a uranium bomb. He asked the President of the National Academy of Science (NAS), the American counterpart of Britain's Royal Society, to assemble a 'blue-ribbon committee' of academy members to conduct 'an energetic but dispassionate review' of potential military applications of fission. The NAS committee was chaired by Arthur Compton, the forty-nine-year-old Professor of Physics in Chicago (Nobel Prize, 1927) and included Ernest Lawrence, with whom Bush had recently crossed swords.[27]

Compton's panel delivered its report to Bush on 17 May 1941: uranium fission might eventually propel ships and generate radioactive poisons, but atomic bombs were only 'a very remote possibility'.[28] James Conant was also extremely sceptical about a uranium fission bomb being produced within five years; this was 'a development for the *next* war, not this one'.[29] Nonetheless, Bush commissioned a further NAS review of the military uses of fission, this time with input from engineers. The second report, on 11 July 1941, again dismissed the atomic bomb as 'not likely to be of decisive importance in the present war' – but with dissent in the NAS's ranks. Ernest Lawrence had insisted on adding an appendix which identified plutonium as a dramatic new portal into the exploitation of nuclear fission. Lawrence recommended intensive research to develop industrial-scale production of plutonium for both reactors and atomic bombs.[30]

His plea was ignored, as were the testimony of Charles Lauritsen, who had sat in on the final M.A.U.D. Technical Committee meeting, and the draft M.A.U.D. report which the uncomprehending Briggs locked away in his safe.

Within three months, however, everything changed. Suddenly, everyone from the S-1 Section to the President believed in the U-235 super-bomb and was desperate to cooperate with the British. The turning-point, and arguably the catalytic event that led eventually to one of the greatest landmarks in the war and in the history of humanity, was Mark Oliphant's showdown with Lyman Briggs, when, incensed by finding the M.A.U.D. report languishing unread in Briggs's safe, Oliphant charged off to Berkeley and argued the case for the U-235 super-bomb so powerfully that Ernest Lawrence became a believer. At last, the American chain reaction began to ignite. Lawrence set up a meeting with Conant and Arthur Compton, and the three met at Compton's house in Chicago a few days later. Lawrence's new-found evangelism converted Conant and Compton, and Vannevar Bush followed suit soon after.[31]

On 3 October 1941, George Thomson travelled down from Ottawa (where he now directed the British scientific liaison office in Canada) and personally presented copies of the final M.A.U.D. report to Bush and Conant, for their eyes only.[32] That theatrical touch – just days after the report reached the British War Cabinet – may well have helped. Six days later, Bush took momentous news to the White House: the British had conceived an atomic bomb which seemed credible but couldn't be achieved without American cooperation to build a massive U-235 factory on American soil. Roosevelt immediately decided that the USA would mount an 'all out' assault on the bomb in close and complete collaboration with the British. The American effort would be led by a 'Top Policy Group' comprising himself, Vice President Henry Wallace, Secretary of War Henry Stimson, Bush and Conant.[33]

Two days later, on 11 October 1941, Roosevelt sent Churchill a secret letter, which looked personal but had been carefully drafted by Bush. The President thought it 'desirable' to discuss 'the subject covered by M.A.U.D. and Dr Bush's organisation', so that 'any extended efforts may be coordinated or even jointly conducted'. Roosevelt had even concocted a codename for the secret enterprise: 'MAYSON'. The letter was personally delivered to Churchill at Chequers by Frederick Hovde, who now headed the NDRC's office in London.[34] Back in Washington, Roosevelt returned his mind to urgent priorities – mounting tension with Japan in the Pacific – and waited for Churchill's reply.

The M.A.U.D. Committee and Lord Hankey's Defence Services Panel would have been pleased and relieved to learn that Roosevelt had embraced Anglo-American collaboration so positively. Churchill, however, thought otherwise. He'd happily cadged destroyers off Roosevelt but couldn't countenance Britain becoming dependent on the USA and was opposed to helping the Americans to develop military technology.[35] And his personal scientific adviser had told him almost exactly what he wanted to hear.

On 27 August 1941, the day when Hankey formally received the M.A.U.D. report, Frederick Lindemann leapfrogged all the little people and sent Churchill his own eight-page memo about the super-bomb. He had been persuaded to go straight to the top by Francis Simon and Rudi Peierls. Lindemann wouldn't bet more than even money on success within two years, but it would be 'unforgiveable' to let the Nazis have the bomb first. It was 'quite clear that we must go forward' and that Sir John Anderson should be the minister in command. A factory to test Simon's separation process must be built in England, not in the USA. Lindemann was prepared to exchange information with the Americans, but nothing more: 'However much I may trust my neighbour and depend on him, I am very much averse to putting myself completely at his mercy.' The U-235 super-bomb was a British invention and must be made in Britain.[36]

Churchill scribbled 'Good' in the margin beside Lindemann's conclusion that 'it seems almost certain that the bomb could be built' and forwarded the memo to his Chiefs of Staff, the commanders of the army, navy and RAF. His covering note contained black humour and a hard steer. Although 'quite content with existing explosives', Churchill felt that 'we must not stand in the path of improvement' and that 'action should be taken in the sense proposed'.[37]

The Prime Minister met his three Chiefs of Staff in private on 3 September 1941. All agreed that the bomb must have top priority and they went even further than Lindemann: Britain would build its own atomic bomb independently, without any American involvement whatsoever. This top-secret meeting wasn't minuted – 'the less there is on paper, the better', as one Chief remarked. Churchill's decision to exclude the Americans completely was shared only with Lindemann and Anderson. Nobody else was told, not even the Deputy Prime Minister or the War Cabinet, and certainly not the M.A.U.D. Committee, Hankey's panel or anyone who would actually make the weapon.[38]

Realism, not defeatism, had forced the M.A.U.D. Committee to conclude that the bomb couldn't be created without active partnership with the USA. The snap decision by four men ignorant of the full facts now left the bomb project in an impossible mess. British scientists and their American counterparts would soon begin designing the US-based U-235 separation factory and the super-bomb itself, supported by Roosevelt's Top Policy Group and the British War Cabinet. But even as the Anglo-American enterprise gained momentum, Churchill and his secret agents – Lindemann and Anderson – would block any practical move to cooperate with the Americans.

Which is why Churchill didn't reply to Roosevelt's letter of 11 October 1941, confirming America's commitment to work with Britain on the top-secret atomic project which the President had tentatively named MAYSON. In early September – three weeks before Sir John Anderson and the War Cabinet received the positive report of Lord Hankey's panel – Churchill had seized control of the bomb's destiny.

Inspectors Call

In late October 1941, Vannevar Bush decided to send two trusted observers to England to slice through the hyperbole and check out the British super-bomb. He chose George Pegram, the Chairman of Physics at Columbia who had given Leo Szilard lab space two years earlier, and Harold Urey, Nobel Prize winner in Chemistry for discovering deuterium and heavy water. Both were 'affable, balanced Anglophiles', charming but ruthless if required.[39]

The pair arrived in London on Monday 27 October 1941, complete with US Army-issue steel helmets and gas-masks. After a briefing by Akers, Perrin and Appleton, they were packed off on a magical mystery tour of Liverpool, Bradford, Birmingham, Oxford and Cambridge (Map 2). They were shown everything and encouraged to ask anybody about anything. In Liverpool, Chadwick levelled with them: 'I wish I could tell you that the bomb is not going to work, but I am 90 per cent certain that it will.' In Cambridge, Lew Kowarski handed a 5-litre canister of heavy water to its discoverer, who had only ever seen tiny samples.[40] In return, Pegram and Urey confirmed that America wanted to pool all knowledge about fission and do together whatever was needed to make the bomb. After seeing the copper diffusion membranes in Bradford, they shared a secret American alternative: an ingenious zinc–silver alloy from which the zinc could be dissolved by acid, leaving a spongy silver matrix riddled with sub-micron channels that might just separate [U-235]F_6 from its U-238 lookalike.[41] Urey also explained what had gone wrong back home; in his diary, Michael Perrin drew a flow-chart of the S-1 Uranium Committee which included a dashed line ending in a square box labelled 'Briggs's safe'.[42]

Pegram and Urey spent a month in England and returned to Washington on 5 November 1941. They told Bush that their visit had 'swept aside all doubts' and that the super-bomb was credible. The Brits were doing great work under horrendous conditions and were desperate for American collaboration.[43] The next day, Arthur Compton handed Bush a third NAS report on the military applications of nuclear fission. This had been updated without direct reference to the M.A.U.D. report (which Thomson shared only with Bush and Conant), but with input from Ernest Lawrence who had been fully briefed by Mark Oliphant in Berkeley. The NAS Committee had made a dramatic U-turn. It might quibble with details but not its conclusion that the U-235 fission bomb was entirely plausible, and 'within a few years' could determine military superiority.[44]

On 27 November 1941, Bush sent Roosevelt the NAS report with his strong endorsement. He didn't expect an instant reply, because the President was preoccupied with Japan, now racking up its threats over American trade restrictions in the Pacific. During the first week of December, Bush activated the American uranium bomb programme. He put Compton in overall charge and set up a two-pronged attack on purifying U-235, with Urey responsible for gaseous diffusion and Ernest Lawrence spearheading an ingenious electromagnetic method based on the cyclotron. Roosevelt signalled his approval on the evening of 6 December 1941, during the last few hours of calm before the storm.[45] Early the next afternoon came the shocking news that Japan had attacked the US fleet at Pearl Harbor.

Churchill normally replied rapidly to Roosevelt – by his own admission, as if responding to 'every whim of his mistress'.[46] But after almost two months, Churchill still hadn't answered Roosevelt's letter of 11 October, offering American collaboration on the bomb. When it finally came, Churchill's 'perfunctory' cable was dated only 'December' and reached Washington after Pearl Harbor. Churchill was evasive and lied about 'our readiness to collaborate in this matter' but arranged for Sir John Anderson and Lord Cherwell to meet Frederick Hovde, the NDRC's liaison officer in London.[47]

Roosevelt didn't reply immediately to Churchill, because America was now at war – not only with Japan but also with Germany and Italy, which had declared hostilities following the attack on Pearl Harbor. When Hovde met Anderson and Lindemann, he was left confused.[48] They insisted that they wanted to collaborate but couldn't until the Americans could ensure 'maximum security' to prevent 'the possibility of leakage of information to the enemy'.

Unaware of Churchill's decision to exclude the Americans completely, Tube Alloys and 'Dr Bush's organisation' continued to build their collaboration. In January 1942, Bush invited Akers to send a team to visit the American investigators working on the bomb and the boiler, with unfettered access to everyone and everything.[49]

James Chadwick headed Bush's guest list, but he refused to go. He hated leaving his lab and thought the trip a waste of time, telling Mark Oliphant that 'we are some way ahead and will remain ahead'.[50]

Oliphant was in regular contact with Ernest Lawrence and knew otherwise. He warned Chadwick, 'You are the only man in the country who really understands the problem ... especially as the Americans are getting ahead.'[51] But Chadwick's mind was made up. The Americans were still lagging behind and had nothing to teach the British.

15

ÜBER ALLES

March 1940 (reprise)–May 1942

In mid-January 1942, Michael Perrin put aside planning the Tube Alloys visit to America to draft a MOST SECRET memorandum to the Secretary of the War Cabinet.[1] The subject hung overhead like the Sword of Damocles, even more ominous because the British had no idea what was going on. Perrin's memo was an update on the German atomic bomb.

The German Army's *Uranverein* had celebrated its second birthday four months earlier. Centred on the KWI for Physics in Berlin, the *Uranverein* radiated out to Leipzig, Hamburg, Heidelberg, Munich and Vienna, with tentacles in Joliot's lab in Paris, Bohr's Institute in Copenhagen and the heavy-water plant at Vemork (Map 3). It was led by Werner Heisenberg, considered by the Nobel Prize committee and others to be more brilliant than Otto Frisch and Rudi Peierls – who had worried that the idea of the U-235 super-bomb could occur to any competent nuclear physicist.

Beyond those bare facts lay uncertainty. Paul Rosbaud had been incommunicado ever since giving John Cockcroft the shocking news of the *Uranverein*'s birth. The only live source was Jomar Brun, the chief engineer and co-designer of the Vemork electrolytic plant, who was now obliged to work closely with two *Uranverein* heavy-water experts, Karl Wirtz in Berlin and Paul Harteck from Hamburg. The information which Brun relayed via the Norwegian 'XU' underground Resistance network was troubling but inconclusive. After

Vemork fell into German hands, Harteck installed new catalytic convertors to boost heavy-water production to an unprecedented 420 litres per month. When monthly output stalled at under 100 litres, Brun was summoned to Berlin to explain why. Wirtz and Harteck treated him respectfully and showed him around the KWI for Physics, but the nearby Virus House and questions about the use of all that heavy water were strictly *verboten*.[2]

Meanwhile, Klaus Clusius, the wizard of isotope extraction, had pulled off another coup by separating two nearly identical xenon isotopes with his thermal diffusion tubes. The next target for this 'extraordinary' technique must surely be the grand challenge of teasing the 0.7 per cent of U-235 out of native uranium – an obvious discussion point which Clusius didn't mention in his paper.[3] Why not? Werner Heisenberg was another enigma, commuting up from Leipzig to spend two days each week at the KWI for Physics. He'd left Niels Bohr profoundly disturbed after they met in Copenhagen in September 1941 and was clearly 'up to something', but what?[4]

Perrin's memo to the War Cabinet explained what was known and surmised about the German nuclear programme.[5] Massive quantities of heavy water were being used for fission research, probably for a nuclear energy generator like the Halban–Kowarski boiler. Perrin dismissed the hypothetical possibility that the Nazis might use a nuclear reactor to generate plutonium, an alternative to U-235 for an atomic bomb. He judged the overall situation 'fairly reassuring'. Despite its head start, the German bomb must have fallen behind as Tube Alloys surged ahead with the promise of American collaboration.

If Perrin's letter had fallen into enemy hands, the Germans would have been delighted to see how blind the British were. By January 1942, the Nazi nuclear programme was ahead in every domain except for one. So far, nobody in Germany had realised that a relatively tiny mass of pure U-235 could make an atomic super-bomb.

Under Cover

The ultimate goal in Heisenberg's first *Uranverein* report to the German Army in December 1939 was a uranium fission bomb 'several orders more powerful than the strongest explosives yet known'. To maximise its destructive power, Heisenberg explained that the bomb's uranium fuel would have to be enriched in U-235 – a task that he delegated to Paul Harteck in Hamburg.[6] In early 1940, Harteck began as Otto Frisch would do several months later, with gaseous hex and Klaus Clusius's thermal diffusion tubes. While Frisch struggled with homemade diffusion tubes a couple of feet long, no lab and no hex, Harteck had top-class facilities, unlimited hex (made by ICI's arch rival, IG Farben) and access to Clusius himself. Within a few months, Harteck succeeded in doubling the U-235 concentration in a sample of hex, a feat worthy of adulation in *Science*. However, this had taken seventeen days of cooking hex in a 25-foot Clusius tube to concentrate a few milligrammes of U-235 – far too little and too slow.[7]

Alternative separation methods were explored after a brainstorming conference which Heisenberg hosted in Leipzig in October 1940. Clusius tried liquid diffusion with solutions of uranium salts, hoping that the lighter U-235 ions would pass more easily between water and ether; pilot trials with rare-earth isotopes appeared promising. Electromagnetic separation, applying a powerful magnetic field across a beam of ionised uranium to deflect the lighter U-235 more than U-238, looked promising but was beyond current technology.[8] A fourth method was an American invention, inspired by an idea that had originally come to Frederick Lindemann in 1919. Jesse Beams at the University of Virginia had separated two neon isotopes using a 'gas ultracentrifuge', a vertically mounted cylinder that span around its axis at 15,000 revolutions per minute; the heavier isotope tended to gravitate outwards, leaving the central zone enriched in the lighter. Harteck and Wilhelm Groth had designed an ultracentrifuge to spin gaseous hex at a colossal 60,000 rpm – forty times faster than a Spitfire's propellor at full throttle – which was currently in blueprint with a gyroscope manufacturer in Kiel.[9] The first gramme of U-235, let alone a militarily useful quantity, wouldn't be ready in time for the *Uranverein*'s third birthday in September. But with four different separation methods being explored by some of the world's greatest experts, surely it was only a matter of time?

The energy-generating *Uranbrenner* reactors were evolving quickly. The search had narrowed for a moderator to slow neutrons to the best speed for sustaining a chain reaction. Ordinary water and paraffin had been crossed off; they stopped neutrons so well that they were used to shield lab workers against neutron radiation. Another potential moderator was examined under conditions that doomed it to failure. In May 1940, Harteck tried 10 tons of dry-ice bricks; unfortunately, Heisenberg was hoarding Substance 38 for his own reactor, and the dry ice had vanished into thin air before Harteck could obtain enough uranium for an experiment.[10] Graphite – the choice of Fermi and Szilard – was abandoned when Walther Bothe reported that even the purest '*Elektrografit*' from Siemens mopped up neutrons. Bothe hadn't done his experiments carefully; he may have been distracted by a beautiful lady scientist (and spy), to whom he wrote: 'I talk physics all day but think only of you.'[11] This left heavy water as the only option. The entire German nuclear programme hung on a heavy-water factory in the middle of Norwegian nowhere, recently upgraded but only producing a quarter of its required output. And even Jomar Brun was baffled by froth that mysteriously appeared inside the electrolytic cells, spoiling whole batches.[12]

By autumn 1940, two lineages of experimental *Uranbrenner* reactors had been established, prefixed 'B' for Berlin and 'L' for Leipzig. 'L-I', designed by Heisenberg's colleague Robert Döpel, and Karl Wirtz's 'B-I' were both cylinders over a metre high, containing alternating layers of Substance 38 and paraffin. Neither produced any hint of a chain reaction, and neither did their successors, moderated by heavy water from Vemork.[13] Heisenberg now pinned his hopes on L-IV, a hollow aluminium ball containing sintered uranium that sat inside a half-metre aluminium sphere filled with heavy water. It would be tested later in the spring; even if it didn't support a chain reaction, it should indicate how much uranium and heavy water would be needed. Sintered uranium was highly toxic – necessitating goggles, masks and protective overalls – and could catch fire on contacting air, skin and especially water. L-IV was an accident waiting to happen.[14]

What about the German atomic bomb? For 'bomb', read 'bombs', because there was now an alternative to Heisenberg's U-235-enriched weapon. Michael Perrin would have been horrified to discover that it was plutonium, which he'd dismissed in his memo to the War Cabinet. In April 1940 – two months before Abelson and McMillan's paper describing elements 93 and element 94 – Josef Schintlmeister in Vienna identified traces of 'a previously unknown chemical element' in metal ores which emitted alpha-rays with a distinctive energy. Schintlmeister tentatively placed the new element in position 94 of the periodic table and predicted that it could be synthesised by bombarding U-238 with neutrons in a *Uranbrenner* reactor and would be highly fissionable.[15] Meanwhile, Otto Hahn's team at the KWI for Chemistry had detected and partially characterised element 93. Hahn's papers, devoid of any reference to its fissionable daughter product, were published in *Naturwissenschaften*,[16] but Schintlmeister's reports were top secret; a copy was handed personally to each member of the *Uranverein*, who read it and then returned it for immediate destruction. It was forbidden to make notes or talk about the papers, and even Schintlmeister wasn't allowed to keep copies.[17]

Carl von Weizsäcker, now Professor of Astrophysics at the Reichsuniversität (Reich University) in Strasbourg, was the first to propose making a German atomic bomb from a transuranic element. In mid-July 1940, von Weizsäcker sent a five-page document (copied to Heisenberg) to Kolonel Erich Schumann, claiming that element 93 (not 94) was fissionable and could make a nuclear explosive: 93 could be made in an *Uranbrenner* reactor and easily separated from its parent uranium.[18] Von Weizsäcker bragged that 'the road to the bomb is clear', but ignored practicalities such as the bomb's design and how to build it using self-destructing explosive (with its half-life of 2.3 days, over 90 per cent of element 93 would have decayed within ten days).

In early 1942, the *Uranverein* comprised forty researchers on nine sites and considered its work to be world-leading,[19] but Erich Schumann warned Heisenberg repeatedly that it would be closed down if he failed to deliver. During the grim winter of 1941–2, the realisation sank in that Hitler's invasion of Russia had been a catastrophe which had cost hundreds of thousands of German lives. Nazi High Command had ordered all new weapons research – no exceptions, not even for

Heisenberg – to be axed if the inventions couldn't be ready for battle within a year.[20]

Heisenberg sailed on, oblivious of the cracks which he'd opened up in the *Uranverein*. Harteck, Bothe and others moaned in private that Heisenberg was selfish, hogged uranium and money, and didn't share his results. He'd won a Nobel Prize but wasn't fit to be in charge of this project because he'd never done any experiments and (according to Harteck) 'had not contributed any basic ideas to the uranium fission problem'.[21] Kurt Diebner, Schumann's Director of Atomic Weapons, took matters into his own hands, pulling his Nazi strings to get everything he needed to build his own reactor in the army's rocket research centre at Gottow, near Berlin. It looked nothing like the others: a chandelier of dozens of uranium cubes strung on wires, to be lowered into a tank of heavy water. And although Heisenberg despised Diebner as an also-ran, Diebner's prototype chandelier design was showing more promise than Heisenberg's own L-reactors.[22]

In February 1942, Schumann received a top-secret, 200-page report which he'd commissioned on the *Uranverein*. The authors were anonymous but included *Uranverein* members as well as outsiders; perhaps unsurprisingly, progress on the reactors and bombs was judged impressive. The report included the estimate that the critical mass of uranium – sufficient to perpetuate a chain reaction – lay between 10 and 100 kg.[23]

Schumann decided to give the *Uranverein* one last chance to prove itself. He organised a three-day technical symposium with twenty-five lectures to bring *Uranverein* researchers up to date with progress in isotope separation, reactor design and heavy-water production. Bernhardt Rust, the Nazi ideologue and head of the Reich Research Council, spotted an important propaganda opportunity and commandeered the first morning of Schumann's meeting for military and industrial VIPs to receive a non-technical summary of how nuclear research would win the war for Germany. Those invited included Himmler, Reichsmarschall Hermann Goering and Wilhelm Keitel, chief of the Armed Forces High Command. Heisenberg picked the *Uranverein*'s most powerful communicators – including himself and von Weizsäcker – and gave them each ten minutes to win over powerful, hardened, scientifically illiterate men who routinely bargained with thousands of other people's lives. The lay presentations were scheduled for the morning of 26 February 1942. Obviously, the VIPs weren't meant to attend the in-house technical session that afternoon, because it would have been incomprehensible to them. Himmler, Keitel and several others politely declined to attend, because they'd mistakenly been invited to the scientific symposium. When the error was discovered, it was too late: a golden opportunity to save the uranium programme had been thrown away by a careless secretary.[24]

Heisenberg's lecture described the *Uranbrenner* reactors in detail, with diagrams and photos of the prototypes. These power units consumed no oxygen and could revolutionise naval warfare, as nuclear-powered submarines could remain underwater for long periods. Atomic bombs would be weapons of 'unimaginable force ... about 100 million times greater' than a chemical high explosive. These could be made of uranium enriched in U-235 – work in progress – or the man-made element 94, created inside and harvested from a functioning reactor.[25] All very impressive – but with no information about the size, shape and internal anatomy of the *Uranverein*'s atomic bombs.

Schumann saw through the obfuscation and pulled the plug on the *Uranverein* when Heisenberg refused to set deadlines for producing functional reactors or bombs. However, Bernhardt Rust had been impressed by Heisenberg, and his Reich Research Council now seized control of the entire nuclear-research programme.[26] So this wasn't Heisenberg's last chance. That would come four months later.

Resistance Round-Up

On top of its struggle for survival within the Nazi war machine, the *Uranverein* faced growing threats from others determined to undermine the German nuclear programme. As yet, those threats were still invisible but would soon begin to declare themselves.

At the Collège de France in Paris, Joliot was stuck in hell with the enemy in the next room and all the rooms beyond that. In June 1941, he'd been arrested by the Gestapo but released when Wolfgang Gentner told the local SS that Joliot was doing secret war work for powerful people in Berlin who would be angered by his detention.[27] Thanks to its devoted German engineers, the cyclotron was at last ready to run – but not for long. As Gentner recounted gleefully years later, Joliot waited until the magnet had powered up before shouting that the vacuum was failing (it wasn't). Joliot's chief technician immediately switched everything off, releasing the magnet's pent-up energy in a blast of heat which melted the copper cooling tubes, started a fire and caused 'a tremendous mess'.[28] *Merde!* Where were those engineers?

The cyclotron continued, mysteriously, to misbehave intermittently. Otto Hahn visited occasionally but apparently failed to obtain anything useful from the machine.[29] Walther Bothe persisted in trying to measure the nuclear constants of U-235. Rumours circulated that Joliot was actively helping Bothe; Joliot bit his tongue and allowed the taint to spread. At the Radium Institute, Irène had also reached an equilibrium with the Germans. After leaving the sanatorium at Clairvivre, she was allowed to continue her work providing she didn't interfere with theirs.[30]

Wolfgang Gentner, planted to spy on Joliot, was collaborating with him in ways invisible to the Nazis. The two became friendly – too friendly for the Joliot-Curies' fifteen-year-old daughter, Hélène, who demanded to know why a German spent an evening in their house, talking to her father.[31] A German visitor to the lab also noticed their closeness and denounced Gentner as a sympathiser. Gentner was promptly recalled to Berlin, but managed to engineer the appointment of his successor, outwardly hard-line but covertly sympathetic to Joliot and his cause.[32]

Later that spring, Joliot's hatred of the Nazis became incandescent. The previous October, Gentner had duped the Nazi authorities into releasing Paul Langevin from prison but not Langevin's son-in-law Jacques Salomon, arrested at the same time. The brilliant Salomon never won his Nobel Prize. On the morning of 23 May 1942, he and several other academics were dragged out of their cells and murdered by firing squad. Joliot joined the French Communist Party, banned by the Nazis, and began planning how he and his laboratory could best help to rid France of the Germans.[33]

Jomar Brun later claimed that the idea of destroying the heavy-water plant at Vemork originally came from Niels Bohr; perhaps meeting Heisenberg in Copenhagen had made the peace-loving Dane consider extreme action. Brun had already mounted his own covert sabotage mission, injecting cod-liver oil into the electrolytic cells to cause the mysterious frothing which destroyed random shipments of heavy water.[34]

Recruited through the XU Norwegian network, a team of apprentice saboteurs was coalescing around Professor/Major Leif Tronstad, now exiled in Britain and chief of Section IV (intelligence, espionage, sabotage) of the Norwegian High Command. Many were from the Telemark region, and some worked for Norsk Hydro. They knew the terrain, from the wilderness of the Hardanger Plateau above Rjukan to the factory at Vemork.[35] Most hitched a ride to the MI6 training camp in northern Scotland with the 'Shetland Bus', the clandestine shuttle service operated by Norwegian fishing boats. The bus came too late for one hard-pressed worker at Rjukan; he hijacked a coastal steamer from Stavanger to Aberdeen, received his training and was parachuted back on to the Hardanger Plateau in time for his scheduled return from 'a holiday'.[36] By early 1942, twenty saboteurs were being trained in parachuting, self-defence, living off cruelly inhospitable land, and blowing things up. Expert tuition was given by Tronstad, co-designer of the Vemork plant, with practical sessions on full-scale plywood mock-ups of the heavy-water cells in the concrete basement of the ammonia factory.[37]

Intelligence from senior Germans (unaware that their offices at Rjukan had been wired for sound) indicated that the tempo in Vemork was accelerating. Production of heavy water was being pushed up relentlessly, while German defences around the works were being strengthened. The longer the saboteurs waited, the harder it would be.[38]

Eric Welsh, Leif Tronstad's commander and chief of MI6's Norwegian Section, was a busy man. Welsh had recently inherited a list of agents from Frank Foley, now masterminding counter-espionage operations at MI5. Of special interest was 'Griffin' – Paul Rosbaud at the Springer offices in Berlin, who had spent over eighteen fruitless months waiting by the radio at 6 p.m. for the BBC to tell him that 'the house stands on the hill'. In spring 1942, Welsh contacted Rosbaud, who was overjoyed to be reactivated. Rosbaud's brief was again to relay intelligence about German military science, including the *Uranverein*.[39]

Welsh first sent Rosbaud to follow up a curious story about an island in the Baltic, told by a Norwegian engineering student ('Sigurd') from Dresden. After several meetings with Sigurd in a beer tent on Tiergartenstrasse, Rosbaud went alone to observe the island from an inn on the opposite shore. Rosbaud reported to Welsh that Sigurd had put himself at great risk and identified a significant threat. Since the early 1930s, the island had been a testing-range for rockets, whose vapour trails could be admired over a beer at the inn. The ominous changes meticulously documented by Sigurd included heavy fortifications, purposeful buildings thrown up at speed – and extraordinary projectiles, eight times the height of a man, cigar-shaped and with tail fins. The island was named after the river Peene in whose mouth it sat: Peenemünde. Rosbaud's report was considered on high in MI6 – and shelved, because it just wasn't believable.[40]

Meanwhile, the German atomic bomb programme had gone ominously quiet. The *Uranverein* had tightened its security; former contacts, including the once loquacious Mattauch, had become paragons of discretion. Rosbaud kept his ear to the ground but heard nothing.

16

DOUBLE DEALING

December 1941–November 1942

On 12 December 1941, Churchill flew to America. With the Luftwaffe still pounding British cities, it was a questionable moment to leave (and then to spend five days unwinding in Florida), but the timing was impeccable for Roosevelt. Churchill arrived the day after Hitler declared war on the USA and electrified both Houses of Congress with a passionate speech that welcomed the Americans as Britain's comrades in arms. The two leaders got on famously. Dawn-to-dusk meetings with advisers were interspersed with discussions *à deux*, cocktail receptions and switching on the Christmas tree lights at the White House.[1] When Roosevelt stumbled upon Churchill dictating in the nude to a stalwart secretary, Churchill declared, 'You see, Mr President, I have nothing to conceal from you.' At least Roosevelt was wearing something when he told Churchill, 'Trust me to the bitter end.'[2]

Neither man was telling the whole truth. Significantly, they didn't discuss MAYSON – Roosevelt's proposal to unify the British and American bomb programmes. Churchill flew home determined to build Britain's bomb without American interference. And on 19 January 1942, Roosevelt sent a handwritten memo to Vannevar Bush with the order to lock it in his safe. This was presidential approval for the American atomic bomb to go ahead at full speed, and it acted like a starting pistol. Bush gave the S-1 Section Uranium Committee the energy and muscle it needed to 'Build the bomb and build it fast'.

Lyman Briggs, the token chairman, was sidelined by the 'Program Chiefs', Harold Urey and Ernest Lawrence (jointly responsible for separating U-235), the 'Project Chief' Compton (designing and constructing the bomb) and the 'Planning Board Chief' who would control infrastructure and organisation. Echoing Wallace Akers's appointment to lead Tube Alloys, the Planning Board Chief was an industrial engineer experienced in managing big projects, Eger Murphree from Standard Oil.[3]

The 'slow neutron' reactor programme to generate both nuclear energy and element 94 had been split between New York and Berkeley, over 2,500 miles apart. Compton now united them to form the 'Metallurgy Laboratory' (Met Lab) at his home base, the University of Chicago. The Met Lab took shape away from public gaze under the raked seating of the university's disused football stadium. Compton turned abandoned squash courts and changing rooms into state-of-the-art physics and chemistry labs, complete with mass spectrographs and a cyclotron. People came too, so thick and fast that Compton had to block-book a students' recreation hall for welcome parties.[4] There would be plenty to show when the Brits came to visit.

Seeing is Believing

The Tube Alloys delegation to the USA in January 1942 comprised Wallace Akers, Rudi Peierls, Francis Simon and, to Lew Kowarski's predictable disgust, Hans Halban. Akers and Simon took the posh transatlantic flying-boat; Halban, worried about the risks of flying, went by sea and picked up pneumonia en route.[5] Peierls followed later, after tidying his affairs and leaving detailed instructions for his team in case he didn't come back. His journey began on the sleeper to Kilmarnock ('quite romantic'), followed by a no-frills crossing in a USAF Liberator bomber with a flying suit, oxygen mask and the forgotten luxury of bacon and eggs for breakfast. The flight – his first – left him sleepless with excitement. A diversion to Ottawa to catch up with George Thomson stamped the indelible memory of landing at dusk with lights outlining the city. His third flight, to join the others in New York, was his first contact with America.[6]

As with George Pegram and Harold Urey in England three months earlier, the Tube Alloys quartet was given a warm welcome and access to anyone, anywhere and anything. They attended the S-1 Uranium Committee, run efficiently by Ernest Lawrence, and set off on a two-month voyage of discovery that covered several thousand miles. Any misconception that the Americans were lagging behind was swiftly corrected. The British had staked everything on Simon's gaseous diffusion plant for extracting U-235, without knowing whether the process worked. The Americans were developing three different methods for purifying U-235 – and two types of atomic bomb.[7]

At Columbia University, Harold Urey was working on gaseous diffusion. Peierls was dismissive of Urey's progress until he challenged some of the mathematics and was politely contradicted. After exchanging telegrams with Klaus Fuchs in Birmingham, Peierls realised that the Americans were right; back in England, Simon's team returned to their drawing boards.[8] In Charlottesville, Virginia, Jesse Beams – the pioneer of gas centrifugation – was spinning hex at incredible speeds and Westinghouse was already designing two ultra-centrifuges, one towering 11 feet high.[9] In Berkeley, Ernest Lawrence was modifying his 37-inch cyclotron so that its magnetic field could pull U-235 out of a beam of ionised uranium. Lawrence called his electromagnetic separator the 'calutron' ('California University cyclotron'; Figure 7, Plate 19) and desperately wanted Mark Oliphant, who had fine-tuned the original cyclotron, to come from Birmingham to help.[10]

Fermi and Szilard were transferring from Columbia to the Met Lab in Chicago, to build a uranium–graphite pile which, Fermi boasted, would boil an egg by the end of the year.[11] This was catastrophic news for Halban, who had believed that the secrets of the boiler would entice Fermi to rescue him from England. However, Fermi had surged ahead and his welcome was 'cool', with pointed reminders that the FBI now barred all foreigners from working on secret war projects.[12]

The Met Lab threw in another shock. Element 94 was no longer a theoretical entity but the Americans' alternative explosive for an atomic bomb, on equal priority with the purification of U-235.[13] Glenn Seaborg was already isolating sub-milligramme traces and aimed to extract 'ponderable quantities' in the lab and ultimately on an industrial scale from Fermi's pile. Compare and contrast with Britain, where element 94, alias plutonium, had been sacrificed to pour all resources into one high-risk process to extract U-235.

At Compton's request, Peierls rounded off his American tour at the Met Lab in Chicago. Earlier, Peierls had remarked that 'nobody in America seemed to be giving much thought to the weapon', and Compton wanted to pick his brains about the best man for that job. One name came up immediately. Peierls had known him since the late 1920s, when they'd both worked at the Technical University in Zurich, and had just met him again at Berkeley: Robert Oppenheimer, instantly recognisable by his guileless eyes, gaunt profile, pork-pie hat and pipe.[14]

On his way home, Peierls visited Toronto to see his two children and meet their foster-parents, allocated by the university. Back in Birmingham, he reflected on an eye-opening visit and what it had and hadn't achieved. The two-way exchange of nuclear intelligence seemed 'fairly free' but, despite all the warm words, neither side had made any commitment to work alongside the other. Peierls was left frustrated by 'not knowing whether the American and British bomb programmes would be fused'.[15]

Soon after arriving in New York, Wallace Akers had sounded the alarm about the Americans' 'vastly greater resources' and 'enormous number of people' in a letter to Sir John Anderson, Chairman of the Tube Alloys Consultative Council. The Met Lab boasted ten senior scientists, thirty scientific assistants, seventeen technicians and nineteen support staff; at Berkeley, Ernest Lawrence had forty-six scientists, thirty technicians and three times the total Tube Alloys budget.[16] Akers returned to London in March 1942, excited but chastened by the Americans' energy and organisation, and prepared a detailed report explaining why Tube Alloys was doomed without the Americans. The British bomb effort had to be merged with the American programme, while the Halban–Kowarski boiler could only succeed if transplanted into the USA. Akers sent his report to Anderson in late March 1942, and waited for wisdom to filter down from above.[17]

Anderson was particularly busy, having taken on the brief of Lord Hankey, Minister of Aircraft Production, whom Churchill had sacked in a letter of 'exceptional cruelty' for criticising Lindemann.[18] Anderson had told Hankey to delay answering Charles Darwin's handwritten plea for full cooperation with the American bomb project, received seven months earlier. Anderson himself now wrote directly to Vannevar Bush with vague platitudes about progress but – as ordered by Churchill – avoiding mention of full collaboration. Bush replied promptly and politely, looking forward to 'a more explicit interchange of our plans'.[19]

Akers's cry for help was urgent but Anderson kept him waiting for several weeks. He finally wrote back in June 1942, telling Akers that the Tube Alloys Consultative Council had rejected his prophecies of doom. Britain remained ahead of the Americans and must develop its own bomb without them. Lindemann insisted that the full-scale factory must be built in Britain and would provide all the U-235 required.[20] At last, Akers understood the inexplicable delays and breaks in communication, and why everything was such hard work. He and the front-line scientists were pedalling as fast as they could to keep up with the Americans, while Lindemann and Anderson quietly jammed on the brakes. This must mean that Churchill didn't want the Americans involved – and that he hadn't grasped the awful truth that Tube Alloys couldn't succeed without them.

Hier Endet . . .

After the fiasco of the showcase presentations, the *Uranverein* continued to concentrate on the *Uranbrenner*; the bomb was too far over the horizon to waste time on. In May 1942, Döpel and Heisenberg's L-IV reactor re-enacted the pivotal moment that Halban and Kowarski had celebrated thirteen months earlier: an injection of neutrons whipped up enough new neutrons to indicate that a self-perpetuating chain reaction could be achieved. Heisenberg calculated that they would need 5 tons of heavy water and 10 tons of uranium, quantities that were now feasible.[21]

Prompted by Schumann, the Reich High Command offered the *Uranverein* one last chance to pitch for its survival, at a secret meeting of top-rank military commanders held in the KWI for Chemistry on 4 June. Those present included Field Marshal Erhard Milch and Albert Speer, the no-flannel Minister of Armaments and War Production.[22] The *Uranbrenner* (which would keep a submarine under water for months) and the uranium fission bomb (the most destructive weapon in human history) both seemed to be well received. Then came two killer questions. Milch asked how big the bomb would be. 'About the size of a pineapple,' replied Heisenberg.[23] Milch's question was entirely reasonable. Detailed diagrams of the *Uranbrenner* had been shown, but Heisenberg provided no evidence that the bomb had progressed beyond equations and formulae. The bomb's size, shape or internal workings hadn't even been mentioned, and the pineapple could have been the first thing that came into Heisenberg's head.

The second killer question demonstrated that Heisenberg hadn't given any thought to practical planning. Speer enquired how much money the *Uranverein* needed. Heisenberg suggested 150,000 marks – absurdly cheap for such a huge project – and von Weizsäcker chipped in to request another 100,000 marks.[24] It was obvious that they'd never discussed the budget. How could they be trusted to complete such a complicated project in a few months? Heisenberg didn't realise that he'd killed the German bomb. Milch left before the meeting ended, committed to funding weapons that would be ready for this war: the V-1 flying bomb and the V-2 rocket being tested at Peenemünde.[25]

Everything fell apart after that. On 23 June 1942, a month after its moment of glory, L-IV was wrecked by an explosion – nothing nuclear, just a lab accident caused by ordinary water leaking onto sintered uranium. Nobody was injured because Döpel and Heisenberg had spotted flames and taken shelter before the room was showered with gobbets of burning uranium.[26] The destruction of L-IV was a grave omen. That same day, Albert Speer went to brief Hitler, with Heisenberg's uranium bomb as agenda item number 16. Soon after, Bernhardt Rust relayed Hitler's verdict that the *Uranbrenner* project could continue, but the atomic bomb must be scrapped.[27]

That evening, some of those who welcomed the news met in a bar on Kurfürstendamm, and someone invited their old friend Paul Rosbaud to join them. Inebriated by high emotions and alcohol, their tongues were loosened. Rosbaud's contacts in Leipzig had already told him that Heisenberg's latest reactor had exploded. He'd reported this to Eric Welsh, who did nothing other than confirm that he'd received the message. The new intelligence that the Nazis had axed their atomic bomb was so momentous that Rosbaud delivered it personally to the XU network in Oslo, wearing German military uniform for extra impact (he held an honorary position with the Luftwaffe). Expecting an immediate and excited response from London, Rosbaud was stunned when, once more, Welsh merely acknowledged receipt of the intelligence.[28]

Independent corroboration reached Welsh a few days later. Hans Seuss, working with Harteck in Hamburg and a frequent visitor to Vemork, had beckoned Jomar Brun out of his office and onto a balcony beyond any German microphones. According to Seuss, the Germans were now concentrating exclusively on energy-generating reactors, which wouldn't be operational for at least five years. Brun promptly relayed this information to Welsh, and was surprised by his lack of response.[29]

Shortly after that, Hans Jensen from the KWI for Physics turned up in Oslo to give a confidential seminar to selected old friends. After telling them not to take notes, Jensen explained that Heisenberg had been working on an atomic bomb but had now abandoned it. One of Jensen's friends immediately prepared a full report which a special XU courier smuggled out to London. Yet again, Welsh did nothing.[30]

The demise of the Nazis' atomic bomb should have made MI6 relax if not rejoice, but was apparently ignored by Welsh or his superiors. Perhaps they wanted more evidence. Or did they suspect a hoax to throw the Allies off the scent? Whatever the explanation, nobody told Churchill or the War Cabinet – and certainly not the Americans.[31]

Sleepers Awake

In late June 1942, Churchill visited Roosevelt in Hyde Park, the opulent Roosevelt family residence north of New York City. Churchill later remembered the date and place with total clarity – the stifling summer evening of 20 June, Roosevelt's 'tiny little' study which jutted out over the rose garden – but not the discussion. Churchill came alone and made no notes, which meant that he had to consult Harry Hopkins, Roosevelt's aide, several months later to check that he and Roosevelt had agreed to cooperate over the bomb. It was as if Churchill didn't care about the outcome – or was ensuring that there was no British record of what he said.[32]

Michael Perrin was in the USA at the same time. He'd been sent out by Akers to secure uranium from the Eldorado Company (currently monopolised by America), to see if the Canadians would take Halban and Kowarski in case the Americans didn't want them, and to smooth the way for integrating Tube Alloys into the American nuclear programme. Perrin flew out from Bristol on 3 June 1942 via Lisbon and Bermuda, and hit the ground running in New York thirty-five hours later. The six-week trip began well. His goodwill gift – 5 litres of heavy water from Kowarski's stock, which had gurgled loudly inside its diplomatic bag – was well received. Uranium would be forthcoming, thanks to the secret compulsory purchase of Eldorado planned by the Canadian government. And the Canadians were enthusiastic about the Halban–Kowarski boiler, which Fermi had refused to accommodate in Chicago.[33]

Perrin retraced Akers's steps and documented incredible growth in just three months. The Met Lab in Chicago now had over 200 staff. Fermi's latest pile was a three-dimensional lattice of tennis-ball-sized uranium oxide spheres embedded inside an 8-foot cube of gleaming black graphite blocks; confidence was high that its descendants would indeed boil an egg by Christmas. Glenn Seaborg and John Wheeler provided valuable updates for their counterparts back in England.[34] Perrin impressed Vannevar Bush, who wrote approvingly to John Anderson about their 'profitable and interesting discussion', and hoped that Perrin would stay long enough in the USA 'to handle liaison in this field, for which he seems so well qualified'.[35]

Perrin flew home on 10 July 1942 and briefed Akers about the 'enormous progress' in America. With the possible exception of Simon's extraction process, the Americans were out in front and pulling away.

There was very little time – 'probably less than a month' – to nail down the collaboration with the Americans before they 'completely outstrip us ... and quite rightly, will see no reason for our butting in'. It was now or never, so Akers seized the bull by the horns. Lindemann initially dug his heels in – forget the Americans – then made an astonishing tactical error by admitting that he wanted Simon's extraction factory in England for 'sentimental reasons'. Charming as always, Akers went for the jugular. Anderson assisted, having been persuaded by Perrin's evidence. Lindemann gave way, leaving Anderson to break the news to their commander-in-chief.[36]

On 30 July 1942, Anderson sent Churchill an urgent note. 'With some reluctance', he'd had to revise his views on the bomb. Trying to build a U-235 factory in England would cause a 'major dislocation of the war effort' and was now 'out of the question'. Even the pilot plant could 'only be erected in the United States'. Given American 'enthusiasm and lavish expenditure over the whole field of Tube Alloys', the British team should move to the USA so that 'work on the bomb project would be pursued as a combined Anglo-American effort'. Time was running out: 'We now have a real contribution to make to a merger. Soon we shall have little or none.'[37]

Churchill had just survived a vote of no confidence over his handling of the war, mounted by Lord Hankey's 'CHURCHILL MUST GO' campaign in revenge for his ignominious sacking. When Anderson's note reached him on 31 July, Churchill immediately accepted Anderson's recommendation without comment. Twelve months of Churchillian duplicity, unsuspected by the Tube Alloys scientists and their American colleagues, ended as silently and abruptly as it had started. As far as Churchill was concerned, everything was back on track as though it had never fallen off. The little people could now sort out the mess and build the Anglo-American bomb which would help to win the war.[38]

The Shape of Things to Come

The summer of 1942 saw the brakes released in America. Thirty-year-old Glenn Seaborg had established his thirty-strong team (all even younger) in the Met Lab to chase element 94; unaware that this had already been christened in Cambridge, Seaborg now called it 'plutonium'. Blasting uranium salts with high-energy deuterons from the cyclotron was routinely producing plutonium but in minute amounts – just a fraction of a milligramme from hundreds of kilos of uranium. The invisible samples were characterised by 'ultramicrochemistry', using specially designed instruments so delicate that they could be crippled by the draught from opening a door. On 18 August 1942, Seaborg secretly made history at his weekly lab meeting by revealing the first ever glimpse of plutonium: tiny pink crystals, photographed down the microscope. Somebody asked which plutonium salt this was. 'I'm not allowed to tell you,' replied Seaborg. The crystals, the weight of three grains of table salt, comprised less than one-millionth of the amount needed to make a plutonium bomb – which is why they were already working up plans for harvesting plutonium from an operating pile, even before Fermi had built one.[39]

Two momentous decisions were made that summer. In June 1942, James Conant put Robert Oppenheimer (Plate 18) in charge of fast neutron research as the 'Coordinator of Rapid Rupture', and Oppenheimer began assembling a team to work on the nuts and bolts of the bomb.[40] Meanwhile, Vannevar Bush had sized up the gargantuan task ahead and sought Roosevelt's approval to bring in the professionals. Roosevelt agreed and, on 13 August 1942, the US Army Corps of Engineering assumed responsibility for planning, building and maintaining all the weapon production facilities. It was a logical choice: the corps had recently completed the Pentagon complex, the world's largest ever engineering project. The corps traditionally assigned each new project to its geographical district; this one would be administered by the corps headquarters at 270 Broadway, in Manhattan. Accordingly, and for reasons of obfuscation, the American atomic bomb programme was known as the 'Manhattan Engineer District' project, or 'Manhattan' for short.[41]

A senior colonel in the corps was appointed to lead the project but was pushed aside by the man who had built the Pentagon. The usurper was forty-six years old and considered himself ideally qualified to run a huge multidisciplinary enterprise that had to be driven by military men, not scientists. To give himself extra clout to slap down self-important Nobel Prize winners, he demanded and was granted a general's star. Meet Brigadier General Leslie Groves, appointed Director of the Manhattan Project in September 1942.[42]

Action and Inaction

If MI6 had told the British War Cabinet that the Nazis had killed off their own atomic bomb, a blind eye might have been turned to their increasing demands for heavy water. However, that information was withheld, and in July 1942 the Cabinet approved direct action against the Norsk Hydro electrolytic plant at Vemork. The US Air Force had offered to bomb the plant; Leif Tronstad patiently explained the near-impossibility of pin-pointing the reinforced basement of an eight-storey concrete building stuck on the side of a deep ravine.[43] The task therefore fell to the Special Operations Executive, which was already training a team of Norwegian saboteurs in Scotland for that purpose.[44]

The ground was laid by 'Operation Grouse', parachuting four Norwegians with local knowledge onto the Hardanger Plateau above Rjukan. They would lie low and radio through updates to inform 'Operation Freshman', when the Grouse pathfinders would guide in gliders carrying a team of British Army sappers (explosives experts), then lead the sappers to the plant and help them to escape.[45]

Operation Grouse began auspiciously on 18 October 1942, with a full moon and reasonable parachuting conditions over the Plateau. Unfortunately, the four men were 35 miles off course when they jumped. It took them two days to find most of their equipment (but none of their food), and another two weeks of arduous cross-country skiing to drag it to the stone shelter where they should have landed. On 9 November, after local Resistance fighters provided a battery to replace the one that had died inside their radio, they finally contacted London where, after three weeks of silence, everyone had assumed the worst.[46]

On the same day, Jomar Brun and his wife flew from Stockholm to Scotland. Brun had taken ever greater risks in passing information to London, mostly using the well-known (and therefore 'most foolish') microfilm-inside-the-toothpaste-tube trick. It was now time to get him out. A 'student from Oslo' brought a secret 'leave now' message – and was treated to excellent wine and a slap-up meal by Mrs Brun, who until then hadn't known that her husband was working for the XU network and the British. The Bruns slipped away from Vemork with two litres of heavy water and, just in case, two cyanide capsules from 'a friendly Rjukan pharmacist'. They reached London on 11 November. Brun was debriefed by Leif Tronstad, Eric Welsh and R. V. Jones, Welsh's boss in MI6 – and learned that the plant which he and Tronstad had built at Vemork was to be blown up in just over a week.[47]

The thirty-four sappers from the Royal Engineers prepared thoroughly for Operation Freshman: hard physical training in the north Welsh mountains, familiarisation with hydroelectric plants in Scotland and the plywood mock-up of the basement at Vemork, and – as they would escape through Sweden, 100 miles away – winter survival skills. They didn't learn to parachute, as they would be delivered in a Horsa glider towed across the North Sea by a Halifax bomber. The Horsa, a big wooden brute with a 90-foot wingspan, was new to the RAF. The Halifax pilots were given a crash course in pulling a glider, something none of them had done before. In retrospect, other risk factors were obvious. Nobody had tried landing a Horsa in the depths of winter, or had calculated the strength of the tow-rope needed if a fully loaded Horsa iced up. And the experienced Swedish meteorologist keeping vigil on the night was ignored when he told them not to go.[48]

On the evening of 19 November 1942, two Halifaxes, each towing a Horsa containing seventeen sappers and two pilots, took off from Wick on the north-eastern tip of Scotland. Light drizzle turned into heavy wind and freezing rain when the aircraft crossed the Norwegian coast and headed for the Hardanger Plateau and the landing site which the Grouse team had outlined with lights. The telephone lines connecting each glider to its Halifax failed, leaving the Horsa pilots deaf as well as blind in appalling visibility. Only one Halifax came home, minus its Horsa; the tow-rope had iced up and snapped. The other Halifax flew into a mountain near Stavanger, killing all the crew. Its Horsa crashed some distance away, killing half of those on board. German soldiers were soon on the scene; a doctor dealt with those seriously injured, and the rest were taken away. The survivors from the other crashed glider were quickly rounded up.[49]

On 21 November 1942, German radio reported that saboteurs on a failed British glider mission to southern Norway had been 'wiped out to the last man'. Their fate was no surprise, as Hitler had recently decreed that all 'sabotage troops' were to be killed on capture. The badly injured crash victims who survived the doctor's lethal injection were strangled and their bodies dumped at sea. The rest were interrogated by the Gestapo and then shot. The sabotage equipment in the wrecked Horsas should have told the Germans everything they needed to know, especially as one of the sappers had taken a blue pen to a map printed on a silk handkerchief and drawn a neat circle around Vemork.[50]

17

CRITICAL MASSES
September 1942–March 1943

Fear was tightly woven into the fabric of the Manhattan Project from the start; as a researcher at the Met Lab said, 'When the Nazis get the bomb, they'll drop it on us first.'[1] After their welcoming party at the Met Lab, new recruits and their spouses were summoned to a private screening of *The Next of Kin*, a British propaganda film which hammered home the message that 'careless talk costs lives'. Laura Fermi later recalled, 'After the film, there was no need for words.' They were constantly reminded that Chicago was infested with 'known agents of the German government', perhaps the stranger striking up an innocent conversation over a burger. Asking nothing was as vital as saying nothing. Enrico Fermi shared only one secret with Laura, on condition she told nobody: there were no metallurgists at the Metallurgical Laboratory. She assumed that he was still working on the pile, and didn't probe.[2]

The Manhattan Project was driven by the belief that the German atomic bomb was three years ahead. The Americans had no reliable agents in Europe and their occasional intelligence included a third-hand report in October 1941 that an unnamed German physicist believed Heisenberg to be close to perfecting a bomb.[3] Everything they knew about the German nuclear programme came from British Military Intelligence and was pre-digested by Eric Welsh, now working closely with Michael Perrin of Tube Alloys. The British had passed on everything about heavy water and the failed sabotage mission at Vemork, but not the news that the Nazis' bomb had been axed – which could have pulled the plug on Manhattan.[4]

Under New Management

Leslie Groves (Plate 18) was the son of an army chaplain who instilled in him 'a strong moral sense of right and wrong ... and of duty to God and country'. Shaped by West Point and two years at MIT, Groves developed into a megalomaniac workaholic. Welsh ancestry and Anglophobia were embedded in his genes, and a visit to England during his twenties convinced him that Britain served no useful purpose. Following his promotion to run the Manhattan Project – 'on September 23, I became a Brigadier General and officially took charge' – power and paranoia about security went to his head.[5] Groves built up Manhattan as an array of 'cells', stuck together like a wasps' nest. Everyone inside a given cell worked exclusively on their allocated tasks; only the leader of the team was allowed to peep into adjacent cells whose contents were crucial to his own mission. 'Compartmentalisation' was routine for sensitive US Army projects, but Groves turned secrecy into a fetish. Only he and his right-hand man at the apex of the Manhattan command pyramid would ever see how the cells and their self-contained projects fitted together into the whole picture.[6]

The British felt the first Grovesian tremors during autumn 1942. An English nuclear physicist was shaken when an old American friend told him, 'Sorry, but I'm not allowed to talk about that.' Charles Darwin and George Thomson felt 'shut out' of discussion about the bomb, and were informed by a suddenly brusque Vannevar Bush that the US War Department had taken over 'certain phases of the project'. Extreme secrecy was now paramount, and it was no longer 'necessary for members of Tube Alloys, other than your Executive Committee, to know of the entire plans'.[7]

Bush was even blunter in a long and maddeningly vague letter sent to Sir John Anderson on 10 October 1942.[8] The Manhattan Project had already cost $55 million and couldn't afford to build Simon's separation plant unless the British proved that it worked.

Regrettably, by Anderson's own admission, a prototype hadn't yet been tested. Bush wholeheartedly supported the transfer of Halban's boiler group to Canada but rejected Anderson's requests for freely exchanging all information and fusing Tube Alloys with Conant's S-1 committee to create a joint controlling organisation. Hard on the heels of Bush's letter came an urgent message from Francis Simon, in New York for discussion with Harold Urey: 'The Army want to run the thing 100% American.' The normally imperturbable Anderson was 'alarmed' and especially infuriated by the Americans' refusal to house Simon's separation plant because he'd carefully explained to Bush that it was impossible to build the pilot plant in war-ravaged Britain.[9]

In early November 1942, Anderson sent his smoothest operator, Wallace Akers, to pat down the Americans' ruffled feathers. Akers set off, ready to deploy his charm, clear thinking and communication skills.[10] Up his sleeve was the latest British design of a gaseous diffusion membrane for the U-235 separation units. This had been shown to Michael Perrin when he visited the Mond Company's factory near Swansea, and he'd sellotaped a strip of it into his diary. It was made of sintered nickel and, unlike all the American versions, it should work.[11]

Akers's natural optimism might have been dented if he'd seen the note from James Conant congratulating Vannevar Bush on the 'masterly' evasion in his obfuscating letters ('masterpieces worthy of a lawyer or a diplomat') to Anderson.[12] And if Akers had known about the secret treaty that Churchill had recently signed with Stalin, he might have stayed at home.

On meeting Groves in Washington, Akers was immediately told that the Manhattan Project demanded complete secrecy. Oliphant and Lawrence had been talking together, as had Peierls and Oppenheimer. They must stop, because the British weren't in the Americans' cells. Akers pointed out that both pairs of researchers were long-term collaborators and entirely trustworthy. Groves slapped him down. If Oppenheimer really needed Peierls, then he could have him – but Peierls would remain locked inside Oppenheimer's cell, incommunicado from everyone outside until the end of the war. Akers protested that scientists of Peierls's calibre wouldn't accept such terms, and the first Akers–Groves encounter ended with no meeting of minds.[13]

On 1 December 1942, Akers visited Chicago and was staggered to see the massive graphite ziggurat towering over the ex-squash court under the deserted football stadium. The pile that Fermi called 'CP-1' was the descendant of thirty experimental piles, tested and discarded during the year. CP-1 resembled a huge doorknob, 20 feet high and 25 feet across its middle, propped up by wooden scaffolding like a gigantic square egg cup. Inside were 50 tons of uranium oxide and uranium metal spheres, embedded like fossils within 40 strata of ultrapure graphite. The graphite came in blocks 4 inches square by 16 inches long that weighed 19 pounds; there were 45,000 of them, totalling 380 tons.[14] Since mid-November, gangs of volunteers had worked twelve-hour shifts, day and night, adding new layers. Graphite dust filled the air and coated everything and everyone. Ignoring the sign that read 'POSITIVELY NO SMOKING IN THIS ROOM' could have had explosive consequences, while the craftsmen who fashioned the graphite blocks oozed black sweat for hours after showering themselves clean.[15] As CP-1 grew, its ability to sustain a nuclear chain reaction was checked periodically. When Akers visited, they had just added the final stratum which should make the pile 'go critical'; they would find out the next day. Awed by the sheer size of the pile and the hundreds of people serving it, Akers realised that he'd glimpsed the future.[16]

When he flew on to Montreal that evening, it was like stepping into the past. After Fermi rejected the boiler, Canada had welcomed in Halban's group to bolster their own interest in nuclear energy. Halban and five colleagues had moved there from Cambridge in September but were now stuck in a limbo of Halban's making. He'd appointed himself chief of the Canadian operation, with an inferior post for Kowarski and salaries to match. Belittled, Kowarski stayed in Cambridge with the others who loathed Halban. Unsurprisingly, Kowarski ignored Halban's demand to inform the Canadians in writing that they'd always worked happily together.[17] There was some good news: Tube Alloys would now get uranium because the Canadian government had just bought the Eldorado Mining Company in a stealthy grab that even their Cabinet didn't know about.[18] And with the boiler safely off their soil, the Americans were happy to give Halban heavy water and even invited him to a secret workshop on heavy-water reactors in Washington in March 1943.[19]

From Montreal, Akers returned to the fray in Washington, and a lunch on 11 December 1942 that would forever stick in his throat. Groves and Conant, both seemingly determined to crack his carapace of English urbanity, dictated their new rules of engagement to him. The British could continue sharing all their secrets, but the Americans would only give them knowledge that would help the USA to win the war. Akers politely argued the injustice, but they talked over him. He threw in his only trump – the revolutionary nickel diffusion membrane – and Groves ignored it. Akers's diplomatic skills deserted him, and what should have been a pivotal discussion about the future of the most destructive weapon in human history descended into a spat between three angry, opinionated men who were all used to having their own way. Back in England, Akers's report about the meeting caused dismay and foreboding. Chadwick was 'extremely anxious' about the future of Tube Alloys but refused to visit the USA on Groves's terms. So did Rudi Peierls.[20]

After seeing off Akers, Vannevar Bush prepared an update for Roosevelt to consider over Christmas 1942. To complicate an already delicate situation, the US ambassador to London had just dropped a grenade on Roosevelt's desk. Churchill had been at it again, signing a secret Anglo-Russian Treaty with Stalin which would enable Britain and Russia to share all their secrets about existing and future weapons.[21] In September 1942, facing the common enemy of Nazism, this seemed a jolly good plan to Churchill, and to Stalin. A tripartite American–Russian–British agreement also appealed to the ambassador. Roosevelt, who hadn't been consulted at all, thought it was a lousy idea. So did Henry Stimson, Secretary of War. The Russians were de facto allies now, but could they be trusted after the war, when the stakes of conflict would be vastly higher than ever? And no way could American atomic bomb secrets shared with the British ever be passed to Russia.[22]

The Americans' verdict was relayed on 13 January 1943, in a brutal memorandum from Conant. Akers, still in Washington, was handed his copy by Conant and Bush. Henceforth, the British and Canadians were barred from a long list of no-go research areas: fast neutron reactions, atomic bomb design, plutonium, U-235 separation methods other than gaseous diffusion, graphite-moderated piles, and the production of heavy water and hex. The embargo would only be broken if sharing the information would assist the American war effort. Tube Alloys was locked out of the Garden of Manhattan.[23]

John Anderson described Conant's memorandum as 'a bombshell ... quite intolerable'.[24] The temperature plummeted and animosity snowballed. On 16 January 1943, Groves reissued his invitation to Chadwick and Peierls. Both said no. 'Most unfortunate,' mused Groves. In March, the British retaliated with self-harm, when Anderson refused to let Halban attend the heavy-water reactor meeting in Washington. American revenge was swift, cancelling all heavy-water deliveries from their new factory at Trail, British Columbia.[25] And with those acts of petulance, it looked as though a joint British–American atomic bomb was dead in the water.

Nuovo Mondo

Estimates of the age of the Earth vary somewhat, from a snappy 6,500 years claimed by Creationists to the 3.4 billion years calculated by Rutherford. By contrast, there is no doubt when the Nuclear Age was born: just after 3.52 p.m. Chicago time on Wednesday, 2 December 1942. Its birth certificate was the paper-chart recording of neutron release inside Fermi's pile, CP-1.[26]

The morning was bitterly cold when Fermi crunched his way through fresh snow to the squash court with Leona Marshall – 'the only girl' in his team, and a close friend of his wife, Laura. CP-1 was waiting for them, held in check by a 15-foot cadmium control rod pushed sideways into its heart to kill off errant neutrons. Fermi's team squeezed onto the squash court's viewing gallery, with the 'big wheels' – Fermi, Compton, Leo Szilard and Eugene Wigner – in front and the others packed behind. To determine when CP-1 reached critical mass, they injected neutrons from the classic source – radon and beryllium – into the core of the pile, and measured neutron production inside the structure. Each new neutron registered as a click on a counter – and if too many poured out, the control rod of neutron-killing cadmium, poised above a vertical channel in the top of the pile, would automatically drop down to snuff out the reaction.[27]

Throughout the morning, the horizontal control rod was gradually withdrawn 6 inches at a time, checking for an exponential rise in neutron production. The proceedings were interrupted by the automatic control rod slamming down (the failsafe had been set too sensitively) and lunch ('I'm hungry,' said Fermi). At 3.52 p.m., the control rod was pulled out another 6 inches and the sedate clickety-clack of the neutron counter – 'like a fast train over the rails' – accelerated to a constant roar because it was overloaded and couldn't respond quickly enough. The chart recorder and Fermi's broad grin confirmed that the Nuclear Age had begun. CP-1 was put safely back to sleep, and they all savoured the moment. Eugene Wigner felt the thrill of achievement and then an eerie premonition of 'very far-reaching consequences' which could not be foreseen.[28]

Photographs had been taken of CP-1 during construction, but no cameras were there that afternoon; the scene was eventually reconstructed by an artist in 1947. Wigner had brought a raffia-bound bottle of Chianti, just in case. Fermi uncorked the bottle, divided the wine into paper cups, and they drank to success and being ahead of the Germans. Most of them autographed the label on the wine bottle, which is how they later worked out who'd been there.[29] After the others had gone, Leo Szilard, Fermi's partner and occasional bugbear for the past three years, shook his hand and – echoing his thoughts on confirming that fission generated excess neutrons – said that this was 'a black day in the history of humanity'.[30] A pre-coded phone call to Harvard was more positive. Arthur Compton rang James Conant and told him that 'The Italian navigator has just landed in the New World'. Conant enquired if the natives were friendly, and Compton said yes.[31]

Laura Fermi had spent the day preparing a dinner party for her husband's team and their wives. When Enrico returned from the Met Lab, he wore 'a steady grin' but gave no hint of anything unusual – even when one guest after another shook him by the hand and said

'Congratulations'. To Laura's consternation, nobody explained why. Eventually, her friend Leona Marshall turned up but said only that 'Enrico has sunk a Japanese admiral'. It wasn't until the second week of August 1945 that Fermi could finally tell his wife what had happened before their dinner party on 2 December 1942.[32]

Breaking Ground

James Conant was extremely relieved to hear of the Italian navigator's safe landing in the New World, because a great deal was riding on CP-1. Weeks before they laid its foundations, a huge contract had been drawn up to build its successors – on Fermi's assurance that it would work. CP-1 couldn't have boiled the promised egg. It generated half a watt, just enough to light a torch bulb, and only for a few minutes; it couldn't operate at full throttle because there was no means of dissipating the heat of the chain reaction, potentially ending in a non-nuclear but intensely radioactive explosion.[33] CP-1's much bigger progeny would be cooled by water and safely generate 'huge amounts of energy'. That energy was a waste product. The new reactors weren't being built to boil eggs or illuminate cities, but to make plutonium for atomic bombs.

Soon after taking up post, Groves decided to push plutonium as a nuclear explosive. He instructed the Du Pont Company to design a plutonium factory and drew up a shortlist of possible sites: covering at least 200 square miles, remote from population centres (in case of 'catastrophic explosions' or the 'release of intensely radioactive gases'), and with easy access to road, rail and unlimited supplies of cold water and electricity. The winner – 'Perfect' – was chosen in late December 1942 during a flight from the Rattlesnake Mountains to a wide bend in the Columbia River in the south-east of Washington State. Soon to be camouflaged as the 'Hanford Engineering District', the 625 square miles of arid desert lay midway between two large hydroelectric power plants and had a population of just 1,500, centred on the town of Richland at its northern end.[34]

In March 1943, residents received letters from the government, giving them thirty days to leave 'for the war effort'. They all went. 'We were so patriotic,' lamented one of those expelled, 'but it was still a terrible blow.'[35] Repopulation began while the desert was being cleared, weeks before the first prefabricated barracks were put up. Applicants to work on a 'Construction Project' were lured by 'attractive wages' (twice the going rate) and 'immediate living facilities'. Retention was a problem from the start, due to the claustrophobic restrictions of secrecy, the 'bleak' surroundings and the fact that the 'immediate living facilities' were tents. But enough men – and a few women, housed separately – stayed to found the new community of 'Hanford' and to break the ground for whatever was going to be built there.[36] Hanford, codename 'Site W', was a void on the map for several months while Du Pont and Manhattan engineers worked up plans for three massive, water-cooled, graphite-moderated nuclear reactors, together with three plutonium extraction plants. The only guidance was that they must be close to the Columbia River, as far as possible from Hanford, with the reactors at least a mile apart and over 4 miles from the three extraction plants, each also separated by a mile.[37]

Site W was Groves's first project within Manhattan, and his first test of competence. Vannevar Bush began with grave misgivings about Groves's bluntness and intolerance, which could only alienate scientists. Groves wore his prejudices as proudly as his general's star and never hid his contempt for the non-military, no matter what they'd discovered or whichever Nobel Prizes they'd won. After first sitting down with Groves, Bush sent an agitated note to Conant ('We are in the soup') and began looking for a replacement.[38] But the obtuse general quickly proved his worth, bulldozing through War Department red tape to get his Manhattan Project the top war-effort priority of 'AAA'. His snap decisions to give the plutonium project to Du Pont and to locate it at Hanford were soon vindicated by blueprints of the plants and boots on the ground, belonging to the thousands of workmen who began shifting 640 million cubic yards of earth.[39]

Groves played Monopoly on a colossal scale. Hanford was one of three vast Engineer Corps projects, retrospectively christened 'Atomic Cities' after the War (Map 4). Oak Ridge (Site X), known to the general public as 'Clinton Engineer Works', was near Knoxville in Eastern Tennessee, 1,250 miles to the south-east of Hanford. Its 90 square miles of poor, thinly populated agricultural land became the centre for purifying U-235. Initially, the focus was on three processes: Ernest Lawrence's electromagnetic separation using the calutron, Jesse Beams's ultracentrifuge, and the gaseous diffusion method which Harold Urey was developing in parallel (and in competition) with Francis Simon's group in Oxford. All used gaseous hex, to be synthesised on site. Another option was being explored: Phil Abelson was experimenting with thermal diffusion, exploiting a heat gradient to pull out the marginally more restless U-235 from liquid hex, rather than the gas. Groves later said: 'We could not afford to wait ... we had to abandon all normal, orderly processes'. This was evident at Oak Ridge in spring 1943, when space was cleared for factories that would be huge, ravenous for electricity, potentially hazardous and a gigantic security risk, but whose structure and footprint were entirely unknown.[40]

Groves played God as well as Monopoly. He ran his project and ruled his people with 'a fist of iron', riding roughshod over anyone who questioned him or got in his way. He followed US government policy of paying contracting companies just $1 more than the work's actual cost and refused them any share of patents or royalties.[41] His own budget existed to be stretched, and blown if it didn't stretch far enough. In June 1942, Bush gave Roosevelt an estimate of $131 million for the coming year. By New Year 1943, Groves wanted $500 million, and his first billion a few months later. Manhattan's great secrecy meant keeping even Congress in the dark, so huge sums of money had to be buried in an unmarked grave. This was easy for Groves, because the army's Engineer Corps had a vast budget which Congress was programmed not to challenge. The Pentagon had cost almost $50 million; in July 1942, a few weeks before taking on Manhattan, Groves was personally responsible for projects totalling $740 million, or fifteen Pentagons. The neatest way to hide Manhattan's astronomical expenditure was to drop it into a black box containing other obscenely expensive projects.[42] When Congress finally demanded answers, it would be too late to question the accounting.

Groves's third Atomic City followed the nomenclature of the other two. 'Site Y' was the metropolis on which converged the roads from sites W and X – the place where nuclear explosive, either U-235 from Oak Ridge or plutonium from Hanford, would be turned into atomic bombs. This time, Groves's finger settled on the map of New Mexico, and a volcanic plateau that housed an exclusive boys' school. It was called Los Alamos (Maps 4 and 5).[43]

On 25 February 1943, another crucial piece of the Manhattan jigsaw clicked into place. Groves and Bush appointed the Scientific Director of a 'special laboratory' in New Mexico, 'concerned with the development and final manufacture of an instrument of war which we may designate as Projectile S-1-T'. This would involve 'certain experiments in science, engineering and ordnance' and eventually 'difficult ordnance procedures and the handling of highly dangerous material'. Groves and Bush selected their man against stiff opposition from colleagues suspicious of his Communist leanings. He was the current Coordinator of Rapid Rupture, Robert Oppenheimer.[44]

18

BREAKDOWN AND REPAIR
March–November 1943

The monumental projects at Hanford and Oak Ridge during early 1943 were completely invisible to the British. Rudi Peierls's worst fears had been realised: the collaboration was breaking down, forcing Tube Alloys and the Manhattan Project apart.[1] The scientists on both sides of the Atlantic were still trying to push ahead, but against increasing resistance from their political masters.

Groves, Conant and Bush, the three men driving the billion-dollar Manhattan bulldozer, no longer wanted British involvement. Groves was scientifically illiterate and openly despised anything served on a plate stamped 'Made in Britain'. Conant and Bush had also turned against the British. Conant's 'no entry' memorandum barring Britain and Canada from most of Manhattan's activities hadn't come 'from the top' as he'd claimed, but from himself and Bush, and they'd egged each other on to squeeze the British out.[2] None of them trusted the British, with good reason. Back in autumn 1941, the Americans had been desperate to cooperate but were kept dancing on a string for months by Sir John Anderson's stalling letters to Bush. They didn't know – and neither did Tube Alloys or even the British War Cabinet – that Churchill had secretly decided to build a British bomb without the Americans, and had ordered Anderson to fob them off. The Americans also believed that the British were lying about their true interest in nuclear energy. ICI – at the time being sued in US courts for violating commercial law – seemed unhealthily close to Tube Alloys. Two ICI men, Akers (who still worked part-time for the company) and Perrin, fronted Tube Alloys, while Anderson, leading the British negotiations, was an ICI director.[3] American suspicions crystallised when – inexplicably – Frederick Lindemann told Bush and Conant that the top British priority wasn't to build the bomb, but to exploit nuclear power after the war. Whatever the reason for this massive blunder, Lindemann didn't correct it and the damage was done.[4] The last straw was Churchill's secret treaty with Stalin, which could potentially feed American bomb intelligence to the Russians.[5] Conclusion: the British were collaborating only to plunder American science and industry, and were a huge security risk. Conant's memorandum was therefore logical and proportionate.

The British were also victims of suspicion and paranoia. They didn't realise that Bush's 'evasive' correspondence was simply tit-for-tat, because nobody in Tube Alloys ever saw Anderson's prevaricating letters to Bush. Tube Alloys viewed Manhattan as a predator, stealing British ideas and inventions just as they'd done with the cavity magnetron. Conant's memorandum was an outrage, giving the USA free access to everything developed in Britain or Canada while denying their 'partners' any reciprocal rights. They'd stolen Halban's vision of the heavy-water-moderated reactor and then killed his research by cutting off his supply of heavy water.[6] British perceptions of American foul play peaked in May 1943 when the Eldorado Mining Company confessed that the USA had bought up all the uranium it would produce for the next three years. Churchill exploded and accused the Canadian government, which had recently acquired the company, of 'selling the British Empire down the river'. The Canadians admitted that there had been a high-level 'slip-up' but couldn't give the British any uranium.[7]

In Britain, morale hit rock bottom in May 1943. Rudi Peierls noted gloomily that 'we may as well give up now, as the work will never get done in any reasonable time.'[8] In fact, reconciliation was imminent – and brokered by the man who had set about wrecking the Anglo-American collaboration to build an atomic bomb.

1. Otto Hahn (1879–1968) and Lise Meitner (1878–1968) in their laboratory in Berlin, around 1909. They discovered protactinium (element 91) in 1918. In 1938, Hahn discovered nuclear fission (Nobel Prize, 1945), while Meitner and Frisch worked out that fission released a massive burst of energy. Meitner fled to Sweden in 1938.

2. James Chadwick (1891–1974). Discovered the neutron (Nobel Prize, 1932) and masterminded the British Mission to the Manhattan Project. Chadwick was instrumental in driving the Anglo-American collaboration which delivered both the atomic bombs used against Japan in August 1945.

3. Werner Heisenberg (1901–1979) and Niels Bohr (1885–1962), at Bohr's Institute in Copenhagen, 1934. Both won Nobel Prizes in Physics (1926 and 1922, respectively). Bohr was Heisenberg's long-standing mentor and personal friend until they fell out in 1942 over the development of atomic bombs.

4. Marcus (Mark) Oliphant (1901–2000). Hot-tempered Australian-born physicist who promoted the Frisch–Peierls memorandum and eventually persuaded the Americans to take the British U-235 bomb seriously. At Berkeley, Oliphant was Ernest Lawrence's right-hand man, and his group was crucial in enabling Lawrence's calutron to enrich U-235 for Little Boy.

5. John Cockcroft (1897–1967), listening for atomic disintegrations in the Cavendish Laboratory, late 1930s. With Ernest Walton, Cockcroft split the lithium atom (Nobel Prize, 1951). He played crucial roles in the Tizard mission and the M.A.U.D. Committee, and in establishing Britain's defensive radar chain and the Anglo-Franco-Canadian nuclear reactor.

6. Otto Robert Frisch (1904–1979). Jewish refugee physicist who coined the term 'fission' and demonstrated that this released a huge amount of energy. He conceived the U-235 bomb detailed in the Frisch–Peierls memorandum. At Los Alamos, Frisch was crucial to the development of Little Boy, dropped on Hiroshima.

7. Attendees at Niels Bohr's annual symposium, Copenhagen, July 1937. Bohr, Heisenberg and Meitner are first, second and fifth from left in the front row. Others present include Peierls and Szilard (fourth and seventh in second row), von Weizsäcker and Oliphant (second and eighth third row), and Frisch (two rows behind Oliphant).

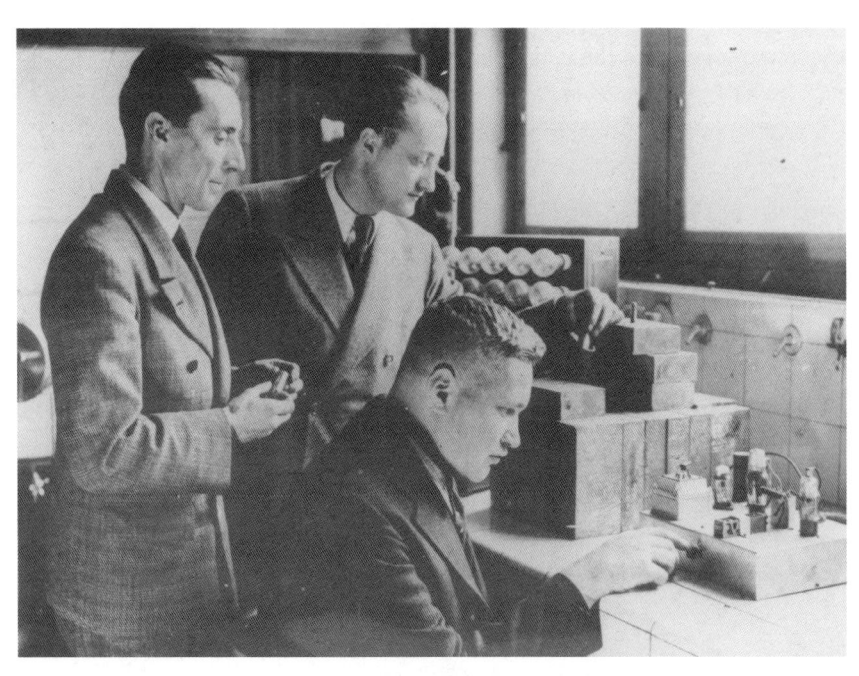

8. Frédéric Joliot (1900–1958, left) with Hans von Halban (1908–1964) and Lew Kowarski (1907–1979) in their laboratory at the Collège de France, Paris. The photograph was taken in 1947 during the filming of *La Bataille de l'Eau Lourde* (*The Battle of the Heavy Water*).

9. Rudolf (Rudi) Peierls (1907–1995). German Jewish theoretical physicist who worked in Birmingham with Oliphant and Frisch. Co-wrote the Frisch–Peierls memorandum which foresaw a U-235 atomic bomb. At Los Alamos, Peierls made vital contributions to both Little Boy and Fat Man, dropped on Hiroshima and Nagasaki.

10. Sir Henry Tizard (1885–1959). Chair of scientific advisory committees in the Ministry of Aircraft Production which promoted radar and the Frisch–Peierls U-235 atomic bomb. In autumn 1940, Tizard led his 'Technological and Scientific Mission to the United States' to encourage the exchange of military scientific secrets between Britain and America.

11. Jacques Allier (1900–1979, left), banker and French military intelligence agent, with Frédéric Joliot. In March 1940, Allier removed all the heavy-water stock from Vemork to prevent it from falling into German hands. The photograph was taken in 1947 in Joliot's laboratory during the filming of *La Bataille de l'Eau Lourde*.

12. George Thomson (1892–1975). English physicist and Nobel Prize winner (1937), who chaired the top-secret M.A.U.D. Committee. In 1941, the Committee concluded that the Frisch–Peierls U-235 atomic bomb was viable and, with American collaboration, could be built in time to influence the outcome of the war.

13. Franz (Francis) Simon (1893–1956). German Jewish physicist who moved to Oxford in 1933. With Peierls, Simon devised gaseous diffusion units for concentrating U-235 from vapourised uranium hexafluoride. These laid the foundations for the factory at Oak Ridge, Tennessee, which enriched U-235 for the Little Boy bomb dropped on Hiroshima.

14. Paul Rosbaud (1893–1963). Austrian-born scientific editor with Springer publishers in Berlin, and undercover agent for the British Secret Intelligence Service, MI6. Rosbaud relayed vital intelligence about the V-2 rockets and the *Uranverein* ('Uranium Club'), the German nuclear fission research collective which set out to build atomic bombs and reactors.

15. Klaus Fuchs (1911–1988). German theoretical physicist who fled to England because he was a Communist. Fuchs worked closely with Peierls in Birmingham and at Los Alamos to optimise gaseous diffusion enrichment of U-235 and the implosion device to detonate Fat Man. He was imprisoned in 1952 for spying for Russia.

16. Vannevar Bush (1890–1974, second from right) and James Conant (1893–1978, right). All-powerful scientific administrators in the USA, who led American atomic policy despite knowing little about nuclear physics. Both were sceptical of the Frisch–Peierls U-235 bomb and obstructed collaboration with the British. Also pictured are Ernest Lawrence on the left and Arthur Compton next to him.

17. Wallace Akers (1888–1954). English industrial chemist and research director at Imperial Chemical Industries (ICI). Director of Tube Alloys from its inception in late 1941 but, at American insistence, replaced by James Chadwick as leader of the British Mission to the Manhattan Project.

18. General Leslie R. Groves (1896–1970) and J. Robert Oppenheimer (1904–1967). The US Army engineer who built the Pentagon and ran the Manhattan Project; and the Professor of Theoretical Physics at Berkeley who directed the atomic bomb factory at Los Alamos. Two men who ensured the project's eventual success.

19. The 'calutron girls' in the Y-12 electromagnetic U-235 separation plant at Oak Ridge, 1944. Each calutron pulled U-235 out of a beam of vaporised uranium and required constant monitoring and adjustment. Dozens of calutrons were arranged in pairs around oval 'racetracks' 122 feet long; the Y-12 factory was over a mile in length.

20. Mark Oliphant's research group from Birmingham University, photographed in 1944 in Ernest Lawrence's laboratory in Berkeley. Oliphant served as Lawrence's deputy and his group played a crucial role in optimising the recovery of U-235 by the electromagnetic separation method.

21. Otto Frisch playing the piano for the Los Alamos radio station, April 1945.

22. The Alsos team removing the lining of graphite blocks from the *Uranverein*'s reactor shell, Haigerloch, April 1945. The Germans had already extracted the heavy water and uranium cubes. Eric Welsh, in uniform without a helmet, is standing in front of the circular cover of the reactor vessel.

23. The Alsos team digging up a buried cache of uranium cubes from the Haigerloch reactor, April 1945. Michael Perrin is standing at far left, and Samuel Goudsmit is second from the right.

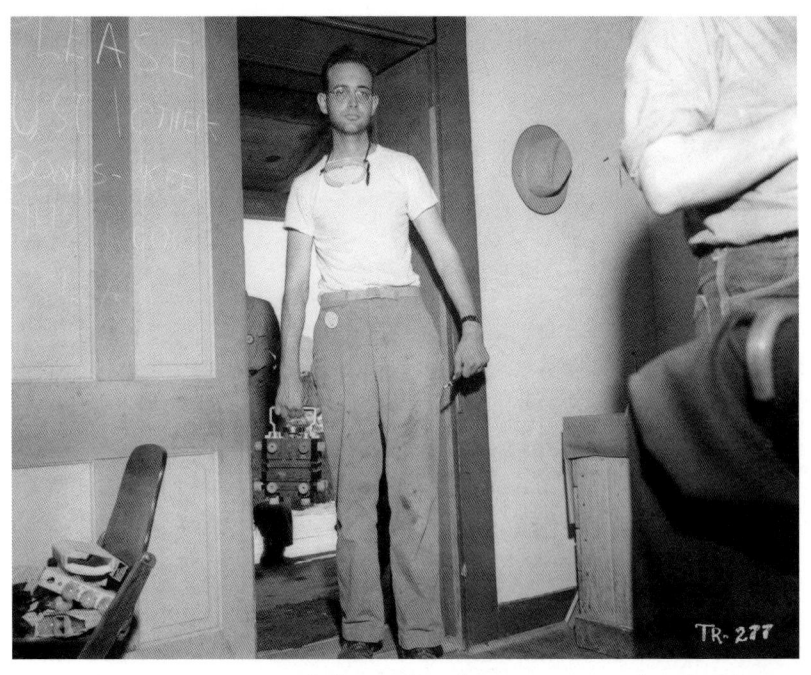

24. A US Army sergeant delivers the plutonium core for the Trinity bomb, 14 July 1945. The two plutonium hemispheres weighed just over 6 kilos and exploded with the force of 20,000 tons of TNT.

25. The Trinity explosion, photographed by ultra-high-speed cameras, 16 July 1945. The top two images were taken at 6 and 25 milliseconds (thousandths of a second) after detonation, and the bottom images after 4 and 9 seconds. Note the change in magnification: the scale bar is 100 metres.

26. Little Boy in the loading pit on Tinian Island, about to be lifted into the bomb bay of the B-29 Superfortress *Enola Gay*, 5 August 1945.

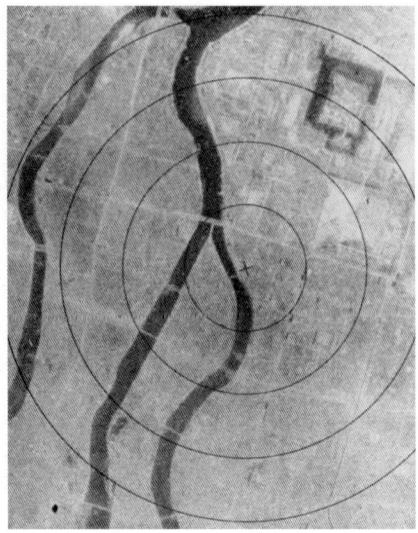

27. The centre of Hiroshima, before (left) and after (right) the detonation of Little Boy, 6 August 1945. Ground zero is indicated by the cross at the centre of the concentric circles, spaced at 500-yard intervals.

28. Under the mushroom cloud. Street scene near the docks in Nagasaki shortly after the detonation of Fat Man, 9 August 1945.

29. The British Mission's farewell production of *Babes in the Wood*, Los Alamos, 22 September 1945. The Trinity tower was represented by the ladder and the bomb's detonation by a cigarette end thrown into the bucket on top. Genia Peierls is standing to the right of the ladder.

30. James Chadwick and General Leslie Groves, with Richard Tolman (Groves's scientific adviser) and Henry Smyth (author of the Smyth Report), photographed in 1945.

31. Otto Frisch and Rudi Peierls (centre) in Washington DC, 1946, after the presentation of the American Medal of Freedom to them, William Penney (left) and John Cockcroft (right).

32. Albert Einstein and Leo Szilard at Einstein's retreat in Peconic, Long Island. They are pictured in 1947, re-enacting the drafting of the letter that Einstein wrote to Roosevelt in August 1939, warning of recent developments in fission research and the possibility of a German atomic bomb.

Action Men

In mid-January 1943, MI6 asked James Chadwick to help bring Niels Bohr to Britain, because the Danish underground believed that the Great Dane's immunity with the Nazis was wearing thin.[9] On 25 January, Chadwick wrote to 'Dear Bohr', explaining 'how delighted I myself should be to see you again', guaranteeing 'a very warm welcome' and requesting Bohr's assistance with 'a particular problem'. Chadwick's letter, handwritten on University of Liverpool headed paper, was reduced to a 1-millimetre square microdot which was hidden inside the hollowed-out stem of a key, sent to Danish agents via Stockholm. In Copenhagen, the microdot was liberated, enlarged, read by Bohr and buried in a metal tube in his garden.[10]

Bohr wrote back regretfully, declining Chadwick's 'tempting invitation' because his duty was to protect Danish scientists, and stating that 'any immediate use of the latest marvellous discoveries of atomic physics is impracticable'. However, he left the door open: 'If there comes a moment when things look different, I shall make an effort to join my friends.' Bohr's letter was also shrunk to a microdot, which was wrapped in metal foil and taken to England inside the drilled-out tooth of a courier.[11]

At this time, six Norwegian ski commandos were completing winter survival and parachute training at the SOE's 'Special Training School' (STS) No. 12, near Aviemore in the Cairngorm mountains of Scotland. The saboteurs of 'Operation Gunnerside' were determined to finish the job in Vemork after the disaster of Operation Freshman two months earlier. They were led by twenty-three-year-old Second Lieutenant Joachim Rønneberg. Leif Tronstad, Professor of Chemistry and a former Swedish relay-running champion, applied to join them but Norwegian High Command refused; he was too valuable to risk losing, and Rønneberg needed endurance skiers, not retired track athletes.[12]

After Aviemore came STS 17 at Brickendonbury Manor in Hertfordshire, for intensive familiarisation with plastic explosives and – locked in a guarded barn – the plywood reconstruction of the electrolysis room at Vemork. Finishing school was STS 61, a country house near St Neots in Cambridgeshire, where SOE agents learned hand-to-hand fighting and wrote difficult letters to their next of kin before being flown low across the moonlit English Channel and dropped into Nazi-occupied Europe.[13] During a trip to the cinema in Cambridge, Rønneberg spotted an ironmonger's shop and, on inspiration, bought a heavy-duty bolt-cutter.[14] After STS 61, the commandos were sent back to Scotland to wait for a full moon and clear skies over the Hardanger Plateau.

The other half of Operation Gunnerside were the five men of Operation Grouse, now optimistically renamed 'Swallow'. Deprived of all their fuel and much of their food by their shambolic air drop, they had seen out a cruel winter of blizzards in a stone cabin high on the plateau. Their pitiful rations of oatmeal were eked out with lichen and eventually a reindeer (with a blood appetiser served hot from the jugular). They gleaned information from insiders at Vemork and by hair-raising sorties to the target: the heavy-water factory stuck on a promontory high above a gorge which sees no direct sunlight during the six months of winter. They found heavy German reinforcements in anticipation of another attack, with barbed wire, mines and booby traps all around the installation, and armed guards on the narrow suspension bridge that carried the only road across the gorge. There was a potential way in along the single-track railway that served the factory – but at the top of a 600-foot cliff face overlooking the river.[15]

Swallow finally met Gunnerside in late February 1943, a week after the new arrivals had been parachuted into the wrong place. They voted on which route to take and chose to climb the cliff to the railway.[16] On the evening of 27 February 1943, Rønneberg's team trekked on skis from their staging cabin 5 miles west of Vemork, left their skis and survival kit in trees across the gorge from the plant, waded through the half-frozen river and climbed the cliff, reaching the railway – luckily unguarded – shortly before midnight. They were wearing British Army uniforms, to deflect any German reprisals away from the locals; and emulating Jomar Brun and his wife, each carried a cyanide capsule in case of capture.[17]

Rønneberg's new bolt-cutter made quick and quiet work of the lock on the metal gate through which the railway entered the compound; the

SOE-issue hacksaw might well have alerted the German soldiers billeted a few yards away. The saboteurs crawled into the plant through a cable duct which Brun had identified, and ran through their well-rehearsed routine with the plastic explosive and the eighteen electrolytic cells in the basement. They were briefly delayed – finding the spectacles belonging to the night watchman whom they'd surprised – then lit the fuses with thirty seconds to get clear. The explosion was muffled but quickly brought out the soldiers. By the time all hell was let loose, Rønneberg's team were scrambling back down to the river. They reclaimed their skis and kit and headed north, zigzagging up a steep escarpment beneath a ski-lift that in happier times had provided a scenic ride to the plateau. They reached the top under 'a mackerel sky, a wonderful sunrise' and dispersed, some to escape through Sweden (200 miles and fourteen days away) and the rest to remain in hiding locally.[18]

They all escaped, despite almost 3,000 German troops, including a crack Alpine ski regiment, sent to hunt them down. One of the Swallows – Claus Helberg, who had scouted the cliff-railway route – lurched to freedom across a tightrope of near disasters. In late March, Helberg ran into three German Alpine soldiers. He outskied two of them and was left in a life-or-death biathlon, gasping for breath on his skis and exchanging shots with the last man, whose pistol jammed at the crucial moment. Helberg was later captured after falling into a ravine and breaking his arm – which didn't prevent him from jumping out of the bus taking him to a concentration camp, then dodging bullets as he disappeared into the trees.[19]

The morning after the raid, the XU underground network was fizzing with rumours that the electrolytic plant and a large stock of heavy water had been destroyed. It was several days before the radio operator hidden high on the Hardanger Plateau could transmit confirmation of success to London. Everyone from Churchill down was delighted, and General von Falkenhorst, the senior German Army commander in Norway, paid a grudging tribute: 'The best coup I have ever seen.'[20] His admiration wouldn't have stopped von Falkenhorst from having Rønneberg's team shot if they'd been captured, just as he'd ordered the murder of the brave but unlucky men brought to earth by Operation Freshman.

Reconciliation

It was Winston Churchill who finally brought an end to the Anglo-American impasse which, through ignorance and pigheadedness, he'd been instrumental in creating. In April 1943, he demanded an updated plan and costs for an exclusively British bomb. Akers and Perrin gritted their teeth, dropped everything else and gave him the facts: 20,000 workers, £70 million, half a million tons of now-scarce steel and – even if rushed through at highest priority – three to five years to make the first bomb.[21] Even then, Churchill wasn't ready to turn his back on Lindemann's absurd fantasy that Britain could go it alone; soon after, he threw away the chance to corner Roosevelt for straight talking about the bomb at the Trident Conference in Washington in May 1943.[22]

Across the Atlantic, harsh realities had also been confronted. Groves, Conant and Bush had been forced to accept that the Manhattan Project could fail unless some of the British were brought back on board. As prophesied by Peierls and Simon, the American method for separating U-235 by gaseous diffusion wasn't working. And Chadwick had expressed grave misgivings about plutonium, insisting that it wouldn't explode in the type of bomb being designed for U-235.[23] The near-simultaneous realisation that neither Britain nor America could build the bomb alone set the scene for rapprochement.

In mid-July 1943, Bush and Henry Stimson came to London to discuss other aspects of the joint Anglo-American war effort. Churchill seized the initiative, bringing them together with Anderson and Lindemann to thrash out the reasons that the bomb collaboration had foundered.[24] Disturbing home truths were revealed. Each side was paranoid that the other would run off with all the nuclear booty after the war. The Americans could have exposed Lindemann as the 'high up' source who had 'forcibly' insisted that Britain's nuclear priorities were 'commercial and after the war'; instead, they dropped a diplomatic veil over his identity.[25] The British resented the Americans' piracy of Canadian uranium.[26] Recently, Bush had taken 'serious offence' at Churchill's apparent attempt to undermine him by approaching Roosevelt through his aide, Harry Hopkins.[27]

Magically, the air cleared. A draft agreement between Roosevelt and Churchill was re-written 'in more majestic language' by Churchill and despatched to Washington with John Anderson to deploy his 'judicious diplomacy'. The telegram waiting for Anderson in Washington – 'BEST OF LUCK – WINSTON' – might have helped him to face last-ditch resistance by Groves and Conant.[28] Within a week, a secret Tube Alloys–Manhattan accord was presented to Roosevelt and Churchill, who were both in Quebec for the Quadrant Conference of 17–24 August 1943. Bigger things were on the table, notably Roosevelt's enthusiasm for an Allied invasion of Normandy, a notion that Churchill actively opposed. Churchill softened his stance, and this may have helped the Anglo-American atomic partnership to get off the ground. After a brief delay 'for fear of criticism by members of Congress', the Quebec Agreement was signed on 19 August 1943.[29]

For a couple of weeks, Conant refused to accept that the Quebec Agreement had been ratified, but it was too late. Akers, still *persona non grata* because of his ICI links, had been waiting in Ottawa and now flew down to Washington. There, he met Chadwick, Oliphant, Peierls and Simon, and the 'British Mission to the Manhattan Project' was born.[30]

A Difficult Extraction

It took months to persuade Niels Bohr to 'join his friends' in England. Throughout spring and summer 1943, the Great Dane remained deaf to Chadwick's repeated pleas – until the imperative 'WAR CABINET PRIORITY HAS BEEN GRANTED FOR IMMEDIATE JOURNEY', cabled urgently by Chadwick and Lindemann in late September, made Bohr realise that the Gestapo were about to arrest him. Eric Welsh swung into action, and the Danish underground spirited away Niels and Margrethe Bohr. They crossed the Kattegat to Sweden by fishing boat on the night of 29 September 1943, followed a couple of days later by their sons.[31]

Bohr was a celebrity in Sweden – which, although neutral, was thick with German spies and informers. The star-struck Swedish officer who met the Bohrs off the boat introduced him to everyone

THE IMPOSSIBLE BOMB

when they stopped for lunch en route to Stockholm. Bohr then wandered the streets for a week, visiting Lise Meitner and other friends. Eric Welsh fumed and finally snapped one evening in early October, when Bohr wanted a further delay to have dinner with the King of Sweden.[32] At the time, Welsh was out on the North Sea in a surfaced submarine, delivering two secret agents to Norway. Before submerging, Welsh ordered Bohr to lie low and whistled up help to fly him to Britain.[33]

Bohr's saviours arrived in a Mosquito fighter-bomber at Stockholm airport on 6 October 1943. Because Sweden was neutral, the aircraft was unarmed; as there were only seats for the crew, Bohr had to sit on the bomb-doors in the empty bomb bay in a flying suit, ill-fitting helmet and parachute. The plane took off after dark and headed south-west over Nazi-occupied Norway towards Scotland, flying high and fast beyond the reach of German fighters. In case of attack, the bomb-doors would be opened and Professor Bohr would drop into Norway or the North Sea; even if his parachute behaved, his chances of live rescue were essentially zero. As they climbed through 10,000 feet, the pilot told Bohr over the intercom to put on his oxygen mask. Bohr didn't hear him and didn't respond, leaving the crew – who couldn't access the bomb bay in flight – wondering if he'd died from hypoxia. Luckily, Bohr recovered consciousness as the plane descended towards the Scottish coast and was back to his version of normality when released from the bomb bay.[34]

Next came a dizzying merry-go-round of meetings and briefings in London. Bohr believed that the bomb was theoretically feasible but impossible in practice because both Britain and the USA were bled dry by their existing war efforts; even America, he argued, would have to be turned into a gigantic weapons factory.[35] It took a day to stun Bohr into realising how wrong he was. After briefings by Akers, Perrin and Lindemann, he was taken to dinner in a private room at the Savoy on 8 October 1943 and hit with the full force of information about Tube Alloys, Manhattan and Fermi's CP-1 pile. The decisive hammer-blows were administered by Akers, Welsh and Chadwick, who had flown back from his first visit to Los Alamos.[36] Chadwick asked the subdued Bohr if he was prepared to go to Los Alamos. Yes, of course. Lindemann enquired if the bomb would work. 'Yes,' Bohr replied, 'and that's what I'm determined to stop.'[37]

Bohr was reunited with his son Aage, flown out less excitingly soon after his father. Bohr Senior's new-found freedoms included the cinema, where he left his briefcase (fortunately, this contained no secret papers, just his wallet and all his money). Thereafter, each of the six pockets in Bohr's suit contained a copy of the day's appointments and he was always accompanied by a minder; Eric Welsh took turns to do this and struck up a lasting friendship with his charge, who appreciated his 'directness and humour'. The Great Dane couldn't visit Maud Ray in Kent but spent a weekend at Michael Perrin's house, where he entranced Perrin's three-year-old son Charles by whittling a stick into two intertwined spirals.[38]

At the end of November 1943, 'Nicholas Baker' and his son 'James', whose Danish accent was much less pronounced than his father's, were flown out to Washington and then on to New Mexico, to join the British Mission to Los Alamos.[39]

Coming Clean

The dinner at the Savoy on 8 October 1943 was another landmark, because Eric Welsh finally told the whole truth about the Nazis' non-existent atomic bomb.[40]

This was a further massive shock to Bohr. He believed that Heisenberg had engineered that bizarre meeting in Copenhagen in September 1941 to discover what Bohr knew about an Allied atomic bomb, and was shocked when Heisenberg told him that the Germans were working on their own weapon. Bohr held to his view that the bomb was technologically unachievable, but now regarded Heisenberg as a palpable threat. Chadwick felt the same, describing Heisenberg as 'the most dangerous German in this field, because of his brainpower'. Lise Meitner even rated him above Chadwick: 'I'm afraid that the Allies have no such man as Heisenberg.'[41] Heisenberg's shadow also stretched across the Atlantic. Viktor Weisskopf, a German refugee and one-time associate of Heisenberg and von Weizsäcker, wrote to Robert Oppenheimer in late October 1942, urging him to have

Heisenberg kidnapped. No immediate action followed, but American military intelligence began to consider ways of deleting Heisenberg from the Nazi equation.[42]

Eric Welsh, spymaster and chief of the intelligence operation against the Nazis' nuclear programme, hadn't believed Paul Rosbaud's first reports in June 1942 that the *Uranverein* had been ordered to stop working on the bomb. He'd inherited Rosbaud from Frank Foley, who was convinced that Rosbaud was reliable, but Welsh remained sceptical, suspecting an elaborate smokescreen spread by Rosbaud or his informants to throw the British off the scent. However, more supportive evidence filtered in during the winter of 1942–3, and by mid-March 1943 Welsh finally accepted that the Nazis' atomic bomb was no longer a cause for concern.[43] For some months, he kept this information to himself. And why not? After the long deadlock that spring, Tube Alloys and Manhattan were finally working together, and an Anglo-American atomic bomb now seemed achievable. Why destabilise a potentially war-winning enterprise that had nearly collapsed and could still fall apart?

Welsh finally let the Americans into the secret in September 1943, when Groves sent one of his brightest young men, Major Robert Furman, to London to extract up-to-date information about the German bomb from British Intelligence. Welsh levelled with Furman, who was impressed by the evidence and returned to America to tell Groves the good news that there was no German bomb and that Manhattan had an open field.[44]

Groves's reaction reflected his mission statement, 'Build an atomic bomb, and build it before the enemy can.' He silenced Furman before the revelation could contaminate the Manhattan Project.[45] Fear of a Nazi bomb wasn't the only thing that kept Groves awake at night. If Manhattan failed, he would be dragged in front of Congress to explain away billions of wasted dollars and why he'd kept them in the dark for so long.[46]

Blood and Water

Professor/Major Leif Tronstad, kicking his heels with Norwegian soldiers at the SOE's Special Training School in the Cairngorm mountains, was restless and frustrated. The spectacular success of Operation Gunnerside had been swiftly recognised by command of Churchill himself. As the architects of the mission to destroy their brainchild at Vemork, Tronstad and Jomar Brun received the honorary Order of the British Empire, while Joachim Rønneberg and Jens Poulsson, leader of the Grouse/Swallow team, were both awarded the highly coveted Military Cross. Fair enough: Tronstad and Brun had been with the saboteurs every step of the way, but only in spirit and from the safety of Britain.[47]

It was now two and a half years since Tronstad had escaped from Norway, and he was homesick and aching to play his part in clearing out the Nazis. His wife and two sons were still stuck in Trondheim; coded messages from the XU underground network brought only limited comfort, telling Tronstad (codename, 'Mikkel', or 'Fox') that 'Mrs Fox and the Cubs' were well. Tronstad was fiercely patriotic and shared the Americans' suspicions about ICI. Determined that Norsk Hydro wouldn't fall victim to British commercial imperialism after the war, he explained that 'blood is thicker than even heavy water.'[48]

Tronstad was justifiably worried about the future. The wrecked electrolytic cells were quickly replaced and upgraded with impressive German efficiency, and heavy-water production restarted in early May 1943.[49] As before, it was stored on site and shipped out on the railway line which had given the saboteurs access to their target. The track ran 20 miles along the side of the valley to the terminal at the northern end of Lake Tinn, where the ferry *Norsk Hydro* shuttled cargo and up to 120 passengers to the Oslo-bound railhead at the lake's southern tip.[50] By September, heavy-water deliveries to Berlin had almost returned to pre-Gunnerside levels. But what were Heisenberg and the *Uranverein* doing with it all? If the German bomb project really had been scrapped, the *Uranbrenner* reactors were being used only to produce energy. This could eventually lead to a revolutionary and game-changing engine for a submarine or a battleship, but not the super-weapon to end the war with a German victory.

Meanwhile, Groves suppressed the intelligence from Norway and Britain and perpetuated the myth that the Germans were still developing their bomb and must be stopped at all costs.[51] In September

1943, Groves informed Bush that Norsk Hydro was now supplying Heisenberg with 120 kg of heavy water each month. US military planners had already interrogated Tronstad about bombing Vemork, and did so again. Tronstad simply repeated the truth. The electrolytic plant was unassailable from the air; an air raid couldn't succeed and would kill civilians.[52]

Tronstad was correct on all counts. On 15 November 1943, Groves ordered a 'pin-point' bombing raid on the electrolytic plant at Vemork. The pin was a lamentably blunt instrument. The raid by the US Eighth Air Force, just after lunchtime, wrecked parts of the ammonia factory. The electrolytic plant sustained trivial damage and was back in operation almost immediately. Twenty-one Norwegian civilians died.[53]

Surprisingly, the Germans reacted by ordering Norsk Hydro to stop making heavy water. Next came the bad news. The heavy-water stocks were to be kept under guard for transfer to Berlin, together with the electrolytic equipment from Vemork and a smaller plant at Såhem. The Nazis intended to make their own heavy water in Germany.[54]

19

OVER THERE

September 1943–February 1944

As an eminent American historian once explained, the 'British Mission to the Manhattan Project' had nothing to do with the Church of England. The mission comprised over sixty scientists from England, some British by adoption, who went to Los Alamos and Berkeley from autumn 1943 onwards to enable the Americans to wreak devastation on a biblical scale. Some stayed for a few months, others until after the war ended, and a couple never left. While in America, members of the mission blinked out of existence in Britain; not even the War Cabinet knew where they were or what they were doing.[1]

The mission was born in late September 1943, a month after Churchill and Roosevelt signed the Quebec Agreement. Rudi Peierls, Mark Oliphant and Francis Simon led the way, flying to New York on the Pan Am clipper from Foynes in Southern Ireland; killing time in Limerick, they watched a lurid documentary about nuclear fission and atomic bombs.[2] Chadwick followed a day later. Thanks to food poisoning from the oyster bar at New York's Grand Central Station, he spent the night writhing on the bathroom floor and looked 'like death' for his first meeting with Groves the next morning. Their relationship didn't start well.[3] Chadwick attacked Groves's obsession with compartmentalisation, and Groves reacted like 'a dictator'. Luckily, Groves respected Chadwick's plain speaking, and the two ill-matched men agreed to do business because they had to. Groves later boasted that he'd often dragged his feet because he 'detested the interchange', but Chadwick found him 'friendly and cordial'.⁴

Groves wanted Chadwick to lead the British Mission to work with Oppenheimer, who had already amassed over 300 scientists at Los Alamos, codenamed Site Y and nicknamed 'Shangri-La'. Over dinner that evening, James Conant put the knife into ICI. With Chadwick now running the British Mission, Conant explained, Wallace Akers no longer had any role in America. Akers fell on his sword without fuss and retreated to London.⁵

From Washington, Chadwick began assembling his dream team in a flurry of top-secret telegrams to Akers and Perrin. Within 'two or three weeks', he needed eight researchers from Birmingham to help with Lawrence's electromagnetic separation in Berkeley and with bomb design at Los Alamos. After haggling with the Air Ministry, who wanted some of them for radar, Chadwick got his men.⁶ Returning briefly to England in early December 1943, he nailed Otto Frisch with two typically blunt questions: 'Do you want to work in America?' and 'Are you prepared to become a British citizen?' Frisch jumped at both offers and, assisted by a bemused policeman, things moved 'with bewildering rapidity': citizenship fast-tracked, call-up papers issued and immediately rescinded, and a brand-new passport stamped with an American visa handed to Frisch as he waited to sail from Liverpool. His fellow passengers on the *Andes* included Rudi Peierls, who had returned to collect Genia, and Ernest Titterton, Mark Oliphant's first PhD student at Birmingham and now an expert in electronics. Akers waved them off; he couldn't find taxis to take them from the Adelphi Hotel to the docks, so conjured up a cortège of shiny black funeral limousines and a hearse for their luggage.⁷

The *Andes*, a luxury cruise liner converted into a troopship, had brought 4,000 American soldiers to England but was thinly populated on its return trip. The crossing was rough, with right-angle zigzags to confuse U-boats, and particularly grim for Genia Peierls who was recovering from a miscarriage. Frisch had a dormitory to himself and relaxed by playing an elderly upright piano lashed to a pillar, hanging on to the keyboard whenever the stool started to roll away.⁸

On arrival in Newport, Virginia, Rudi Peierls went to New York, soon to be joined by Francis Simon and Klaus Fuchs to advise the Kellex Corporation, which was building the huge gaseous diffusion separation factories at Oak Ridge. Genia rescued their children from Toronto, the first time she'd seen them in nearly four years, and took them to find a family apartment in New York. Frisch suffered a series of culture shocks. Seeing fresh oranges piled high in a brightly lit street made him laugh hysterically.[9] Then he met Groves, who was long-winded and security-obsessed and told him that he'd be working at a place called Los Alamos.[10]

Site Unseen

Los Alamos lies in a wild and remote corner of New Mexico, 60 miles north-east of Albuquerque and a few miles west of the Rio Grande river (Map 5). It is beautiful and invigorating: 7,500 feet above sea level, with crystal-clear air and breathtaking sunrises over the Sangre de Cristo mountains. The area is dotted with ancient settlements of the Native Americans whose descendants are the local Pueblo Indians. Site Y was centred on 'the Hill', an elongated flat-topped mesa chiselled out of the Pajarito Plateau between two roughly parallel canyons running east–west. The Hill was served by a single rutted road; Santa Fe, the nearest town, lies 35 miles and a visceral two-hour drive to the south-east.[11] Since 1917, the Hill had been occupied by the exclusive Los Alamos Ranch School which provided the sons of the wealthy with 'health, strength and self-confidence' through learning outdoor skills and sleeping on open verandas. The school's last cohort graduated hurriedly in January 1943, when the US Army commandeered the mesa and the surrounding 50,000 acres at $225 per acre, over thirty times the take-it-or-leave-it rate offered to the Hispanic homesteaders whom the army evicted.[12]

Otto Frisch and Ernest Titterton were the first British pioneers on site, two weeks before Christmas 1943. Frisch was enthralled by the landscape – 'like the Lost World of Conan Doyle' – and the cliffs that he (an experienced alpinist) judged 'unclimbable'. The mesa was a mess, with green-painted prefabricated housing for over 3,000 people being thrown up alongside the 'Big House', dormitories and workshops of the Ranch School. Orientation was challenging: there were no street names and the buildings were numbered in order of completion, irrespective of location. The best landmark, and a homing beacon for anyone lost in the vicinity, was the cylindrical wooden water-tower that dominated the skyline.[13]

James Chadwick arrived in mid-January 1944, exhausted by a tour of the other Manhattan sites which hammered home the realisation that the Americans had 'scope and vigour far beyond anything we could do'.[14] He'd flown into Albuquerque, where a welcome from Native Americans dancing in warpaint left him unmoved. The Chadwicks were allocated a VIP two-bedroom log cabin on 'Bathtub Row' with a bathroom (a rarity) and Appolonia, a Pueblo cleaning lady complete with traditional blanket-shawl and turquoise jewellery. Aileen Chadwick came straight from an emotional reunion with their twin daughters, who had been sent to safety in Halifax, Nova Scotia. James had visited them and either hadn't noticed their transformation or didn't think it worth mentioning. Aileen was shaken by the two forthright seventeen-year-olds with make-up and Canadian accents, and was still in shock when she joined her husband on the Hill. James was miserable company, having made himself ill haggling with Groves and Halban over the 'unhappy and disorganised' Parisian boiler group. Chadwick's health and the group's fortunes improved markedly when he replaced the kiss-of-death Halban with John Cockcroft, but then Washington needed him. He left Aileen behind to chain-smoke on her own in Bathtub Row. Aileen invited English and American wives over for high tea, but won little sympathy when she told them how much she resented her 'primitive' lifestyle. Isolated and lonely, she pined for summer and the arrival of her daughters.[15]

By contrast, the Peierlses instantly took to life on the Hill. In late January 1944, Rudi was summoned urgently from New York to Los Alamos to tackle a completely new problem: creating an explosive shockwave powerful enough to crush a plutonium sphere into half its volume. The Peierls family travelled down to Santa Fe by slow train and were driven to Los Alamos, stunned by the 'always breathtaking' ascent from desert to the sheer-sided volcanic plateau. Their flat was across town from the Chadwicks, below one belonging to 'Mr and Mrs Henry Farmer', alias Enrico and Laura Fermi. They were immediately among friends, including Frisch and Oppenheimer. Still sporting the pork-pie hat of fifteen years earlier, Oppenheimer invited them over for a welcome supper and trial by dry martini, after which Rudi and Genia could barely walk home.[16]

Niels Bohr had returned to his disciples. Just before New Year 1944, 'Nicholas Baker' arrived at the Hill, having finally shaken off the two US Secret Service agents who had shadowed him and slept outside his hotel bedroom in New York. Mr Baker's brand-new British passport had done the trick, despite his incomprehensible English and 'NIELS BOHR' painted boldly on his luggage. Groves wanted Bohr on his payroll, but Chadwick and Bohr refused. The Great Dane added lustre to the British Mission and, according to Oppenheimer (who christened him 'Uncle Nick'), brought 'hope to the entire macabre enterprise'. At fifty-eight years old, Bohr was twice the average age of the Los Alamos scientists, and old enough to be the grandfather of many. He skied and scattered advice, ideas and inspiration as energetically as always – and remained everyone's 'father confessor', even when they weren't sure what he was saying.[17]

Bohr, who 'thought about politics and other human matters almost as deeply as he did about physics', had impaled himself on the horns of a dilemma. He was absolutely determined that the bomb must be built, but also that it must never be used to commit mass murder. The only way he saw to prevent the nuclear annihilation of humanity was for Britain and America to cooperate with Russia after the war – and to share the secrets of the atomic bomb. To test the temperature, he sounded out friends who were confidants of Roosevelt and Churchill: Supreme Court Justice Felix Frankfurter in Washington, and Frederick Lindemann in London. They responded favourably, which encouraged Bohr to plan a meeting with Churchill to argue his case in person.[18]

Bohr's ambivalence over the bomb infected another member of the British Mission. Joseph Rotblat initially refused Chadwick's invitation to America; the offer of British citizenship fell flat because Rotblat was proudly Polish and desperate to return home as soon as the Germans were defeated, to find his wife and resume his career. Chadwick eventually persuaded Rotblat to come to Los Alamos, having strong-armed Groves into accepting him as the only non-American, non-British member of Manhattan. Rotblat arrived in late March 1944 and stayed with the Chadwicks for a couple of weeks, reliving good times in Liverpool. Already fretting about how the bomb would be used, Rotblat became depressed after discussing nuclear physics and the future of humanity with Bohr. He was derailed when, over supper at the Chadwicks' house, Groves bragged that the bomb wasn't being built to defeat Germany but to intimidate and if necessary crush the Russians after the war. Perhaps Groves was baiting Rotblat, whom he still regarded as a security risk; whatever his motive, he left Rotblat distressed by thinking of the hundreds of thousands of Russians who had died fighting the Nazis.[19]

All Change

Nineteen forty-three was the year in which the tide of the war turned. For ordinary Germans, it was a procession of catastrophic defeats: Stalingrad in February, North Africa in May, Sicily in July, the Italian mainland in September. And the war came home to the Fatherland, with thousand-bomber air raids hammering cities and the steel-making crucible of the Ruhr, then the Dam Busters and the firestorm that incinerated Hamburg and 40,000 people on 24 July. The dreadful year ended with another knife in the guts – the sinking of the *Scharnhorst*, pride of the *Kriegsmarine*, off the Norwegian coast on 26 December 1943. And the New Year of 1944 began with Allied bombers penetrating into the very heart of the Reich – Berlin.

In early December 1943, Groves set up a 'Scientific Intelligence Mission' at the OSRD's Liaison Office in London. The mission was led by Lieutenant Colonel Boris Pash and initially comprised a dozen scientists, intelligence officers and interpreters. Pash was an American Army intelligence officer who arrived fresh from an undercover operation in Berkeley, where he'd failed to find Soviet spies in Ernest Lawrence's lab.[20]

The Scientific Intelligence Mission gave itself the cryptic title 'Alsos'. Even if the enemy had realised that this was classical Greek for 'grove' they might not have made the connection with an American general of a similar name (who was furious when he learned how Alsos came to be christened). Groves created Alsos to plunder scientific war booty after the defeat of Germany. Alsos would hunt down scientists working for the Nazi war effort – on everything from antibiotics and vaccines to rockets and bombs – to extract their secrets for sole exploitation by the USA. The primary target of Alsos was the *Uranverein*.[21]

Within two weeks of being established, the Alsos Mission swapped desks in London for the Allied front lines in Algiers and then Italy.[22] The sixteen-strong group now included Major Robert Furman, Groves's most trusted intelligence officer, and Morris (Moe) Berg, whose unique curriculum vitae featured both Princeton and the American Baseball League. In Rome, Alsos rounded up some of Fermi's ex-colleagues who were convinced that a German bomb, if it existed at all, was years from completion. Afterwards, Moe Berg hung around in Italy while the others returned to London to await the invasion of France.[23]

The *Uranverein* was also high on the agenda for Eric Welsh. Just after Christmas, he'd planted an item of disinformation through the 'XX' (Double-Cross) sub-committee of MI5. This made a big splash in the *Sunday Express* on 26 December 1943 – a world-renowned Swedish nuclear physicist ridiculing the idea that atomic bombs could play any role in the war – but might not have grabbed the enemy's attention because it coincided with the sinking of the *Scharnhorst*.[24]

Welsh now accepted that the Nazis had killed off their own atomic bomb, but intelligence pointed to continuing fission experiments with an unquenchable thirst for heavy water at the Virus House in Berlin and Heisenberg's lab in Leipzig. Perhaps they weren't just trying to build a nuclear power generator? There was another, more troubling possibility: using nuclear reactors to synthesise lethally radioactive isotopes to be scattered by conventional explosives. These 'dirty' bombs would be cheap and easy to produce and could block the advance of massed soldiers. Anticipating an Allied invasion of northern France, the top-secret Operation Peppermint was rushed through to provide radiation detectors for front-line troops.[25]

Two pieces of intelligence in February 1944 made Welsh smell enough danger to take decisive action. The first was another gem which Paul Rosbaud smuggled out of Berlin. Working alongside concentration camp prisoners (and flirting with disaster by slipping them bread rolls), Rosbaud had helped to clear rubble after an RAF air raid on the night of 14 February which badly damaged the KWI for Chemistry. He learned that key members of the *Uranverein* were being relocated from Berlin and the heavily bombed Leipzig to a cluster of sites 30 to 40 miles south-west of Stuttgart (Map 6). These included a textile factory in Tailfingen and, close to the Black Forest and the Swiss border, the Hohenzollern Castle in Hechingen and the riverside village of Haigerloch.[26]

A few days later, sources inside Vemork revealed that Norsk Hydro's remaining heavy-water stocks and electrolytic plant machinery would be shipped to Germany on 20 February 1944. Welsh agonised, decided that the risk couldn't be taken, and instructed the Norwegian XU underground. The target – the ferry *Norsk Hydro* as it crossed Lake Tinn to the rail terminal for Oslo – was softer than Vemork but still represented a formidable challenge. The night before it sailed, members of the Swallow team who had stayed in hiding on the Hardanger Plateau planted a time-bomb in the ferry's hold. The operation went like clockwork. *Norsk Hydro* sank within minutes of the explosion at 10.20 the next morning, consigning the last of Vemork's heavy water and an embryonic electrolytic plant to 1,400 feet of water, together with fourteen Norwegians and four German soldiers.[27]

This was a short-lived setback for the *Uranverein*. Within weeks, Welsh learned that IG Farben's ammonia factory near Leipzig was being modified to produce heavy water – and that the relocated laboratories deep in the Black Forest were up and running.[28]

20

MISSIONARIES
May 1944

In early May 1944, the British Mission to Los Alamos comprised eight congenital Britons, three adopted refugees, one immutable Pole and a badly disguised Dane. During the next twelve months, another dozen would join them on the Hill. Including visitors such as Frederick Lindemann and Mark Oliphant, the mission totalled twenty-eight scientists – a tiny Britannic island in an ocean of over 3,500 Americans.[1] By coincidence, the proportion of 'British' scientists at Los Alamos (0.8 per cent) was near that of U-235 in native uranium (0.7 per cent); and like U-235, they punched far above their weight.

Around them, the town of Los Alamos grew fast, echoing St Joachimsthal during the Bohemian silver boom four centuries earlier. The entire Hill, including the sprawling residential quarters, was fenced off by barbed wire, with armed guards at the two access points. Another barbed-wire enclosure protected the inner sanctum of the Technical Area – laboratories, cyclotron, library and administrative buildings (Map 5). The Hill's outermost defences included armed guards on horseback and 'K-9' dog patrols around the base of the mesa. The town had its own shops, cinema, theatre and barber's shop.[2] A small hospital dealt with emergencies, although predictable illnesses were avoided: Chadwick rejected an otherwise ideal candidate because the high altitude could cause a previously punctured lung to collapse again.[3] The school, largely staffed by scientists' wives, was generally 'excellent' but strangely 'weak' in science.

All scientists on the Hill were brought together for regular seminars, updates and trouble-shooting sessions chaired by Oppenheimer in the cinema, heavily guarded for these special matinee performances.[4] Essential knowledge was distilled into the *Los Alamos Primer*, a twenty-four-page cyclostyled resumé of tutorials given in spring 1943 by Oppenheimer's thirty-four-year-old colleague Robert Serber. Written with clarity, elegance and 'an engaging dryness', the *Primer* was a top-secret do-it-yourself guide to making 'a practical military weapon' from '25' or 'material 49' (U-235 and plutonium were not mentioned by name). Nobody owned a copy; the *Primer* could only be read in the Technical Area's library.[5]

Los Alamos remained one of the best-kept secrets of the war, because Groves regarded it as a gigantic compartmentalised cell and a recurring security nightmare. Materials and equipment – ranging from copies of *Police Gazette* for the barber's shop to meteorological balloons, sewing machines and seventy-five cows for unspecified purposes – were ordered through a non-existent company in Denver. Incoming mail for Americans arrived via an address in Santa Fe, and through a postal box in Washington for the British Mission. Outgoing letters were left unsealed because all mail was scrutinised by army censors – not to catch spies (which it signally failed to do), but to prevent inadvertent leaks of information. The practice began after Groves investigated rumours that letters were being opened; they weren't, but it seemed a neat idea.[6] When an alarmed censor informed Groves that Peierls regularly sent detailed updates to Chadwick in Washington, Groves merely asked Peierls for copies to keep himself informed.[7]

Some thought security an absurd imposition – enough, Edward Teller later claimed, for Groves to win 'almost any unpopularity contest'. Richard Feynman, a future Nobel Prize winner but then a twenty-three-year-old post-doc from Princeton, was a grand master of mischief. Holes in the outer security fence were well-known shortcuts for residents on the Hill, but not for the army guards – one of whom was bewildered when Feynman re-entered the compound without having checked out. Feynman also baited the censor by exchanging letters with his wife in Albuquerque, in which they encrypted gossip and shopping lists into impenetrable code. When she sent Feynman a letter cut into a jigsaw puzzle, the censor inserted a plaintive note: 'I don't have time to play games.' He might have reacted differently if he'd known the backstory. Feynman's wife Arline, his childhood sweetheart and soulmate, was in a sanatorium and losing her battle with tuberculosis.[8]

Despite the grimness of its purpose, life on the Hill was good. Most put aside their ferocious work ethic and followed Oppenheimer's directive not to labour on the Sabbath. Diversions included barn-dancing, the cinema and the theatre, where the corpses of poison victims in *Arsenic and Old Lace* received a rousing curtain call. Frisch and others provided music for dances, the Hill's radio station or a restful supper after a hard day's skiing; one listener remarked that Frisch was 'wasting his talents' on physics. Outside the fence, there were Indian archaeological sites; shopping, sightseeing and the La Fonda restaurant in Santa Fe; riding for the 'horsey set'; 14,000-foot peaks in the Sangre de Cristo mountains where Frisch and Egon Bretscher relived happier times in the Alps; and, rarely, trout fishing for Chadwick.[9]

Peierls found the lack of hierarchies and boundaries at Los Alamos 'an enormous pleasure', while Frisch had 'never seen a small town with such a variety of intelligent and cultured people'. Oppenheimer had skimmed off 'the cream of the universities' – not just physicists, chemists, mathematicians and engineers, but also a philosopher and a painter.[10] He'd also transformed himself from a left-wing hazard into 'a superb leader'. Wearing his trademark pork-pie hat, now sweat-stained, 'Oppie' would greet newcomers with 'Welcome to Los Alamos, and who the devil are you?' – and, with very few exceptions, the relationship blossomed from there. The British Mission also fell under his spell. Ernest Titterton's wife, Peggy, said, 'We just worshipped him ... we would have done anything for that man,' but the Chadwicks' twin daughters found him 'odd ... not as clever as our father'.[11]

The setting was idyllic and the atmosphere and camaraderie exhilarating, like a never-ending scientific conference permeated with the clandestine thrill of Marie Curie's 'floating university' in Warsaw. But this idyll had to deliver the 'instrument of war' that Groves and Bush had codenamed 'Projectile S–1–T' in their letter of February 1943, commissioning Oppenheimer to lead Manhattan. Now, fifteen months and over a billion dollars later, the 'extraordinary galaxy of scientific stars' was yet to bring that instrument of war any closer. Rudi Peierls had made Groves's 'blood boil' by telling him that 'only a miracle' would create enough pure U-235 for a bomb by mid-1945[12] – shortly before a depressing meeting with Conant convinced Groves that the U-235 bomb, even if ready in time, might only match a paltry 500 tons of TNT.[13] Even worse, suspicions were deepening that plutonium wasn't a clever shortcut but a liability which could never explode in the weapon taking shape to detonate U-235.

Old World

The Tube Alloys office in Old Queen Street still seemed frantically busy but was just a sideshow now that the centre of gravity had slid across the Atlantic. Wallace Akers, severed from the British Mission, kept in regular contact with Chadwick, while Michael Perrin was working ever closer with Eric Welsh and MI6.[14] Heroic efforts were being made to secure essential supplies for when (if?) Britain could pursue its independent nuclear ambitions. ICI had investigated potential sources of heavy water and found that the legendarily dense water of the Dead Sea was stiff with NaCl, but not D_2O.[15] There was better news about uranium. The dastardly Canadians were still giving all their ore to the Americans, but supplies were now reaching England from the Belgian Congo – personally arranged by Edgar Sengier, President of Union Minière, who was exiled in London.[16]

Manhattan had torn the guts out of British nuclear physics research as well as Tube Alloys. Those left behind in Chadwick's laboratory in Liverpool filled their days with experiments that would never set the world alight. Morale plummeted after Mark Oliphant flew in, cocky with excitement about the astonishing facilities in America, and told Chadwick's team that they were wasting their time in Britain. On receiving anguished messages from Liverpool, Chadwick gritted his teeth and reminded everyone that their role now was to ensure that the Americans made the bomb before the war ended. He added stoically, 'I cannot hesitate to make any sacrifice which will help ... including the future of my own laboratory.'[17]

In Oxford, Francis Simon's team continued to experiment with gaseous diffusion, but without Simon, Peierls or Fuchs. The 'Special Products' factory at Rhydymwyn in North Wales at last housed three prototype Simon separation units, inside which massive fans forced gaseous hex through a thin skin of sintered nickel. The results were promising but the programme would have to be axed without the next step – the pilot separation factory which would never happen because it could only be built in North America.[18] Rhydymwyn was a metaphor for the entire British atomic bomb programme: marking time while the general British war effort and the Manhattan Project were frenetically busy. This incongruity permeated high up the British chain of command. One VIP confessed in a memo that he couldn't remember the codename of 'Anderson's project'. His special adviser prompted him by heading his reply 'TUBE ALLOYS'. The special adviser was Frederick Lindemann, and the man with the memory lapse was Winston Churchill, co-architect of the Quebec Agreement which had supposedly welded together Manhattan and Tube Alloys to make the Anglo-American atomic bomb.[19]

An Innocent Abroad

Roosevelt, Churchill and Stalin had presented a unified front at the pivotal Tehran Conference in December 1943, but the Americans and British harboured grave doubts about Russia's atomic intentions after the war. During the 1930s, Russian nuclear physicists had fallen behind, but a few serious players emerged before the war shut off all communication. There was no hard evidence, but Secretary of War Stimson and others believed that Russian spies had infiltrated the Manhattan Project and were feeding its secrets back to Moscow. In Britain, hysteria about the Russians was still embryonic, although Churchill now realised the dangers of telling them about new weapons, as it had been imagined he would in the now-abandoned Anglo-Russian Treaty.[20]

This was the minefield, heavily sown with suspicion and paranoia, into which Niels Bohr blundered in spring 1944. He foresaw America and Russia as the biggest and most dangerous players after the war, and argued that the world would be safer if Russia also had the bomb; America and Britain should therefore tell Russia about the bomb now, and demonstrate trust by divulging its secrets to the Soviets before they made their own weapon.[21] Frederick Lindemann and Sir John Anderson agreed about the perils of a nuclear-armed confrontation between post-war superpowers. On 21 March 1944, Anderson tested Churchill's reaction, without mentioning Bohr's role. He suggested in a briefing document that the Russians should be informed about the 'devastating weapon' and invited to collaborate on planning international nuclear control. Churchill's response was immediate and unequivocal – 'No' and 'On no account', scribbled in the margin.[22] Churchill's veto wasn't relayed to Bohr, who sent Churchill a rambling letter about his utopian vision of a balanced nuclear future. Churchill didn't bother to reply but Bohr flew to London in early May, expecting a sympathetic hearing. Churchill was ferociously busy but Lindemann winkled Bohr into the PM's diary on the afternoon of 16 May 1944.[23]

Bohr's courtiers prepared him carefully. He was tutored in concise Churchillian English by R.V. Jones, Assistant Director of Intelligence (Science). Sir Henry Dale, President of the Royal Society, sang Bohr's praises in a strongly supportive letter to Churchill. Lindemann attended on the day, to make the introductions and smooth over any Bohrisms that might mystify or irritate the great man.[24] Unfortunately, the meeting of minds powerful enough to change the world was a complete disaster. Bohr ignored his English tutors and did it his way. To physicists worldwide, Bohr was God Almighty made human by endearing idiosyncrasies. To Churchill – exhausted, ill and crushed by conflicting priorities – he was a bumbling, incoherent foreigner who couldn't string a sentence together, fiddled incessantly with his pipe and hadn't even bothered to spruce himself up to meet the Prime Minister. When Lindemann chipped in to support Bohr, Churchill slapped him down and 'scolded us like a pair of schoolboys'.[25]

The Great Dane was branded a serious security risk and had to be vetted by Groves and Chadwick before returning to Los Alamos.

Luckily, even Groves was satisfied that Bohr hadn't broken any confidences.[26] When the dust settled, Bohr tried to mend bridges and sent Churchill another long, clumsy letter which elicited no reply. Churchill was now highly allergic to Bohr and reprimanded Lindemann: 'I did not like him, when you showed him to me with hair all over his face.'[27]

To be fair, the Prime Minister had weightier matters on his mind. Roosevelt and later Stalin had chiselled away at Churchill's resistance to a mass Allied attack on Nazi-occupied France, and finally coerced him into agreeing to the Normandy Invasion.[28] The final planning meeting for 'Operation Overlord' had taken place late on 15 May 1944, the evening before he met Bohr. Ready or not – and Churchill feared that Britain wasn't – 'D-Day' was set for 5 June, less than three weeks away.

21

LIBERATION
June–August 1944

The Allied invasion of Normandy on 6 June 1944 had been delayed by foul weather but quickly proved its strategic value. It also heralded blacker times for Germany. A few days later, the Russians began pushing the Eastern Front back into Poland, leaving hundreds of thousands of German troops killed, wounded or captured. And on 20 July 1944, just six weeks after D-Day, a heavy oak table in Hitler's 'Wolf's Lair' in eastern Prussia saved the Führer from a briefcase bomb deposited a few paces from his feet. It killed three others and left Hitler with cuts and a burst eardrum, and hell-bent on revenge. The failed assassins, all senior army officers, were rounded up and executed, the lucky ones by firing squad and the others garrotted with piano wire strung over a meat hook.

The US Secretary of War Henry Stimson had directed the US Army to provide 'every facility and assistance' to Boris Pash's mission, which was 'of the highest importance for the war effort'. While Allied troops flushed the Germans out of Normandy and Brittany, Pash's Alsos team waited in London. His chief scientist was Samuel Goudsmit, a forty-one-year-old physicist from MIT and an old friend of Werner Heisenberg, whom he'd tried to tempt to Michigan just before the War. Like Peierls and Fuchs, Goudsmit had drawn up a list of serious players in the *Uranverein*; as a Dutch national whose parents had been killed in a concentration camp, he had an additional motive for hastening the Nazis' demise.[1] Eric Welsh and Michael Perrin, desperate to be there when the *Uranverein* were hunted down, shared everything from their agents in Germany, but were pointedly left behind when the all-American Alsos team went to France.[2]

On 9 August 1944, Pash landed on Omaha beach with ten men in two jeeps and set off for Ploubazlanec on the north Brittany coast where their first target, Frédéric Joliot, had reportedly been seen after vanishing from the Collège de France. The two calm, hard American soldiers riding shotgun soon saw action because enemy fire encircled their destination. It was a stone cottage overlooking a beach nicknamed 'Sorbonne-Plage' because it used to be awash with intellectual friends of Marie and Pierre Curie, who had bought the house as a holiday refuge. Nobody had been there for ages.[3]

On the way to Paris, they picked up a stray puppy which they named 'Alsos', possibly in deference to Groves, who was widely known as 'the greatest son of a bitch'. They started 300 miles behind the Allied front line but reached the outskirts of the capital on Liberation Day, 25 August 1944, at the same time as the Free French Army. They found a schizophrenic city, with streets swept clean and tables laden with champagne to welcome the liberators, while German soldiers fought their way out nearby. Joliot couldn't be allowed to fall into anyone else's hands, German or Allied, so Pash bluffed his way to the head of the Free French Army cavalcade forming up for General de Gaulle's victory parade. The Alsos jeeps were among the first ten vehicles to enter Paris, then headed for the Collège de France on the Boulevard Saint-Michel. Hazards en route included a crossroads guarded by a German tank and snipers blocking access to the Collège. On the fifth attempt, they broke through and acquired their target at 4 p.m., three hours before de Gaulle reclaimed Paris.[4]

Frédéric Joliot added little to existing intelligence about the German occupiers. Wolfgang Gentner, Joliot's co-saboteur of the cyclotron, and Otto Hahn were both good men and anti-Nazi. The bad guys, pro-Nazi and untrustworthy, included Schumann, the thuggish musicologist and chief of Army Ordnance; Walther Bothe, a 'hostile visitor' from Heidelberg; and Kurt Diebner, a 'key man' with 'thorough knowledge' of nuclear fission and its military potential.

Heisenberg had never come to Paris, while von Weizsäcker had only visited once to lecture. Joliot reckoned that Diebner and Bothe would be closely involved in building an atomic bomb, but that Germany was 'not remotely close' to a workable weapon.[5]

Joliot had walked a perilous tightrope. Some French patriots accused him of collaborating with the Germans, while the Gestapo had questioned him twice on suspicion of supporting the Resistance. Jacques Salomon's execution had converted Joliot into a fervent Communist and coordinator of the *Front National*, an underground collective of academics, writers and clerics dedicated to destroying the Nazis. His lab quietly diversified into producing radios, explosives and incendiaries (recycled from unexploded Allied bombs appropriated by the chief of the Paris police laboratory). In late June 1944, with the Gestapo preparing to pounce again, Frédéric Joliot disappeared from the Collège de France and 'Jean-Pierre Gaumont' materialised in a northern suburb. As the liberating forces closed in, Joliot had dodged bullets alongside his fellow Resistance fighters and cursed his luck for being 'reduced to throwing Molotov cocktails at German tanks while the Americans are probably making atomic bombs'.[6]

Pash and Goudsmit grilled Joliot for five days and then flew him to London for further questioning by Eric Welsh and Michael Perrin. He met James Chadwick, their first encounter since that awkward Nobel Prize ceremony in December 1935. The Quebec Agreement prohibited Chadwick from disclosing nuclear information, but he told Joliot to keep quiet about being a Communist – which, for Groves, was on a par with being a Nazi.[7]

After Paris, the Alsos Mission's next target was 100 tons of uranium oxide belonging to Union Minière – essential nutrition for the Manhattan Project, which was devouring every last kilo that Eldorado produced. A 1,600-mile paperchase through active combat zones led from the Union's offices in Antwerp to a warehouse in Toulouse. The uranium oxide, bright yellow and heavy enough to trigger rumours that it was gold, was transported to Marseille by the 'Red Ball Express' convoy of all-Black US Army truckers and then shipped across the Atlantic. When British officers complained about the 'Wild West' tactics of the Alsos Mission, their concerns were logged and ignored.[8]

LIBERATION

Choose Your Weapon

In June 1944, the Manhattan Project employed 160,000 people and was well into its second billion dollars. The first few grammes of pure U-235 and plutonium would be among the most expensive commodities in human history – if they ever materialised. The deadline for delivering 'a workable weapon of war' had been set in stone for midsummer 1945. Halfway through the project's allotted lifespan, U-235 and plutonium were still in the running, but only just.[9]

Assuming that they could obtain enough of it, a U-235 bomb seemed straightforward. It could be detonated as the Frisch–Peierls memorandum had envisaged four years earlier: smashing together two small pieces of U-235 to create a whole that exceeded the critical mass. Nobody knew how much U-235 was needed. Frisch and Peierls had estimated 5–10 kg. Enrico Fermi suggested anywhere between 20 kg and 1.2 tons, while the *Los Alamos Primer* plumped for 200 kg. For now, a mass of 50 kg was assumed, corresponding to a 5-inch sphere of the metal.[10]

Getting this right was crucial for designing the bomb and determining how long it would take to produce. While waiting, Frisch devised a 'flamboyant' (and potentially deadly) experiment to calculate the critical mass. Two pieces of U-235, each safely below the likely critical value, would be brought close together 'for a split second' to provoke increases in neutron production. Frisch suggested dropping a U-235 cylinder through a slightly larger ring of U-235. This wasn't for the faint-hearted or clumsy, because a non-explosive but furious chain reaction could flare up if the two pieces didn't separate quickly; the experimenter would have received a lethal radiation dose by the time he noticed that the invisible inferno of neutrons had turned the air blue. Richard Feynman, on the Los Alamos steering committee, likened the experiment to 'tickling the tail of a sleeping dragon', and Frisch was surprised to be given permission to proceed.[11]

Even though U-235 was still a promissory note, the 'gun' to detonate it was already being tested at Los Alamos. This was essentially the device which the Royal Arsenal in Woolwich had started developing for Tube Alloys in 1941: the breech and sawn-off barrel of an artillery gun that would fire a hollow U-235 cylinder onto a snugly fitting plug of U-235.[12] The concept was simple but timing was everything: a one-millisecond lapse could turn the biggest bang ever into a damp squib. The chain reaction would rip through 50 kg of U-235 in well under a *micro*second. It must not be triggered before the two pieces of U-235 had mated fully, as part of the U-235 could then 'predetonate' and so prevent the entire mass from exploding. Moreover, the chain reaction would instantly heat the U-235, making it expand and physically separating the U-235 atoms; it was calculated that an immediate expansion of just 1 cm could freeze the chain reaction by preventing neutrons from reaching the next U-235 nucleus. Most people had confidence in the choreography of the chain reaction for a U-235 slug blasted in at 1,000 feet per second, but others were racked by doubts. In theory, 1 kg of U-235 should explode with the force of 20,000 tons of TNT. The worst-case estimates downgraded the power of a 50-kg uranium bomb to just 200 tons of TNT.[13]

The top-secret gun that, with a single shot, would change warfare for all time was built by the Naval Gun Factory in Washington. One-ton prototypes, cut down from 3-inch anti-aircraft guns, were first fired in March 1944 in a deep, echoing canyon at Anchor Ranch, 2 miles south-west of the Hill (Map 5). Improved models with higher-tensile steel, 6-inch bore and four primers to ignite the cordite propellant were ordered from the Naval Gun Factory, for urgent delivery with their custom-built testing mounts.[14]

Two other ways of enhancing the explosion were intensely researched. Surrounding the U-235 explosive with a heavy shell of non-explosive 'tamper' would reflect escaping neutrons back in and stoke up the chain reaction. The tamper would also be a physical barrier, preventing the slug from smashing through the target. Ten-inch discs of potential tampers were tested, including lead, tungsten carbide, gold and platinum. The precious metals – 'vulgar wonders in a nuclear physics lab' – attracted many visitors fascinated to see what several years' salary looked like, even though a few milligrammes of plutonium would cost hundreds of times more.[15] The second refinement, the 'initiator', was still hypothetical. This would be a powerful neutron source, safely tucked away in the heart of the bomb and waiting to be activated when the two pieces of U-235 mated. It would then pour out a torrent of neutrons to ignite the chain reaction at the critical moment.[16]

By early June 1944, Los Alamos had received less than 1 per cent of the U-235 likely to be needed for a bomb. This wasn't for lack of trying. Site X at Oak Ridge was as big as Inner London and home to 44,000 people.[17] Groves had thrown hundreds of millions of dollars at two U-235 enrichment processes – gaseous diffusion and electromagnetic separation – but neither was yet operational.

The gaseous diffusion plant (K-25) was a vast U-shaped building served by workers on bicycles because it was a mile long. It covered the same area as the Pentagon, but at half a billion dollars, cost eight times more. Like the rest of Site X, K-25 had been thrown up at speed without checking that it functioned. With its shell half-finished, only two of the 2,000 extraction units in the blueprints had been delivered; because of poor materials and workmanship, neither worked. Without the British – whom Groves barred from Site X – it would have been even worse. Peierls, Fuchs and Simon had pushed through improvements against resistance by Groves, but the eventual game-changer was the flat diffusion membrane made of sintered nickel (codenamed 'K1'), developed in Swansea and initially supplied gratis by the reviled ICI. Even so, the process could only double the U-235 concentration from the original level of 0.7 per cent to 1.4 per cent, far short of the goal of 50 per cent.[18]

The electromagnetic separation factory (Y-12) was Ernest Lawrence's stroke of genius but had also been flogged into galloping when it wasn't ready to walk. Lawrence had turned the 184-inch metal ring of the Berkeley cyclotron into a series of U-235 collecting units (calutrons), mounted along a massive electromagnetic coil that stretched around an oval 'racetrack' 122 feet long by 77 feet wide. The prototype racetrack carried forty-eight pairs of calutrons, each spraying vaporised uranium into a magnetic field designed to guide U-235 atoms precisely into a collector (Figure 7). Each separation unit was sealed inside a high-vacuum tank – in total, the biggest vacuum ever created.[19] Vast amounts of copper wire were needed for the electromagnet coils; when

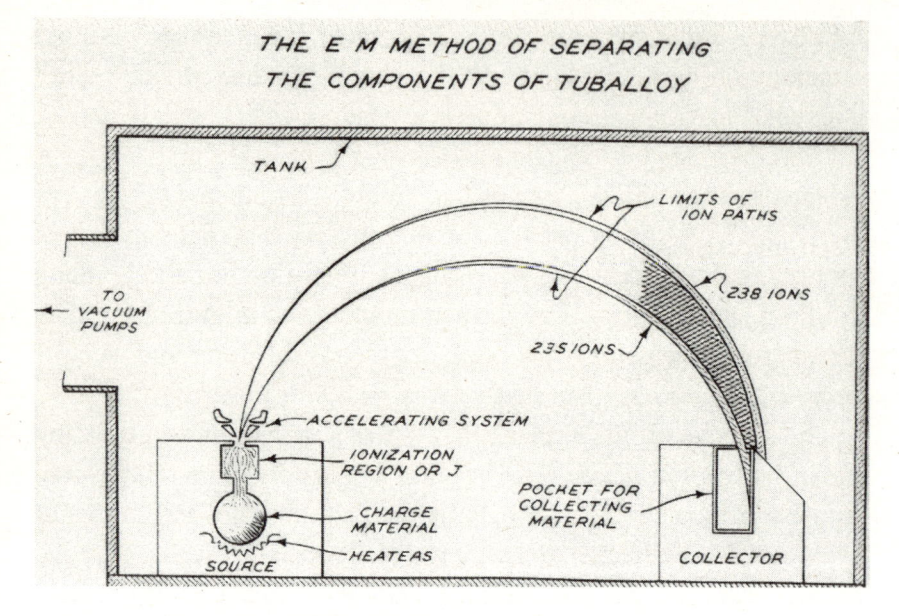

THE E M METHOD OF SEPARATING THE COMPONENTS OF TUBALLOY

7. The 'calutron', used for the electromagnetic separation of U-235 from U-238. Note use of 'Tuballoy' to denote uranium. Original drawing by Robert Hile, Westinghouse.

copper ran out, the US Treasury quietly stepped in with 15,000 tons of silver bullion – as suggested by Mark Oliphant, who had brought his twenty-strong research group from Birmingham to Berkeley in early 1944.[20] The extremely high operating voltages were constantly controlled by teams of 'calutron girls' (Plate 19) who worked seven-day rotating shifts and knew only that what they did was vital to the war effort and that 'God have mercy on us all' if the enemy were doing it better.[21] It was originally calculated that 2,000 calutrons would be needed to produce 100 grammes of U-235 per day, but yields were dismally low because U-235 had stuck to the wrong surfaces inside the vacuum tank. Lawrence's and Oliphant's groups redesigned the separation units, which eventually straddled ninety-six racetracks.[22]

Oppenheimer had prompted Groves to evaluate a third enrichment method. Phil Abelson had spent three years at the Philadelphia Naval Shipyard exploring liquid thermal diffusion. Liquid hex (rather than the gas) was pumped through 100 vertically mounted columns, each 48 feet high and comprising concentric tubes of nickel (inner, heated by steam) and copper (outer, cooled by water). The process was dangerous. When a pipe ruptured and sprayed six men with liquid hex, two died horrifically from multiple organ failure. But U-235 hex concentrations were slightly higher at the top of the column, and Groves ordered an urgent review of Abelson's method during summer 1944.[23]

By Saturday 10 June 1944, electromagnetic separation had yielded only 200 grammes of U-235 and gaseous diffusion none at all; accumulating 50–60 kg for a single U-235 fission bomb in the next thirteen months seemed out of reach. On that day, Vannevar Bush and Secretary of War Stimson went to Capitol Hill to justify the astronomical cost – currently $100 million per month – of the clandestine project known only as 'S-1'. Their scrutineers were the Senate Appropriations Committee who, by order of the President, now had to be let into the secret. Stimson and Bush won the day, thanks to their communication skills and assurances that the American superbomb (no details, but small enough to carry in an aircraft) was far ahead of the one that the Nazis had started building first, and that a single device could affect the outcome of the war. Groves breathed a sigh of relief, but not for long. As James Conant said later, it was good to have three horses ahead in the home straight, but there was no guarantee that any of them would clear the final hurdle in time.[24]

Plan B

The Naval Gun Factory had also turned out 'fantastic' gun detonators for the plutonium bomb, assuming that there would be enough plutonium and that it would behave exactly like U-235.[25] Both assumptions fell apart during late spring 1944.

Chadwick and others had worried about plutonium from the start. With its half-life of 24,100 years, the main plutonium isotope (Pu-239) decays 30,000 times faster than U-235 (half-life, 700 million years). Its decay generates alpha-particles which strike neutrons off impurities such as beryllium, even if present in only a few parts per million, potentially triggering a chain reaction. A gramme of plutonium produced in the cyclotron was estimated to trickle out 2 trillion neutrons per second, so that predetonation was inevitable during the

20 microseconds taken by a slug fired at 1,000 feet per second to mate with its target and create a critical mass.[26]

Naive hopes of eliminating impurities and making a bigger gun to fire the slug faster were immediately dashed when the first few milligrammes of plutonium were delivered from an experimental pile (X-10) at Oak Ridge. Their resting neutron counts were even higher, because conditions inside the pile had also created an unstable plutonium isotope, Pu-240, which decays much faster. Pu-240 was baked into the recipe for making plutonium in a pile; the huge reactors at Hanford would produce even more, and it couldn't be removed. This meant that even the most powerful artillery gun couldn't fire a plutonium slug fast enough to create a critical mass before the background neutrons had ignited a predetonation.[27]

On Wednesday, 19 July 1944, Oppenheimer called a crisis meeting in the heavily guarded cinema. With 'an aura of grim desperation', he explained that the plutonium gun could never work and was being scrapped. Work would begin immediately on a completely different detonator which he called the 'gadget'. This would have to be dreamed up, designed, built and tested – and plugged in to fire the plutonium bomb that would be dropped in anger in just over a year. 'Gadget' wasn't just security camouflage; Oppenheimer had no idea what it would be. Its origin was an unpromising idea that had been dumped in the rush to build the plutonium gun, and it would be the product of a shotgun wedding between two high-risk research groups that didn't yet exist. Anyone else would have stood accused of grasping at straws, but Oppenheimer's clear thinking, plain speaking and charm carried everyone with him, including Groves.[28]

The rescue plan now portrayed 'in a most enchanting light' had started life a year earlier as 'a completely untried and undeveloped method that no one wanted to use unless absolutely necessary'.[29] In summer 1943, Seth Neddermeyer had proposed that a high-explosive blast could compress a lump of plutonium into a 'supercritical' mass in which the plutonium nuclei were crammed together so tightly that an explosive chain reaction would immediately follow. The notion seemed crazy, because metals are taken to be incompressible. 'Implosion' – directing an explosive shockwave inwards rather than to blow things apart – was a young and imperfect science, and Neddermeyer, a thirty-six-year-old 'lone wolf', wasn't its best ambassador.[30] Then a serious player came on board. John von Neumann, a brilliant Hungarian émigré at Princeton, visited Los Alamos in autumn 1943. Rudi Peierls had known Neumann in Germany and described him as a mathematician 'of the purest and most abstract kind', who was also dedicated to 'good living' and had been untouched by fifteen of Peierls's full-strength vodka martinis.[31] Neumann agreed that fast implosion was the best way to detonate plutonium, and Neddermeyer began experiments with high explosive wrapped around steel cylinders. The results – sometimes dramatic but always unpredictable – encouraged Neddermeyer but nobody else. Hence the initial incredulity when Oppenheimer picked implosion to save the plutonium bomb and potentially the entire Manhattan Project. Oppenheimer shunted Neddermeyer into an advisory role and handed his project to two hastily assembled teams of hard hitters: the 'Theoretical Group' led by Hans Bethe and with Rudi Peierls in charge of 'implosion dynamics', feeding ideas into the 'Gadget Group' which would invent and build the implosion detonator. The Gadget Group's leading light was George Kistiakowsky, Professor of Chemistry at Harvard and an expert on high explosives.[32]

The common aim was to create a massively powerful spherical shockwave that would shoot inwards, like a high-speed movie of an exploding grenade played backwards. Peierls was relatively new to 'hydrodynamics' – the behaviour of matter subjected to colossal forces – having taught himself the elements to lecture to undergraduates in Birmingham.[33] He and Hans Bethe tried to predict mathematically how a spherical shockwave could rush in symmetrically towards its centre. They failed, because the shortcuts which made the calculations possible also made the results uninterpretable. An IBM electronic calculator donated by the US Treasury enabled them to do cleverer sums but delivered a troubling verdict. Their simplest model – a hollow plutonium sphere enveloped in high explosive – wouldn't implode evenly into a tiny ball of hyperdense, supercritical metal. The tiny fraction of a second taken for all the explosive to ignite was too long: a series of shockwaves would shoot inwards, creating irregular 'jets' that would crush the plutonium erratically.[34]

THE IMPOSSIBLE BOMB

The Gadget Group began by investigating 'shaped' explosive charges, previously researched in England and by the US Navy. The force of an explosion can be 'focused' by tweaking the geometry of the charge or inserting a metal filler. An example was the copper plug just behind the tip of an explosive armour-piercing shell: detonation on impact fires out a thunderbolt of molten copper which can punch a hole through the side of a tank so that the shell finishes exploding inside the vehicle. The uneven jets predicted by Peierls and Bethe were confirmed when attempts to implode hollow steel cylinders wrapped in high explosive produced a mangled mess of metal.[35] It was like trying to squash a fistful of plasticine into a tiny ball, only to find it squeezing out between your fingers.

In mid-summer 1944, with just twelve months left to deliver the 'workable weapon of war', it seemed that the only option left for detonating a plutonium bomb had defeated the Manhattan Project's best brains.

Body Image

Incongruously, others already knew how the plutonium bomb would look in August 1945, hanging in the belly of a B-29 bomber as the bomb-doors opened high above its target. It was ugly and obese: a 4-ton steel egg, 11 feet long and 5 feet across its waist, with an ungainly square tail-unit bolted onto its rear. The tail-unit ensured a clean fall; early prototypes had wobbled during descent, which would have played havoc with the delicate instruments inside that would tell the bomb when its moment had come. Most of the egg's interior was empty, to be filled with equipment yet to be invented: altimeters, the electronic firing circuits and the high-explosive charges that – with luck – would crush a few kilos of plutonium into a supercritical mass. For now, the heart of the bomb was mocked up as a large football-like sphere made of polygonal steel plates bolted together.[36] The plutonium bomb was nicknamed 'Fat Man' – not a snide reference to Churchill, but a character in Dashiell Hammett's *The Maltese Falcon* who inspired Robert Serber, author of the *Los Alamos Primer*.[37] The plutonium bomb had started life as 'Thin Man' (the title of another book by Hammett), built around the ill-fated plutonium gun device. This looked like a torpedo, 2 feet wide and 17 feet in length to accommodate the extended gun-barrel that accelerated the slug to 3,000 feet per second. Despite being too long for the B-29's bomb-doors and misbehaving badly during test drops, Thin Man was the chosen vehicle for the plutonium bomb until 19 July 1944, when Oppenheimer killed the plutonium gun and Fat Man began to take shape around the hypothetical implosion gadget.[38]

By contrast, the U-235 bomb was a cinch. This cut-down version of Thin Man, christened 'Little Boy', was 10 feet long by 28 inches in diameter, weighed over 4 tons and contained the breech and bottom 6 feet of a 6-inch gun-barrel. And it worked, having fired a slug of ordinary uranium metal into a target at 1,000 feet per second.[39]

To wreak the greatest possible devastation, the bombs had to explode at an optimal height above 'ground zero', the centre of the target. This demanded exquisitely accurate detonation systems which couldn't fail under battle conditions. Detonators for ordinary high-explosive bombs were only 90–95 per cent effective; those for the atomic bombs – sitting on the summit of a two-year, $2 billion project – had to operate with 99.99 per cent certainty, allowing just one bomb in 10,000 not to explode within 100 feet of the desired altitude. No existing equipment approached this degree of precision. Highly sensitive barometric gauges were invented and quickly abandoned after test drops of dummy bombs. State-of-the-art electronic 'radio altimeters' appeared 'very promising' in exhaustive trials that included diving an AT-11 trainer aircraft towards the ground at full speed. As a safeguard, the firing circuit included four radio altimeters, each set to trigger detonation at the correct altitude. And in case the fancy electronics failed and the weapon fell intact onto enemy soil, four ordinary nose-impact bomb fuses were embedded in the front of each bomb to ensure self-destruction.[40]

With the Germans fast losing ground in Europe, attention shifted eastwards towards the only enemy potentially left standing in August 1945. The long-range bomber tasked with dropping the first atomic weapons on Japan was the US Air Force's new but accident-prone B-29 Superfortress. By autumn 1944, B-29s were routinely testing full-sized mock-ups of Fat Man and Little Boy at the US Navy's range in Dahlgren, Virginia. The trials went well.[41] The only missing ingredients were U-235, plutonium and an infallible means of detonating them.

Given the dearth of U-235 and plutonium reaching Los Alamos, it became obvious that, at best, only one or two bombs of each kind might be ready by August 1945. Confidence in the gun detonation device for U-235 was high enough for Little Boy to be sent into battle untested. By contrast, the plutonium bomb was so speculative that it had to be tested before use in anger, because the technical, psychological and strategic consequences of an atomic bomb failing to explode on enemy territory were too grim to contemplate.[42]

Plans for 'the first implosion test of active material' were drawn up throughout summer 1944. The shortlist of remote, unpopulated localities included islands off the Californian coast and the deserts of Utah, Arizona and New Mexico. In August 1944, the test site was selected: a desolate lava field on the Alamagordo Bombing Range in New Mexico, 200 miles south of Los Alamos. The area was known as the *Jornada del Muerto*, or the 'Dead Man's Trail'. The provisional date for the test was Independence Day, 4 July 1945.[43]

The implosion test device, not yet born or even conceived, had already been christened: 'Trinity'. It was rumoured that Oppenheimer had chosen the name, inspired by John Donne's seventeenth-century sonnet entitled 'The Trinity', which lauded the three prime forces of power, love and knowledge. Oppenheimer had been introduced to Donne's poetry in 1936 by Jean Tatlock, a literary-minded and Communist-leaning medical student. They had an on–off relationship which she terminated in 1939; Oppenheimer met his future wife, Kitty, soon after. He later denied any involvement in naming the first atomic bomb, but Jean Tatlock may have been in his mind that summer. In early January 1944, Oppenheimer had received the news that she had been found dead in her bath with a suicide note, and he wept.[44]

22

THE GIVING OF ALL HELP
May–December 1944

By spring 1944, James Chadwick had made himself and the British Mission indispensable to the Manhattan Project. He and Groves were as friendly as their chalk-and-cheese personalities allowed, and Groves granted him the same privilege as Oppenheimer – a grandstand view of the entire Project except for the plutonium factory at Hanford. To their great relief, Chadwick had taken his wife and daughters to Washington, leaving Rudi Peierls in charge of the British Mission at Los Alamos.[1]

In mid-May 1944, the Tube Alloys Technical Committee met in Washington for a three-day review of the British Mission's progress towards (in Chadwick's words) 'the giving of all help to the Americans' to ensure that 'TA [Tube Alloys] bombs ... would be a military factor in this War'. Eight months earlier, the first wave of the mission had carried fourteen British scientists to Los Alamos and a further twelve, led by Mark Oliphant, to Ernest Lawrence's U-235 electromagnetic separation lab in Berkeley. Both groups quickly proved their worth, but at a huge cost back home in asset-stripped Britain. The Tube Alloys committee hoped that sending Britain's best researchers to the USA would be an excellent long-term investment; while collaborating with the Americans to finish the bombs, they were acquiring the skills and knowledge to launch Britain's own military and civilian nuclear programmes after the war.[2]

THE IMPOSSIBLE BOMB

Without the British Mission, both the U-235 and plutonium bombs would have lost momentum and ultimately the race against time. Riding through Groves's attempts to exclude them, Rudi Peierls, Francis Simon and Klaus Fuchs had helped to refine the gaseous diffusion process, now at last in operation thanks to the sintered nickel separation screens invented in Britain.[3] In Berkeley, Oliphant's theoreticians and experimenters (Plate 20) were working 'harmoniously' with Lawrence, diagnosing and ironing out problems in the calutrons; Oliphant deputised as chief whenever Lawrence was called to Washington or Oak Ridge.[4] Otto Frisch's hair-raising 'dragon' experiment – on ice until enough U-235 arrived – would determine how much U-235 to put into the bomb and finally confirm that the gun detonation method would work.[5] Peierls was leading the theoretical assault on the implosion gadget – the last hope for the plutonium bomb – while Ernest Titterton was inventing electronic circuits to detonate the gadget's complex array of high-explosive charges.[6]

Chadwick hand-picked his mission to cover shortfalls in American know-how. He'd instructed Oliphant to take the strongest possible team to Berkeley because Lawrence's group was large but 'not all of them are good'.[7] There remained crucial holes at Los Alamos, notably for creating the implosion gadget and maximising the damage done by the bombs. The people chosen by Chadwick to plug these gaps arrived on the Hill between May and August 1944.

The first was James Tuck, a thirty-four-year-old physicist whose PhD had been interrupted by the war. Tuck had worked with Frederick Lindemann in Oxford before joining 'Churchill's toyshop', a secret breeding-ground for off-the-wall weapons which tickled the Prime Minister's fancy for ingenious technology but didn't necessarily work. An expert in shaped explosives, Tuck had been the first to use 'flash' X-ray photographs to show that a metal plug in the nose of a shell was liquefied and fired out ahead of the explosion's shockwave. Tuck's ultra high-speed photography now demonstrated that a steel cylinder wrapped in high explosive was torn apart by irregular jets of molten metal shooting inwards; even with sixteen detonation points, there was no hint of the smooth shockwave needed to crush a plutonium ball evenly from all directions.[8]

THE GIVING OF ALL HELP

Tuck settled easily into the bizarre lifestyle on the Hill, revelling in the work, the place, horse riding and 'the spirit of Athens, of Plato, of an ideal republic'. He noted that 'people here are beginning to get rather frightened' but sailed above the general despondency that dogged the Gadget Group. His inspiration was a secret British bomb that looked (and fell) 'like a turnip' and had been invented to hole battleships. The 'CS' ('Capital Ship') bomb exploited two different explosives, with an outer shell that exploded extremely fast and a slower-burning core. Tuck began exploring the possibility that a hollow 'composite' ball of high- and low-velocity explosives could create the elusive spherical shockwave.[9]

Next came Sir Geoffrey Taylor, Professor of Physics at Cambridge and a world expert in hydrodynamics. His Christmas lectures on 'Ships' at the Royal Institution in 1936 were the first ever televised by the BBC, and he was the guru whom Peierls had consulted while teaching himself hydrodynamics in Birmingham. Taylor came to Los Alamos in late May and rapidly made himself popular, even captivating Groves. His expertise was instability, whether in upper-atmosphere weather systems or bomb blasts in the air or under water. He was initially pessimistic about producing a spherical shockwave free from unstable jets, let alone one powerful enough to squeeze a ball of dense metal into half its volume.[10]

In turn, Taylor brought in William Penney, Assistant Professor of Mathematical Physics at Imperial College in London and an authority on the physical impact of shockwaves. Penney had studied what high-explosive bombs did to buildings and the human body, his experimental material being British cities devastated by the Blitz. Americans listened transfixed in the cinema on the Hill as he told them in graphic detail about a horror that the US homeland had never experienced: being bombed in your own house by an enemy determined to kill you. Despite the gruesome content, Penney maintained his characteristic 'brightly smiling face' throughout – earning himself the ironic nickname 'the Smiling Killer'. Penney's task at Los Alamos was to maximise bomb damage by exploiting the topography of the target area and optimising the height of detonation above ground zero. The Trinity test of an implosion plutonium bomb, scheduled for a year hence in July 1945, would provide the experimental data he needed to plan for the real thing.[11]

At Rudi Peierls's insistence, a world-class mathematical physicist was brought to Los Alamos in mid-August 1944 to help the Theoretical Group with imploding a plutonium sphere. Klaus Fuchs left the gaseous-diffusion team in New York and joined Peierls and Bethe in modelling the shockwaves created by the composite explosive 'lenses' made of fast- and slow-detonating explosives which James Tuck was now investigating. Fuchs opened up somewhat on the Hill. Being unattached and unhistrionic, he was a popular babysitter; unexpectedly, he turned out to be a slick mover on the dancefloor. His neighbour in the male dormitory T-102 was the exuberant and irreverent Richard Feynman, and the two became friends. Fuchs was one of the few at Los Alamos who owned a car; he offered his Buick for whenever Feynman wanted to visit his wife, Arline, steadily wasting away in the sanatorium for incurable tuberculosis in Albuquerque.[12]

Critical Flow

In summer 1944, the Tube Alloys–Manhattan Project partnership 'appeared to be working very well indeed' and Vannevar Bush wrote warmly to Sir John Anderson, praising its 'complete harmony and rapid progress'. At the end of August, Chadwick wrote a reality-checking memo to Anderson about 'the position of the Tube Alloys projects in the USA', warning that neither bomb was in the bag.

Little Boy, the uranium bomb: five months after placing the urgent order with the Naval Gun Factory, the improved gun detonator still hadn't arrived but Chadwick anticipated 'no major difficulties' with it. At the current U-235 production rate of only 30 grammes per day, a 50-kg bomb wouldn't be ready until late 1948, but he hoped that U-235 enrichment would hit its target in spring 1945. And Chadwick was concerned that estimates of Little Boy's power might have been 'far too optimistic'.[13]

Fat Man, the plutonium bomb, should be much more efficient than Little Boy, with 5 kg of plutonium releasing more energy than 50 kg of U-235. The pinch-point was the supply of plutonium, which had to reach 1 kg per day by the spring in order to make 'a few bombs' by midsummer 1945. Chadwick believed that 'the outlook is good' for the implosion gadget, but Fat Man would come to nothing if it didn't work.[14]

A billion dollars' worth of hope rested on the plutonium factory – the huge and untested water-cooled reactors being built at Hanford. Their precursor was the smaller, air-cooled X-10 reactor at Oak Ridge: a 1,500-ton, 24-foot cube of graphite enclosed in concrete walls 7 feet thick. The entire structure was pierced by over 1,200 horizontal channels through which uranium slugs were pushed end to end, like boluses of food traversing the intestine. Each slug was a cylinder of uranium metal 4 inches long by 1 inch across, sheathed in aluminium to prevent corrosion. At the reactor's front face, workers fed the slugs into the horizontal channels; X-10 went critical in November 1943 after 30 tons of uranium had been loaded into 600 of the channels. Inside the pile, the slugs were blasted by neutrons which transformed an innocent baton of heavy metal into a demon that sizzled with lethal radioactivity and would have burned its way down to bone if dropped into an open hand. Slugs emerging from the back wall of X-10 fell with a hiss into deep water, where they took thirty days to cool to room temperature. The tiny yield of plutonium – just half a gramme from 5 tons of uranium – was then extracted in a laborious five-step, week-long process which began by dissolving the slug in concentrated nitric acid.[15]

Hanford, a crucial link in the Manhattan chain, was perilously stretched. Its 45,000 employees made it the fourth largest town in Washington State, but turnover was high: the place was 'bleak', the accommodation 'dismal', the work repetitive and unrewarding, and the security constraints hard to bear. Rumours were rife about the site's purpose: could they be making poison gas, Pepsi-Cola, Kleenex or even campaign badges for Roosevelt's fourth term as President? Hanford's own newspaper, *The Villager*, enhanced the claustrophobia, while the attractions of liquor, gambling and prostitution (all illicit but ignored by the authorities) were no match for the 'fearful termination wind', which periodically blew in sand and despair and lengthened the queue of disenchanted workers signing off because they

THE IMPOSSIBLE BOMB

could take no more.[16] And when the first pile (B Reactor) was finished and loaded with uranium slugs in mid-September 1944, it couldn't be coaxed up to full power or to produce any plutonium.[17]

At Los Alamos, Otto Frisch continued to enjoy life on and around the Hill. Unlike the Americans, who were tailed by security agents whenever they left the compound, the British Mission could wander anywhere unaccompanied. While waiting for U-235 to tickle the dragon's tail, Frisch designed a 'water boiler' reactor, climbed mountains, played the piano (Plate 21), honed his cooking skills and admired the 'intense butter yellow' of the aspens as autumn advanced.[18]

By contrast, Joseph Rotblat felt an outsider. He didn't see the bomb as the biggest, most thrilling physics experiment ever, but as a horrific weapon of mass destruction and murder. It was now common knowledge on the Hill that the Nazis' bomb had foundered. For Rotblat, this tore away the final shred of justification for working on such a hideous device. Chadwick urged him to stay, but Rotblat had made up his mind. By Christmas, he would be heading home to Liverpool.[19]

Niels Bohr looked beyond the bomb to a nuclear-armed world after the war, which scared him into leading a double life. On the Hill, he remained the patriarch and supreme troubleshooter; off scene, he continued lobbying those close to Roosevelt and Churchill to share the bomb's secrets with the Russians. Bohr's scientific brainpower was revered around the world, but he didn't realise that his clumsy sorties into realpolitik were doomed to failure.[20]

Klaus Fuchs shared the fears of Rotblat and Bohr and, in a way that made perfect sense to him, was already taking remedial action. One day, he and Richard Feynman were joking that they both looked suspicious enough to be spies. Feynman identified himself as the more obvious candidate, and Fuchs agreed.[21]

Autumn of Discontent

The 'complete harmony' of the Tube Alloys–Manhattan collaboration was too good to last. The French were initially to blame, provoked by perceived British and American treachery. When Charles de

Gaulle visited Montreal in August 1944, he was briefed about Tube Alloys and the Manhattan Project by Jules Guéron, a Free French fighter/nuclear physicist who had fled Paris to join Halban in Cambridge and then Montreal. Guéron told de Gaulle all about the U-235 and plutonium atomic bombs – and that the Americans had stolen the idea of a heavy-water-moderated boiler from the French and then prevented Halban's team from pursuing their own research. De Gaulle was incensed. So were Groves, Roosevelt and Churchill when they heard that a Frenchman had divulged Anglo-American atomic secrets to a third party.[22]

Next, the vengeful Halban ignored Chadwick's order to stay in Canada and travelled to London and then Paris. Those who knew Halban could see him haemorrhaging nuclear secrets to Joliot, who would be impossible to control in France. American fury was focused through the flint-glass lens of Leslie Groves onto Chadwick, who was deeply affected and remained a miserable wreck who struggled to sleep for many weeks.[23] The storm might have blown over quickly if the Manhattan Project had been running on course and on schedule. But it wasn't.

The dribble of U-235 from Oak Ridge wasn't enough for Frisch's dragon experiment, let alone a functional Little Boy bomb in just ten months' time. In desperation, Groves added in Phil Abelson's thermal diffusion separation method. In late September 1944, he ordered the immediate construction of a new 'S-50' plant at Oak Ridge, based around 2,142 48-foot thermal diffusion tubes, to begin operating in March 1945. Groves told the engineering contractors that the ninety-day deadline for completion was 'reasonable', and 'I think you can beat it.'[24] In October, the long-overdue gun detonators for Little Boy finally reached Los Alamos, but the special mounts needed to test them wouldn't be ready until December.[25]

Fat Man was also stuck in the mud because the massive B Reactor at Hanford still couldn't produce plutonium. The cause – acute poisoning by xenon, an unwanted fission product – was diagnosed but apparently untreatable. The Theoretical and Gadget Groups were running at full tilt, but the spherical implosion shockwave for Fat Man still eluded them. And the 'initiator', which had to spew out neutrons in the microsecond when the plutonium went supercritical, was another insoluble problem.[26]

Those working at Los Alamos were generally cocooned against the hard knocks of the war, but the realities of the outside world sometimes broke through. On the morning of 29 June 1944, three V-1 flying bombs hit Croydon in South London. One demolished two houses in Auckland Road, leaving a thirty-four-year-old woman physically uninjured but so traumatised that she was admitted urgently to a hospital specialising in nervous disorders. She was Adele, the wife of William Penney, whose two young sons had luckily been evacuated to safety. Penney was given a military escort for a flying trip home. He spent less than a day in England, visiting his wife in hospital, retrieving documents from the wreckage of their house and checking that their sons were safe. Then he was in a Mosquito bound for Shannon, followed by a Wellington bomber to Newfoundland and a US Air Force plane to Albuquerque. His wife, terrified and mute, had begun electroconvulsive treatment; the man known for his 'brightly smiling face' said nothing, staring out of the window.[27]

In mid-September 1944, following the Octagon Conference in Quebec, Churchill and Roosevelt met again at Hyde Park. Four years had passed since their last encounter there, when Churchill hadn't bothered to take notes and couldn't remember how they'd agreed to cooperate on the atomic bomb. The defeat of Germany now seemed inevitable, but Japan was still fighting and partly winning a bloody conflict in the Eastern Pacific.

This time, the meeting between President and Prime Minister was fully minuted. 'Tube Alloys ... should continue to be regarded as of the utmost secrecy; but when a "bomb" is finally available, it might *perhaps after mature consideration* [Churchill added these words to the draft] be used against the Japanese, who should be warned that this bombardment will be repeated until they surrender.' America and Britain would continue to collaborate fully on 'developing Tube Alloys for military and commercial purposes until after the defeat of Japan, unless and until terminated by joint agreement'.[28]

Churchill added a barbed coda for a man whom he mistrusted and despised. Professor Bohr must be investigated and 'steps taken to ensure that he is responsible for no leakage of information, particularly to the Russians'. In a follow-up telegram, Churchill told Lindemann that Bohr was causing both leaders to be 'much worried' and should be 'confined, or at any rate made to see that he is very near the edge of mortal crimes'.[29]

Countdown to Christmas 1944

Strasbourg, late November. The next stop for Alsos as they moved east across Europe was the Reich University of Strasbourg. Their top targets were Carl von Weizsäcker, Professor of Astrophysics and Heisenberg's close friend, and Eugen Haagen, a microbiologist working on a vaccine against typhus. The human guinea pigs for the vaccine trials had been supplied from the Dachau concentration camp by Dr Sigmund Rascher – to whom Haagen complained in writing when 76 of a batch of 100 experimental subjects were delivered dead or otherwise 'unusable'.[30]

Street fighting continued as Pash's jeeps reached the university quarter on 24 November. Both targets had fled and cleared their houses; von Weizsäcker had blocked his stove chimney by trying to burn documents. A second-rank name on Goudsmit's list was more productive: Rudolf Fleischmann, another nuclear physicist at the university. A trail of clues worthy of Hercule Poirot – a cagey housekeeper, an uneaten egg – led to a nuclear physics lab hidden inside Strasbourg's main hospital, and a sullen Fleischmann who hadn't had time to destroy his records.[31]

These were a treasure trove which Pash and Goudsmit read during two intense days in Haagen's apartment, drinking his brandy and blanking out the sound of incoming shells from German positions a mile away across the Rhine. The cache included detailed correspondence with Heisenberg, von Weizsäcker and others, with full addresses and even their telephone numbers. Paul Rosbaud was right: the *Uranverein* trail ended south of Stuttgart in the mountains on the edge of the Black Forest, in the baroque villages of Hechingen and Haigerloch (Map 6).[32]

Los Alamos, early December. At last, the efforts to focus an implosion shockwave were bearing some fruit. Composite two-explosive 'lenses' developed by George Kistiakowski, James Tuck and Bruno Rossi gave 'encouraging' results and inspired a suddenly upbeat Oppenheimer to predict that 'implosion will work'.[33] Kistiakowski's own prophesy was more guarded: 'The test of the gadget fails. Project staff resume frantic work. Kistiakowski goes nuts and is locked up.'[34] And unless the B Reactor at Hanford started producing plutonium, Fat Man would never happen. In that case, everything would hinge on the uranium bomb.

At first sight, Little Boy looked good. The test mountings had finally been delivered and the gun detonator performed impeccably, firing a hollow 25-kg cylinder of ordinary uranium to mate with a 25-kg uranium plug.[35] But Groves and Conant had picked up Chadwick's fear that Little Boy's power had been greatly overestimated and might only equate to a few hundred tons of TNT.[36]

Zurich, mid-December. Some action had resulted from Victor Weisskopf's suggestion that Werner Heisenberg should be kidnapped to decapitate the Nazis' atomic bomb project. In late 1943, the Office for Strategic Services (OSS) – the American-intelligence covert-operations organisation which also supported the Alsos Mission – hatched plans to abduct Heisenberg during one of his periodic research visits to Switzerland. The notion was dropped after the Nazis scuppered their bomb, but the OSS continued to simmer.

In November 1944, Paul Scherrer, Professor of Physics in Zurich and an OSS agent, relayed news that Heisenberg would visit the university for a few days before Christmas. The OSS responded by launching 'Project Larson', a secret mission that was both audacious and absurd.[37] They sent in the Alsos operative Moe Berg, who was uniquely qualified but not necessarily for the task ahead. Berg was a military veteran and lawyer, fluent in German, Italian and Japanese (*magna cum laude* from Princeton), and a former professional baseball player ('perhaps the slowest man in the American League'). His understanding of nuclear physics was zero.[38] Berg's mission in Zurich was to listen carefully to everything that Heisenberg said during a research seminar and over a supper hosted by Scherrer for Berg and his target. Kidnapping was too complicated. If Heisenberg let slip any hint that he was still working on an atomic bomb, Berg was to shoot him dead on the spot.[39]

Unsurprisingly, Heisenberg survived to tell the tale. He was legendarily tight-lipped, and even casual enquiry would have told the OSS that he'd never reveal anything about his work in the *Uranverein*.

New York, Christmas Eve. Disillusioned and dispirited, Joseph Rotblat was leaving the 'wondrous place' of Los Alamos with a metal box containing memorabilia, letters and research records.[40] When Chadwick had notified the Los Alamos intelligence chief of Rotblat's impending departure, he was shown a 'thick dossier' about Rotblat, which alleged that he was a Russian spy who would be flown secretly out of England and parachuted into Soviet-occupied Poland to pass atomic bomb secrets to the Russians. Chadwick easily convinced the intelligence chief that the dossier was 'a load of rubbish', but Rotblat remained a marked man.

The final reminder that Rotblat had never really belonged at Los Alamos came en route to New York and the boat home. He spent a few days with the Chadwicks in Washington and, on Christmas Eve 1944, Chadwick took him and the box of documents to the station. When Rotblat reached New York, the box had disappeared.[41]

23

COUNTDOWN
January–May 1945

The Manhattan Project was later praised as 'highly adaptable, catch-as-catch-can and brilliant at improvisation', but it ended 1944 mired in 'grim desperation' and deepening pessimism that the $2 billion investment in 160,000 people, including twelve Nobel Prize winners, might come to nothing.[1]

Time was running out very fast. Groves and Oppenheimer had to deliver 'a workable instrument of war' by August 1945, in just eight months' time. The bombs wouldn't be used in Europe, where the defeat of Germany was now inevitable, but could be decisive in 'bringing the war in the Pacific to a fiery end'. With the only other option being an American invasion of Japan, failure to deliver the bombs would be a major catastrophe 'measured in the lives of American soldiers and sailors'. Groves, supporting Manhattan like Atlas carrying the world on his shoulders, had to justify the 'years of effort, the vast expenditure and the judgement of everyone responsible'. He had to prove that both of the billion-dollar bombs could inflict massive damage on the Japanese and force them to surrender – or risk ending a spectacular career.[2]

Spring in the Air

In his residence on Dupont Circle in Washington DC, James Chadwick saw in 1945 with a telegram from his team in Liverpool:

'CONGRATULATIONS DEAR SIR'. He'd been knighted in the New Year Honours List for his services to science, but Sir James was quick to tell his colleagues that this was a collective reward for them all.[3]

Chadwick's knighthood turned out to be a good omen for the Manhattan Project. The New Year brought the first signs that the logjams paralysing both the bombs could be broken in time to alter the course of the war.

During January and February, Little Boy's internal anatomy took final shape inside its 3-metre steel case (Figure 8). The gun detonator was enclosed in an 11-cm thick, 300-kg shell of tungsten carbide tamper to amplify the power of the explosion. The radio altimeter detonation circuits (codenamed 'Archies') had been perfected, firing the gun at precisely 3,000 feet above the ground in thirty-two test drops of dummy Little Boys; just one test failed, because the Archies hadn't been wired up correctly.[4] The uranium bomb was waiting, like a finely wrought crown with two empty sockets to receive its most precious jewels, the slugs of U-235. And at last, the supply of that missing ingredient was about to be transformed from an impossible dream into a routine event.

Groves's 'last card', the S-50 liquid thermal diffusion plant which he'd seized on in desperation three months earlier, turned out to be a trump. Incredibly, the plant was finished early and by mid-February was producing a steady trickle of liquid hex containing twofold enriched (1.4 per cent) U-235. This unlocked the potential of both gaseous diffusion and electromagnetic separation, with each performing much better when presented with higher starting concentrations of U-235. Enrichment became a three-stage cascade, with U-235 levels first ramped up from 0.7 per cent to 1.4 per cent by liquid thermal diffusion, then to 50 per cent in the K-25 gaseous diffusion plant, and finally to 80 per cent by Lawrence's Y-12 calutrons (Figure 9).[5] This was a massive leap, but short of the 100 per cent U-235 assumed in the Frisch–Peierls memorandum. Was this enough to be 'weapons grade'? The truth would only be revealed when Otto Frisch had enough 80 per cent enriched U-235 to perform his long-awaited dragon experiment.

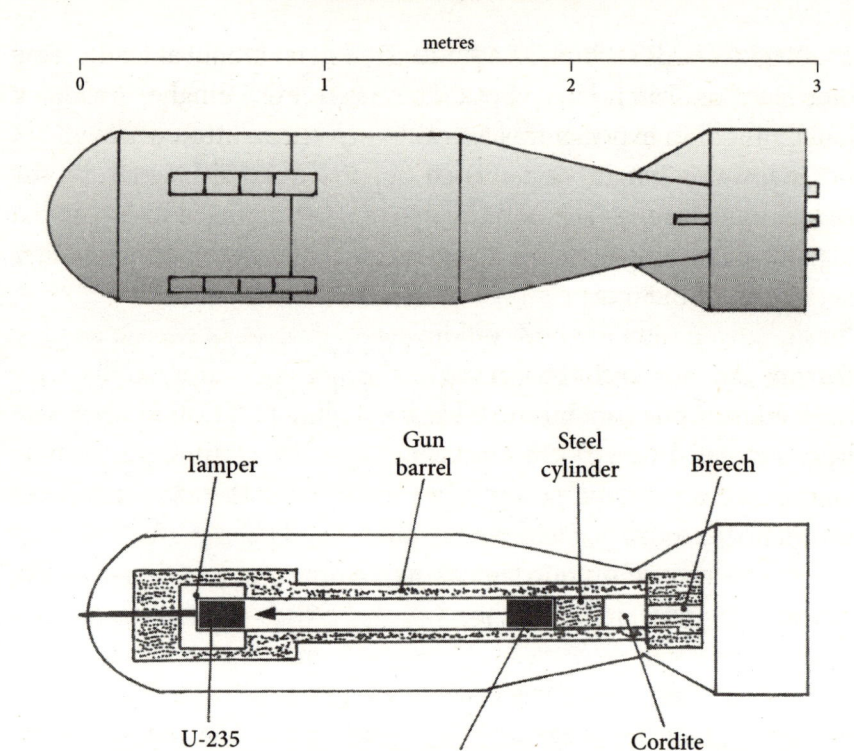

8. Little Boy, the U-235 bomb destined for Hiroshima, showing its internal structure.

From March 1945, 2–3 kg of enriched U-235 were delivered each week to the Hill. The three-day journey from Oak Ridge was designed to confuse the enemy, with armed couriers in civilian clothes, unmarked cars and a reserved compartment on the Santa Fe 'Chief' train to Lamy, a forlorn station in the New Mexico desert 50 miles from Los Alamos. On 11 April, Groves proudly showed off the huge separation factories at Oak Ridge to Secretary of War Henry Stimson, who was suitably 'cheered and braced up'.[6]

Two days later, Otto Frisch was given enough 80 per cent U-235 to tickle the dragon's tail. Frisch had set up his 'Critical Assembly Group' in a canyon codenamed 'Omega', safely remote from the Hill (Map 5). They dropped 6-inch U-235 cylinders through rings made of U-235 blocks and recorded meteoric rises in neutron counts – 'as near as we could go towards starting an atomic explosion without actually being blown up', as Frisch later wrote. They were exceedingly careful but had to abort an experiment urgently when the neutron counter shot off the scale. Just in time, Frisch realised that his own arm was reflecting neutrons back into the reaction, and averted disaster (and his own death) by scattering the U-235 blocks with his hand. Frisch's verdict: 52 kg of 80 per cent enriched U-235 would make a bomb, and the gun detonation method couldn't fail.[7] Subsequent experiments by the Los Alamos cyclotron group on spheres of 80 per cent U-235 reached the same conclusion. It was calculated that Oak Ridge would have produced 64 kg of 80 per cent U-235 by early August – only enough for one uranium bomb, but Little Boy was judged a safe bet for a solo performance without an audition.[8]

Frisch survived his encounters with the dragon, but it was utterly ruthless to those who disrespected it. Later, two researchers who

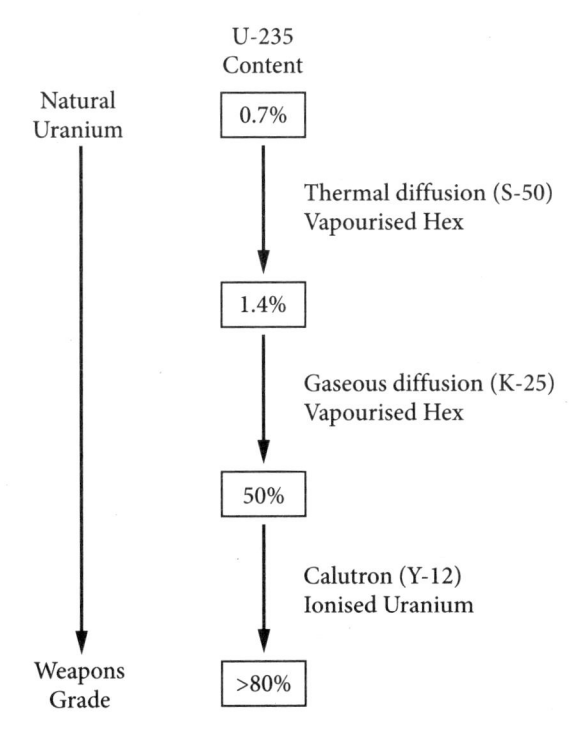

9. Making nuclear explosive for Little Boy: the U-235 enrichment cascade at Oak Ridge.

ignored basic safety rules when following up Frisch's experiments witnessed the blue flash that condemned them to death from neutron irradiation.[9]

Nobody was yet betting on Fat Man, but progress was being made. In late January 1945, the 'Christy solid implosion gadget' was selected to explode the plutonium bomb. Robert Christy, Canadian born and twenty-seven years old, had gone from a PhD with Oppenheimer to work with Fermi in Chicago; he'd autographed the label on the bottle of Chianti when CP-1 went critical in December 1943. The idea of imploding a solid ball of plutonium rather than a hollow sphere had come to him in September 1944 and was supported by Rudi Peierls after a month's hard labour on the formidable IBM calculating machine. Their joint invention of the gadget was acknowledged in the secret US patent (No. SD-3956X), for 'Method and apparatus for explosively releasing nuclear energy', filed by Christy and Peierls in January 1946.[10]

Plutonium finally came online when Hanford cured the xenon poisoning which had blighted the massive B Reactor. Christmas Day 1944 was marked by pushing several hundred tons of heavily irradiated uranium slugs into the plutonium production line.[11] The first shipment of plutonium salts reached the Hill on 2 February 1945, after a 1,300-mile trek from Hanford accompanied by armed guards and radio patrol cars. Plutonium metal 'of excellent purity' was extracted in a special lab where workers were carefully protected against touching or inhaling potentially lethal plutonium dust.[12]

These few grammes marked the start of regular deliveries. Several weeks later, all the plutonium received at Los Alamos was fashioned into a perfect sphere 22 millimetres across and weighing just over 100 grammes. The sphere contained virtually all of the element that had ever existed, except for a few milligrammes in the Met Lab and a few more that had been sprayed into the face of an unlucky researcher when an experiment turned nasty. Bombarding the sphere with neutrons from the Los Alamos cyclotron confirmed that plutonium was more violently fissionable than U-235: its critical mass was calculated as 6.2 kg (13 lb), contained in a sphere just 9 cm (3.5 inches) in diameter. This was enough to make Fat Man explode with greater force than the 52 kg of U-235 in Little Boy. By early May 1945, it was clear that there would be enough plutonium for the Trinity test in July and for a Fat Man to devastate a Japanese city a month later.[13]

The orange-sized plutonium sphere would be imploded down to the volume of a walnut by a huge ball of high explosive, over 1,000 times its volume. Even that mass of explosive couldn't compress metal without focusing its energy precisely. The device's genius lay in the meticulous geometrical arrangement of 'lenses', made of two high explosives with contrasting properties – an idea which the 'CS' battleship-holing bomb had implanted in James Tuck's brain. The surface of the explosive ball was a patchwork of thirty-two hexagons and pentagons, like the panels on a football. Each panel formed the base of a five- or six-sided pyramid that projected inwards, like quartz crystals lining a hollow geode. The thirty-two pyramids were made of Composition B, which detonates extremely fast. The core of the explosive ball – equivalent to the cavity inside the quartz geode – was filled with shaped blocks of a slower-burning explosive, baratol (Figure 10). Peierls and Fuchs predicted that detonating all thirty-two fast-burning pyramids at the same instant would create a spherical, inward-rushing shockwave powerful enough to compress metals.[14]

How to detonate simultaneously thirty-two points on the surface of a 1.5-metre ball of explosive? Even the fastest-exploding detonator cord, which ignited at twenty times the speed of sound, was far too slow. Luis Alvarez invented a novel electric detonator, while Ernest Titterton devised a 6,000-volt electronic pulse that could fire thirty-two detonators within a microsecond.[15] Throughout early 1945, George Kistiakowsky's explosives team blew up 20,000 lb of high explosive in countless configurations. Casting molten high explosive into the geometrical pyramids demanded moulds that were more complicated and exact than ever before; Kistiakowsky sometimes worked through the night with a file, smoothing out imperfections in the interlocking shapes. In February 1945, ultra-high-speed photography showed that detonating a ball of thirty-two explosive lenses generated a converging spherical shockwave, and a sphere of cadmium was successfully imploded into a smaller, hyperdense volume – an

THE IMPOSSIBLE BOMB

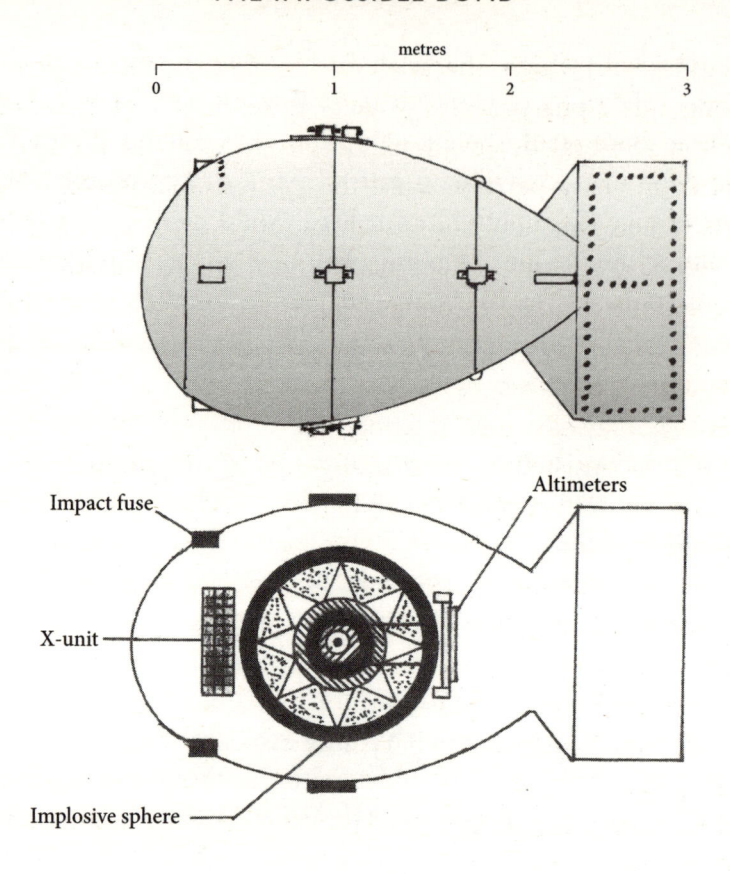

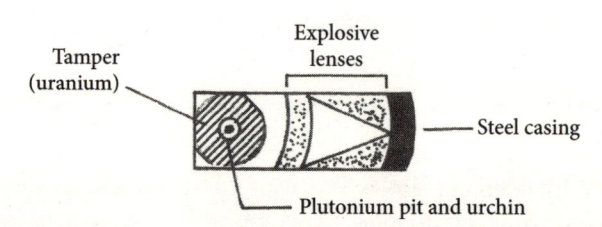

10. Fat Man, the plutonium bomb destined for Nagasaki, showing its internal structure.

astonishing physical 'impossibility' which convinced Oppenheimer that 'lens implosion is going to work'.[16]

The final 'stubborn puzzle', which could yet kill the plutonium bomb, was the initiator that would ignite the chain reaction by spewing out neutrons at the instant of implosion. The ingredients were those with which Chadwick discovered the neutron: polonium, pouring out alpha-particles which strike floods of neutrons off beryllium. Several devices were proposed to shield the beryllium target from alpha-particles until the desired microsecond, but the efforts 'lacked direction'. The first choice of Oppenheimer and Peierls was vetoed by Fermi, whom nobody dared to contradict – until Oppenheimer brought in a man with the wisdom and authority of Solomon. Niels Bohr considered the options and 'clarified what had to be done'. He chose the device favoured by Oppenheimer and Peierls, and may have helped to improve it.[17] The initiator, nicknamed the 'urchin', consisted of a few milligrams of polonium wrapped in aluminium foil to contain the alpha-particles, all enclosed in a thin shell of beryllium. This marble-sized device was inserted into the heart of the plutonium ball, which was bisected and had a small hollow scooped out of the centre of the flat face of each hemisphere (see Figure 10).[18] It might seem foolhardy to trust kitchen foil to hold back the biggest explosion in history – but it prevented alpha-particles from reaching the beryllium eggshell until the implosion tore it apart.

In April 1945, Groves approved the 'Christy solid implosion device with a modulated neutron source'. The design for Fat Man was frozen, clearing the stage to manufacture the components for the Trinity test in just twelve weeks' time.[19]

Testbed

The laboratory for 'Project TR' – Trinity – covered 430 square miles of desolation in the north-west corner of the vast Alamagordo Bombing Range, about 100 miles south of Albuquerque. The 'supreme secrecy' shrouding the site was assisted by its isolation: over 50 miles to the nearest town and 12 miles from any habitation (a ranch, recently emptied by eviction). Every square yard had been surveyed from topographic and grazing maps ('scrounged by devious channels'), supplemented by high-resolution aerial photographs.[20]

From early January 1945, the Trinity site took shape around new roads, settlements and bunkers, strung together by thousands of miles of cables and phone lines. At its centre was Point Zero, where the Trinity plutonium bomb would be detonated on top of a 100-foot metal tower. Base Camp, 10 miles south-west of Point Zero, was the operational centre and 'happy' despite its primitive conditions – water brought in by tanker was stiff with calcium salts and caused major 'sanitary problems'. The Control Station, 6 miles south of Point Zero, was the origin of the wires that ran across the desert to the Trinity tower and would carry Ernest Titterton's electronic signals to detonate the bomb. Observers and banks of monitoring instruments were to be crammed into three ugly concrete bunkers at 10,000 yards north, south and west of Point Zero. These were sunk into the ground like ancient burial chambers, aligned to the momentary sun of Trinity. Closer still were camera bunkers with portholes of bullet-proof glass. Predictions of Trinity's power ranged from a fizzle to a cataclysm, probably between 5,000 and 13,000 tons of TNT. Just in case, the bunkers could withstand a 200,000-ton blast. Safer still would be the viewing point for selected Manhattan–Tube Alloys scientists and VIP guests on Campania Hill, 20 miles to the north-west.[21]

Project TR was launched in tandem with 'Project A' (Alberta), which would drop Little Boy and Fat Man on their still-unspecified targets in Japan shortly afterwards. Trinity would provide proof of concept, a dress rehearsal and a mine of data for final tweaking of the plutonium bomb, the arming and firing procedures, the choice of target, and the infliction of maximal damage. Hundreds of instruments would measure the bomb's force, light, heat and radioactivity. It was estimated that the flash would be visible 200 miles away and that a hypothetical observer a mile distant might live long enough to see a fireball three times brighter than the sun and 350 times as big. The blast would knock a man flat at 2 miles, while the explosion should be audible for up to 100 miles. Sir Geoffrey Taylor predicted that a huge column of incandescent gas would boil up high into the sky until it hit a layer of cold air, which would flatten its top into a mushroom shape. Taylor also calculated that high-speed winds in the upper atmosphere could dump radioactivity on the ground hundreds of miles away.[22]

The instruments to monitor Trinity were state of the art, often invented on the hoof and strictly vetted at weekly meetings in the 'Town Hall' at Base Camp. The explosion would be tracked by ultra-high-speed cameras, some with film roaring through at 100,000 frames per second and potentially able to capture the fireball when still small enough to sit in someone's living room. Its force would be measured by electronic strain gauges and metal 'blast boxes' with holes of various diameters covered with aluminium foil which, like an eardrum, would burst at a critical pressure. Radiation counters would clock up gamma rays and neutrons, while ground tremors would be recorded by seismographs, some carried covertly by intelligence agents to cities 500 miles away. The only biological monitors were eighty-eight cows to be given a grandstand, open-air view 12 miles from Point Zero, blissfully unaware of the command to lie face down, feet pointing towards the Trinity tower, and only to get up after the blast wave had passed.[23]

At the end of March 1945, the no-bullshit 'Cowpuncher Committee', charged with keeping the enterprise moving, decided that 4 July was too early for the Trinity test. It suggested various dates up to two weeks later, the final arbiter being the weather. To rehearse for Trinity and to calibrate the instruments, a test firing of 'ordinary high explosive' was scheduled for 5 May. One hundred tons of TNT would be blown up on top of a 28-foot wooden tower close to Point Zero.[24] This was the biggest intentional explosion in human history, but represented only 1 per cent of Trinity's likely yield – which, if everything ran according to plan, would be demonstrated six weeks later.

Missions Accomplished and Pending

On 30 April 1945, Sir James Chadwick summarised recent progress in a top-secret memo for the Tube Alloys Consultative Council in London. Forty-five British scientists remained in America, at Oak Ridge, Berkeley and Los Alamos. All were making 'significant contributions', especially the fourteen working at Los Alamos on the 'design of the weapon'. All were essential to the Manhattan Project and even though the work was 'unattractive to many physicists', they would stay until 'the crucial period is over', perhaps six months or more. Relations with the USA were excellent again – 'all jam and kippers', in Chadwick's hammed-up Cheshire vernacular.[25]

Maintaining the British Mission also helped to fend off American suspicions that the Brits were setting up rival facilities in the UK – which was actually Chadwick's intention. Mark Oliphant had returned to Birmingham; he had no interest in watching Trinity explode, and anyway wasn't invited. Oliphant was 'very active in progress and discussion', especially with John Cockcroft, laying foundations for Britain's independent nuclear programme after the war. They had already inspected sites near Oxford and Cambridge for a new nuclear research establishment which would cover medicine and biology as well as bombs and reactors. Cockcroft had tied up another loose end. A new British–Canadian heavy-water reactor, staffed by the Montreal group, was nearing completion at Chalk River, north-west of Ottawa. Freed from interference by Halban and the Americans, it was expected to go critical in September 1945.[26]

Signing off his memo, Chadwick effectively wrote the epitaph for Tube Alloys. It and the British Mission had served with distinction, but new organisations would be needed to carry Britain's nuclear ambitions into the brave new world that must follow the war.

Meanwhile, an exhaustive account of the 'achievements of the Manhattan Project' awaited its final flourish. In early 1944, it had struck Henry D. Smyth – the senior Professor of Physics at Princeton – that the story of the bomb must eventually be told to the average American. Having served on the Uranium Committee and worked at the Met Lab, Smyth was ideally placed to write the official US government history of the project, which Groves commissioned him to do in April 1944.[27]

Smyth (see Plate 30) worked part-time on his report in his lab office in Princeton, with a massive safe, bars on the windows and an armed guard in the next room. It was a tricky balancing act: a factual account of how cutting-edge physics had spawned the bomb, to satisfy both scientists and general readers. Restricted to information published openly, Smyth had to omit 'many interesting developments' but would explain who had done what and turn the spotlight on 'those who had worked so long and necessarily so anonymously'. The report would be published soon after the bombs were dropped on Japan, when the world had witnessed the devastation caused and knew the cost in innocent lives. As a bold exercise in public engagement, the report would be unprecedented – even if it was unlikely to be a bestseller.[28]

In February 1945, Smyth had sent the first twelve draft chapters to Groves. Now, with events in the thirteenth and final chapter unrolling with gathering speed, Smyth had begun writing again.

A year earlier, the Hill had witnessed a display of Old World disinhibition when Genia Peierls danced on a table to celebrate the D-Day landings. A week after Chadwick signed his last Tube Alloys memo, there was another 'brief respite' from the daily grind when Rudi Peierls, Otto Frisch, Hans Bethe and others burst into traditional German songs that they'd belted out as students. It was Tuesday, 8 May, eight days after Hitler committed suicide in his bunker in Berlin: VE Day, marking the unconditional surrender of Germany and of Victory in Europe.[29]

The war wasn't yet over, because Japan was being steadily beaten back, but at huge cost and with no sign of capitulation. Japanese resolve to fight to the death had been strengthened by the Americans' horrific firebombing of Tokyo in March, using the newly invented napalm. This stuck to everything and made it burn: buildings, trees, clothes, a horse and then its rider, a baby strapped to its mother's back. Looking down during the raid, an American bomber pilot saw an inferno 'beyond the wildest imagination of Dante', which incinerated 16 square miles of the capital and over 83,000 people.[30] By early May, there was no end in sight for a vicious battle raging on the strategically crucial island of Okinawa, which had already killed 50,000 Japanese and 7,000 Americans.

The President, his military chiefs and Secretary of War Henry Stimson favoured invading the Japanese mainland as the least damaging way to force Japan to surrender; the estimated 250,000 American deaths would be an acceptable price for ending the war in the Pacific. On 25 May, the Chiefs of Staff of the US Armed Forces set 1 November 1945 as the target date for the American-led invasion through Kyushu. For now, the atomic bombs would be held in reserve. The military remained worried that they might not work; the Cowpuncher Committee's most recent recalculation gave Trinity the force of only 5,000 tons of TNT, while just 1,600 tons of napalm had destroyed Tokyo.[31]

Nonetheless, work would continue at full speed on both bombs and how best to use them. They would target major military installations such as armament factories or centres of troop movements and would be dropped on the centre of the target city. Tens of thousands of innocent people would die, but this was a fact of modern warfare, as demonstrated by Warsaw (40,000 civilians killed), Hamburg (37,000), Dresden (25,000) and now Tokyo. Civilian deaths were no longer collateral damage but had become a primary objective and would greatly enhance the psychological impact of the new atomic terror weapons. Indeed, the ideal target would be a strategically important military facility, surrounded by tightly packed housing for the workers.[32]

Groves's chief adviser on target selection was William Penney, a world authority on the physical effects of bombs on buildings and cities. Penney's attention had fallen on Hiroshima, because it was partly encircled by hills which would concentrate the blast of a nuclear explosion. The psychological impact of bombs lay beyond his professional remit but was now close to his heart. Nine months after seeing his wife, Adele, speechless and tormented by demons, in the psychiatric hospital in Surrey, Penney finally received the news that he'd been dreading. Adele died on 15 April 1945.[33]

24

ON THE RUN
April–May 1945

Back in February 1945, the 'Big Three' – Roosevelt, Churchill and Stalin – had met at the Black Sea resort of Yalta to carve up whatever would remain of Germany after the war. Roosevelt looked dreadful, and his permanent attachment to a cigarette augured badly for a sixty-three-year-old polio survivor with uncontrollable hypertension. At a meeting in the White House in late March, he reassured his Secretary of War, Henry Stimson, that he remained 'much interested' in the bomb; although it was noon, Roosevelt was too exhausted to get out of bed.[1] Stimson never saw him again. On 12 April 1945, Roosevelt was felled by a massive stroke and died within the hour. Later that day, Vice President Harry Truman was sworn in as President and Commander-in-Chief. Truman's elevation initially sent anxious tremors across the Hill; he was an unknown quantity and Roosevelt had hidden Manhattan from him. In late April, after intensive briefings by Stimson and Groves, Truman was welcomed safely on board, with his blessing for whatever it would take to deliver atomic bombs for use against Japan in August, just four months hence.[2]

By now, it was obvious that dropping atomic bombs on Japan might solve one acute problem but could stir up others that might never go away. At risk were both America's nuclear supremacy after the war and the survival of humanity. Leo Szilard and other Manhattan scientists had joined Niels Bohr in arguing that America must not have sole possession of nuclear weapons: equipoise with other nuclear-armed nations was the only way to guarantee lasting peace in a world that had lost its nuclear innocence. Bohr's proposal which had outraged Churchill – sharing the secrets of the bomb with the Russians – was the only way to prevent post-war superpowers from becoming 'locked in a secret arms race'.[3]

More immediately, when should Russia be told about the bomb? Yalta had united the Big Three as fighting partners; withholding knowledge of the bomb from Stalin until the whole world had witnessed the nuclear death of a Japanese city would undermine that accord. It was decided that Truman would wait until after the Trinity test and then tell Stalin that the bombs would be used in anger if Japan refused to surrender. Which of course depended on Manhattan producing an infallible instrument of war on schedule.[4]

Russia posed other threats. Suspicions were hardening that Soviet spies had penetrated Manhattan, and it was safest to assume that the Russian atomic bomb programme was already under way. Given Manhattan's intellectual firepower and vast industrial resources, Groves was confident that Russia would take twenty years to make its own bomb. James Conant thought Groves dangerously complacent; his own estimate was just three to four years, and even less if Soviet spies had prised out the secrets of U-235 separation. All the more reason to make Trinity, Little Boy and Fat Man succeed, and then to dictate to the Soviets from a position of strength.[5]

The Russian Army's inexorable advance on Berlin during April caused further concern. Groves had already acted to deprive the Russians of nuclear war booty, by ordering an air strike which obliterated the uranium-producing factory at Oranienburg, north of the capital. However, the Russians would be the first to comb through the remains of the KWIs for Physics and Chemistry in Dahlem. Recent intelligence indicated that they had set up their own version of Alsos to round up *Uranverein* scientists, and they would undoubtedly pick up the trail that led south to Tailfingen, Hechingen and Haigerloch. Which, to add further complexity, lay in the sector of Germany that the Yalta Agreement had allocated to France – and where a French expeditionary force was already operating.[6]

In Washington, Groves lowered himself on his hands and knees in front of a wall map of Germany to scrutinise the relevant area, down near the border with Switzerland. As he later remarked, a camera behind him at that moment could have captured 'one of the more interesting photographs of the War'.[7]

Dead End

From a patchwork of intelligence, the Anglo-American 'Tube Alloys Intelligence Committee' set up by Michael Perrin and Robert Furman had built up a reasonably clear picture of who was where in March 1945. After the air raid on Dahlem on 14 February 1944, Albert Speer had ordered the evacuation of the *Uranverein* groups from Berlin, Leipzig and Hamburg. Consistent with Paul Rosbaud's report, photoreconnaissance had revealed new construction sites tucked away in the woods and valleys around the three towns on the edge of the Black Forest.[8] And the documents which Pash and Goudsmit read over brandy as German shells whistled over Strasbourg had populated those sites with the big names on Alsos's hit-list.

Werner Heisenberg, Director of the *Uranverein*. Carl von Weizsäcker, who left a chimney stuffed with partly burned documents when he fled from Strasbourg. Karl Wirtz, designer of the 'B'-series *Uranbrenner* reactors in Berlin. Paul Harteck from Hamburg, heavy-water expert and leader of the efforts to separate U-235, and his colleagues Erich Bagge and Horst Korschung. Kurt Diebner, one-time Atomic Weapons Director of the Army's Ordnance Research Division, who had covertly diverted uranium and heavy water from Heisenberg for his rival chandelier reactor in Gottow and then Stadtilm, south of Leipzig. Walther Gerlach, a respected physicist from Hamburg, appointed the Reich's 'Plenipotentiary of Nuclear Research' and administrative chief of the *Uranverein* in late 1944.[9] There were also two victims of circumstance: Otto Hahn, discoverer of nuclear fission, and Max von Laue, the anti-Nazi Nobel Prize winner for discovering X-ray crystallography. Neither had anything to do with the *Uranverein* but were evacuated with them to the Black Forest.

The only clue about what Alsos would find in the *Uranverein*'s bolt-holes came from an excited Gerlach, who was in Berlin when Heisenberg phoned him with an update about a rebuilt reactor in Haigerloch. Gerlach was ecstatic and told a confidant that 'The machine works!' – although close questioning indicated that it had generated excess neutrons but failed to go critical. Gerlach's confidant was his old friend Paul Rosbaud, whom he visited in the Springer office on 24 March. Two days later, Rosbaud's report reached Eric Welsh and Michael Perrin, via the XU Norwegian Resistance network.[10]

In late March, Pash set off from Strasbourg on 'Operation Big' to snatch the *Uranverein* scientists, their knowledge and materials before the French could reach them. His group now boasted six jeeps, two armoured cars and a British contingent who flew in on an RAF Dakota: Welsh, Perrin and the flamboyant former SOE commander Sir Charles Hambro, all determined to witness the *Uranverein*'s demise and scavenge what they could from its remains.[11] During their three-week epic trek to the Black Forest, Pash's convoy dodged demoralised remnants of the German Army and, more dangerous, trigger-happy teenagers from the Nazi *Wehrwolfen* ('werewolves'). They squeezed in some profitable diversions. Heidelberg University, where they met Walther Bothe, at last with a working cyclotron but now left alone because they considered him irrelevant. A cavernous salt mine outside Stassfurt with 1,100 tons of uranium oxide. A schoolhouse in Stadtilm containing the gutted remains of Diebner's lab, stripped of its uranium and heavy water by the SS. And finally, after a breakneck dash through the Black Forest on the night of 21–2 April, Haigerloch.[12]

A few days earlier, the town had watched as battered German artillery retreated, and was now decked out in white – sheets, towels, shirts, anything – to signify surrender. Searching revealed nobody of interest, but a padlocked steel door on an ugly brick-and-concrete blockhouse below an 80-foot cliff surmounted by a baroque church. Behind the door, the blockhouse opened into a cave, formerly a storage cellar but now converted into a nuclear research lab, with massive containers – all empty – for heavy water. A circular pit, 12 feet across by 12 feet deep, had been gouged into the limestone floor; nested inside that was a 7-foot diameter metal cylinder lined with graphite bricks, its central cavity empty. A heavy metal cover for the cylinder lay nearby, to be lifted into place by an overhead gantry. It was the carcass of a reactor, minus the heavy water and uranium which would give it life. Pash decided to remove anything valuable and destroy the rest to stymie the French. The next morning, the graphite lining was dismantled for shipment to the USA – photographs show Eric Welsh busy in the Alsos chain gang (Plate 22) – and the empty pit was dynamited. Luckily, there was no damage to the church on the clifftop, or to its exquisite eighteenth-century organ – which, like the piano in Niels Bohr's House of Honour in Copenhagen, bore the fingerprints of Werner Heisenberg.[13]

With French troops perilously close (and skilfully fobbed off by Pash), Tailfingen and Hechingen followed on 24 April and yielded the first clutch of *Uranverein* acquisitions – von Weizsäcker, Wirtz, Hahn and Bagge. Despite his protests that he had nothing to do with them, a bemused Max von Laue was also taken in. Heisenberg and Walther Gerlach had cycled away three days earlier.[14] At first, von Weizsäcker refused to cooperate, but cracked after four days of solitary confinement. He told them all about the reactor in the cellar floor, designated 'B-VIII' as it had been resurrected from the remains of B–VII in Berlin. It was Kurt Diebner's chandelier design (Figure 11). Uranium was lowered into the graphite-lined cavity, heavy water poured in, and the reaction triggered by a neutron source pushed in through the top. The uranium was in a novel configuration: 664 2-inch cubes, with eight or nine of them strung along each of seventy-eight chains hanging from the circular metal cover. Many of the cubes had been sawn out of uranium sheets intended for Heisenberg's own reactor. With 1.5 tons of uranium and 500 litres of heavy water, they had been so close to a self-perpetuating chain reaction, von Weizsäcker said. They could have done it had not supplies been stranded in a goods yard by Allied bombing.[15]

Finally, von Weizsäcker was coaxed into disclosing the real treasure. Back in Haigerloch, a ploughed field was dug up (by a German prisoner, in case of booby traps), unearthing hundreds of

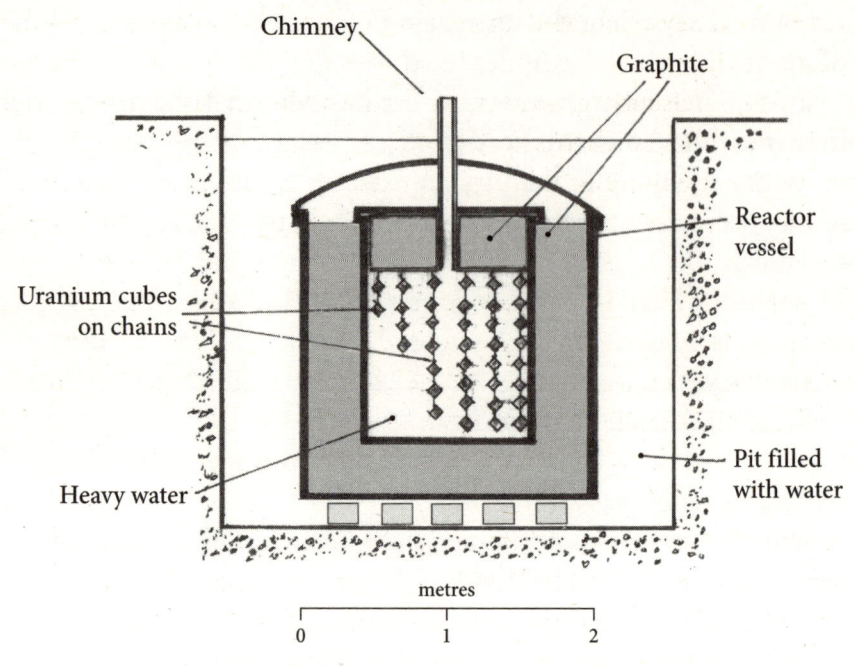

11. Diagram of the *Uranverein*'s B-VIII reactor at Haigerloch.

2-inch uranium cubes; Michael Perrin was among those who retrieved the cache, which weighed over 2 tons (Plate 23). Next came petrol cans full of heavy water, hidden in the cellar of a watermill, and – exhumed by two luckless American soldiers wearing long rubber gloves – a sealed box containing von Weizsäcker's laboratory records at the bottom of a latrine.[16]

Soon after, Gerlach and Diebner were rounded up near Munich, leaving just the biggest fish. Heisenberg's 200-mile pilgrimage to rejoin his wife Li and their six children at their holiday home in Urfeld, high in the Bavarian Alps, took three harrowing days by bike and train. He saw starving prisoners from Dachau packed into a freight train and appeased an unhinged SS officer with a lucky cigarette just in time to avoid being shot. Heisenberg's last few days of freedom epitomised the dying spasms of Hitler's Reich.

29 April. Werner and Li celebrated their eighth wedding anniversary, and 'for the last time, we gave the Hitler salute' at the funerals of two Nazi friends who had killed themselves rather than face defeat.

1 May. They celebrated again, after German radio announced the death of Hitler.

2 May. Heisenberg was taken into custody by Pash, after a brief firefight with SS diehards.[17]

With everything in the bag, an exultant 'EYES ONLY' cable was sent to Secretary of War Stimson in Washington:

ALSOS MISSION HEADED BY BORIS PASH ... HAVE HIT THE JACKPOT ... AND HAVE SECURED PERSONNEL, INFORMATION AND MATERIEL EXCEEDING THEIR WILDEST EXPECTATIONS[18]

The 'materiel' was airlifted to the USA, where the uranium cubes contributed several kilos of U-235 to Little Boy. The eight *Uranverein* nuclear physicists, together with Hahn and von Laue, were confined in a chateau near Paris (nicknamed the 'Dustbin') while their fate was decided. Groves wanted them shot as 'great war criminals'. Chadwick, Welsh and Perrin had other ideas, and 'Operation Big' silently morphed into 'Operation Epsilon'. 'Our German TA detainees' were to be transferred to relative comfort in an old English mansion in Godmanchester, near Cambridge – but not until early July, because certain arrangements had to be made before their arrival.[19]

25

PROOF OF CONCEPT
May–July 1945

What do you expect on an active bombing range when nobody knows you're there? Luckily, both the bombs dropped accidentally near the Trinity site one morning in early May 1945 were dummies, and swift action was taken at the Alamagordo Air Base to 'prevent a recurrence'.[1]

Project Trinity was mostly on course and spirits were buoyant. At 4.37 p.m. on 7 May, the German surrender was marked by a 100-ton salute: the detonation of thousands of crates of dynamite (helpfully stencilled 'THIS WAY UP' and 'DANGEROUS') near Point Zero. The bang was heard at the air base 60 miles away, and a 'highly luminous' fireball rose 1,500 feet and became mushroom-shaped. This rehearsal for Trinity identified many faults to fix – such as vital instruments which overexcited operators forgot to switch on.[2]

Both bombs were now acquiring their last crucial components. Each week brought another 3 kilos of U-235 for Little Boy; in mid-May, they began casting the rings which would be stacked to form the hollow slug to be fired onto the U-235 target. Plutonium deliveries were also on schedule. In early June, the definitive explosive lenses were tested and behaved impeccably. The urchin initiator survived being dropped on a laboratory floor (and was grabbed just before rolling down an open drain). Groves oversaw plans for the emergency evacuation of towns in New Mexico in case radioactivity from Trinity drifted their way (a leak of poison gas would be blamed). He also drafted press releases covering all eventualities, from the explosion of a 'munitions dump' on the Alamagordo Bombing Range to tributes and obituaries for likely victims if Trinity exceeded expectations.[3]

Against this frenetic background, Richard Feynman made time for the comforting ritual of writing to his wife, Arline. Letters were their lifeline while he was trapped on the Hill and she in the sanatorium in Albuquerque. They exchanged jokes, news, puzzles and declarations of love and hope. The last of Feynman's forty letters, written on 16 June 1945 and imploring 'Putzi' to 'keep fighting', never received a reply. On 26 June, responding to an urgent summons from the sanatorium, he jumped into Klaus Fuchs's Buick and drove fast to Albuquerque. He arrived too late to be sure that Arline heard him say that he loved her and always would.[4]

Red-Letter Days

In early May 1945, Truman appointed an 'Interim Committee' to advise him on atomic bombs. Chaired by Henry Stimson, the committee included Compton, Bush, Conant and the Secretary of State; as Groves said, this would show that the decision to drop the bomb wasn't made exclusively by 'men in uniform'. Stimson set the scene: the atomic bomb wasn't just 'a new weapon' but potentially 'a Frankenstein which could eat us up'.[5] After three weeks of deliberation, the Interim Committee concluded that the bomb should be dropped without warning on a large Japanese city containing a major war plant surrounded by workers' homes. Five cities were shortlisted, with Hiroshima and Kyoto designated top 'AA' priority. Hiroshima housed a military headquarters, arms factories and a major convoy port. Stimson vetoed Kyoto, Japan's near-sacred second city, which he'd visited and admired during the 1920s. Kyoto was replaced by Yokohama.[6]

The committee faced a mutiny from within Manhattan. Leo Szilard, Harold Urey, Glen Seaborg and many others at the Met Lab were dismayed to be excluded from discussion about using the apocalyptic bomb that they had helped to create, and wanted its power to be demonstrated on 'a desert or a barren island' without killing innocent people.[7] The committee commissioned James Franck, senior Professor of Physics at Chicago, to write an urgent report on the bomb's 'Political and Social Implications'. Franck, revered as 'a truly wonderful man', had seen physics and warfare at first hand, having fought in the German trenches in the Great War and escaped to America when Nazism took hold. The Franck Report, delivered to Compton in record time on 11 June 1945, reflected extensive canvassing of the Met Lab scientists. It argued that nuclear weapons could never remain secret; in ten years, others would undoubtedly possess them and America's highly concentrated populations would become sitting targets. Tight international control of nuclear weapons was the only way to avoid future conflicts that could wipe out humanity. America could now lead from the front by demonstrating the weapon in an 'appropriate selected uninhabited area'. By contrast, 'an early unannounced attack on Japan' would sacrifice public support for the USA and precipitate a post-war nuclear arms race.[8]

Franck summarised his report in a personal letter to Truman, sent to the White House on 11 June; it got no further than the Secretary of State. The Interim Committee consulted a scientific panel – Oppenheimer, Fermi, Compton and Lawrence – who saw no alternative to direct military use, and rejected the proposal for a harmless demonstration.[9]

Events now moved fast, like a line of dominoes toppling.

21 June 1945. The Interim Committee advised Truman that the bomb should be dropped on a Japanese city, as soon as possible and without warning. The shortlisted targets were Hiroshima, Yokohama and Kokura.

25 June. Truman's military chiefs confirmed 1 November 1945 as the date to invade Japan.

26 June. Truman proposed another Big Three meeting, to persuade Russia to honour its pledge at Yalta to attack Japan. Stalin and Churchill concurred, and Stalin offered Potsdam, near Russian-occupied Berlin, as the venue. Truman accepted but delayed agreeing a date.

30 June. The Los Alamos Cowpuncher Committee told Truman that the earliest possible date for the Trinity test was Monday 16 July.

Truman therefore proposed Sunday 15 July for the start of the Potsdam Conference, to which Stalin and Churchill agreed.

4 July. The Anglo-American Joint Policy Committee met at the Pentagon to rubberstamp an action in the Quebec Agreement: that either country must obtain the other's approval to use a Tube Alloys weapon against a third party.[10]

Big Three

Potsdam, the historic seat of the Kaisers, was steeped in Teutonic make-believe. Truman, Churchill and Stalin met in the half-timbered Cecilienhof Castle, built during the Great War to mimic an English Tudor mansion. The American and English delegations were put up in splendour in nearby Babelsberg, Germany's answer to Hollywood. Babelsberg had produced gems such as Marlene Dietrich's *Blue Angel* (1930) and, latterly, hundreds of hate movies for Josef Goebbels' Ministry of Propaganda.[11]

Henry Stimson crossed the Atlantic on a US troop carrier and reached Marseille on 15 July 1945. He was flown to Gatow airfield outside the remains of Berlin to greet the presidential party, which had steamed into Antwerp on a navy cruiser. The rookie President, Harry Truman, was just ten weeks in post, and Potsdam was his first international fixture. He was chairing the conference and was braced for hard haggling with Stalin to guarantee a Russian attack on Japan soon after the planned American invasion.[12]

Churchill came to Potsdam ill, exhausted and shockingly unprepared. He'd thrown away a golden opportunity to win over Truman ('a man I do not know', he later complained) by declining the President's invitation to attend Roosevelt's funeral and had talked only briefly to him on the phone. Churchill was in deep trouble at home but was punch-drunk with having led Britain to Victory in Europe and oblivious to the omens of doom. He'd ignored hard evidence that he wasn't capable of steering Britain towards post-war recovery – a highly critical letter by Clement Attlee, the Labour Deputy Prime Minister in the creaking Conservative–Labour coalition which necessity had forged in a time of extraordinary danger.

Churchill refused to mend his ways, lost a no-confidence vote in the House of Commons, and was forced to call a General Election. This was scheduled for 6 July, but because many servicemen were still fighting in the Far East, the votes wouldn't be counted until 26 July – midway through the Potsdam Conference, which would break off so that Churchill could return to London to learn his fate.[13]

Churchill had also lost track of the bomb. He hadn't bothered with Sir John Anderson's briefings but had listened to Frederick Lindemann, who had reverted to extreme scepticism after visiting Los Alamos. 'What fools the Americans will look', Lindemann told R.V. Jones, 'having spent all that money'. (On the Hill, Lindemann had impressed everyone with his inability to grasp important stuff.)[14]

Such was the stage set for the Potsdam conference of 17 July–2 August 1945. Henry Stimson hoped that 'America's master card' – a successful Trinity test – would enable Truman to cow Stalin with the world-shattering news that atomic bombs would be dropped on Japanese cities to force Japan to surrender. By removing the need for Russian help to invade Japan, this would sweep Stalin's biggest bargaining chip off the table.[15]

At Los Alamos, the race against time was accelerating. When Stimson boarded his troop carrier on Friday 6 July, Trinity's two silvery plutonium hemispheres were ready for use. When he touched down at Gatow on Saturday afternoon 15 July, Trinity was sitting on top of its firing tower with its plutonium core in place, waiting for 'the biggest and most expensive experiment in scientific history' to take place in twenty-one hours' time.

Don't Look Now

Since early July, a cast of hundreds had gathered at the Trinity site, including 350 scientists and bomb-makers and 125 military policemen to guard them and their weapon. The atmosphere was as febrile as the weather: swelteringly hot by day, thunderstorms boiling up most afternoons, freezing cold at night. Even before the rehearsals began, many were already working eighteen-hour days. Four dry runs for assembling and detonating the bomb and measuring its impact were crammed into successive days from Wednesday, 11 to Saturday, 14 July.[16]

The first rehearsal was cut short by a late afternoon thunderstorm that wiped out some sensitive instruments and hurled a bolt of lightning at the Trinity tower, triggering the detonator circuit which luckily wasn't plugged into anything dangerous. The second and third rehearsals were hastily rescheduled for early morning, and the fourth for just before midnight on Saturday, 14 July. The final rehearsal would test to destruction a near-perfect replica of Trinity, lacking only the plutonium to convert a conventional explosion into the biggest man-made bang ever. It would overlap with the 'Hot Run' – the real thing – which would have started two days earlier. The Hot Run was planned impeccably, with a step-by-step checklist, precise timings (in War Mountain Time) and a dash of whimsicality from Kenneth Bainbridge, the director of the Trinity test:

Friday 13th July, 13.00. Final assembly begins.
Saturday 14th July, 17.00. Gadget complete.
Sunday 15th of July, all day. Look for rabbit's feet and four-leaved clover.
Monday 16th July, 04.00. Bang![17]

Those left on the Hill were also rehearsing, using hollow 6-foot spheres of steel and carborundum – the guts of Fat Man, stripped of its outer casing, high explosive and plutonium. Test spheres survived the 200-mile drive to the Trinity site at 30 mph and being bumped for eight hours along unmade roads; four high-explosive lenses inside the sphere also passed this 'Shake Test' in excellent condition. On Tuesday, 10 July – the eve of the first dry run at Trinity – two full sets of explosive lenses were laid out for minute inspection by Bainbridge and George Kistiakowsky. The 'first-quality' lenses were reserved for Trinity, and the rest set aside for the final dress rehearsal. On Thursday, 12 July, during the second dry run, each set of lenses was packed inside a Fat Man sphere, at two different sites well away from the Hill.[18]

At midnight, each explosive-packed sphere (now gift-wrapped in waterproof fabric) was lifted into a truck and driven south in a military convoy. The Trinity sphere was taken directly to the high altar: a 100-foot steel tower (actually the bottom half of a standard US radar mast), topped by a platform supporting a hoist and a galvanised steel shed which would shelter Trinity for its final hours. The sphere was unloaded into a white canvas tent at the tower's base to await completion.[19] The centre of the 6-foot ball of high explosive was the orange-sized spherical void, to be filled by the 'pit' – the two plutonium hemispheres with the urchin initiator at their heart. The plutonium was already on site, having arrived with a full military escort from Los Alamos the previous evening. It was delivered unsensationally at 18.00 to an abandoned ranch house 2 miles from Point Zero, by a US Army sergeant wearing a metal 'catastrophe' identity badge (in case he ended up unrecognisable) and a faraway expression, as if wondering how such a small carrying case could weigh so much, or what its contents might do (Plate 24).[20]

At 15.00 that afternoon – Friday the 13th – the plutonium core was taken to the white tent where the central cavity in Trinity's explosives had been opened by removing the uppermost lenses. With Oppenheimer watching anxiously, the plutonium ball was assembled around the initiator and lowered carefully into the hole. Even when helped by a hypodermic-needle grease gun and a shoehorn, the ball was a hair's-breadth too big. It had expanded, heated by spontaneous radioactive decay and a blistering afternoon. The plutonium was taken away and cooled down; this time it fitted. The explosives were repacked and the sphere's removable metal cap was bolted back in place.[21]

At 08.00 next morning – Saturday, 14 July – the tent was dismantled and the Trinity sphere hoisted up to the galvanised iron shed on top of the firing tower. The thirty-six detonators were inserted into the high explosive through a geometric array of holes drilled through the sphere, and the detonation circuit was installed. By 17.00, the bomb was ready to go, except for the final connection to the firing cable that crossed the lava field to the Control Center. Trinity was left alone for its last thirty-five hours, except for six-hourly checks of its health and the armed guards constantly patrolling around Point Zero.[22]

PROOF OF CONCEPT

Sunday, 15 July, was not a day of rest, nor conducive to collecting lucky rabbits' feet or four-leaf clover. A steady stream of visitors arrived by plane, train and automobile, from the Met Lab, Berkeley, Washington and Los Alamos, to witness the culmination of all their efforts two hours before dawn on Monday. They were notionally all men; women were banned, but Joan Hinton, a feisty blonde physics graduate working with Fermi, bluffed her way in on the back of a male friend's motorbike. Groves and Fermi would join Oppenheimer and Jack Hubbard, the all-important Chief Meteorologist, in the Control Center. Bainbridge insisted that 'We do not expect danger', but the VIPs – including Chadwick, Lawrence, Conant and Bush – were driven to the safer viewing point of Campania Hill, 20 miles to the north-west. Sir Geoffrey Taylor (who happened to be Joan Hinton's uncle) was there too, at the insistence of Groves. So were the pair who had kicked the whole thing off five years earlier, albeit with a memorandum about an entirely different weapon. As evening fell, Otto Frisch and Rudi Peierls boarded the bus taking observers from Los Alamos to Campania. On arrival at 02.30 – ninety minutes before *Bang!* – Frisch found the surreal scene of the hilltop crowded with people and vehicles, and loudspeakers playing dance music interspersed with countdown announcements.[23]

Even as the VIPs were being met and greeted, Oppenheimer and his team were racked by fears that Trinity would fail disastrously. The fourth and last rehearsal had gone like clockwork until the microsecond early on Saturday morning when the electronic firing pulse should have detonated the huge ball of high explosive inside the plutonium-free Trinity replica. Nothing happened. An urgent postmortem by Hans Bethe concluded that vital components had worn out, possibly through overuse during the rehearsals; these were replaced and extra insulation was added to Trinity's detonators. Human error may have contributed: a later account refers darkly to 'shooting being too good' for anyone making a mistake that late in proceedings. To cheer up the visibly distressed Oppenheimer, George Kistiakowsky bet him $2 that Trinity would work; Oppenheimer countered by wagering a month's salary that it wouldn't.[24]

THE IMPOSSIBLE BOMB

The precise moment of *Bang!* before dawn on Monday, 16 July 1945 is known 'only very poorly', as the Control Center couldn't pick up the local radio station WWP for a time check.[25]

A thunderstorm brewing shortly before midnight pushed the 04.00 test back to 05.00, assuming that conditions improved. The decision was made by the triumvirate in the Control Center who had the final say – Oppenheimer, Groves and meteorologist Jack Hubbard, any of whom could veto the firing. Within the hour, the skies cleared and the Arming Party – Bainbridge, Kistiakowski and two soldiers – drove their jeep to the foot of the tower. At 05.00, Hubbard predicted clear skies ('I'll hang you,' warned Groves) and the final-say group decided on 05.30, to be reviewed at 05.10. The Arming Party connected Trinity's detonator circuit to the firing cable from the Control Center. At 05.10, the time of 05.30 was confirmed and broadcast by loudspeaker and short-wave radio. It had been agreed that Groves and Oppenheimer must be in separate shelters in case of catastrophe; Groves left the Control Center and drove to Base Camp, where he lay on the ground between Conant and Bush. The Arming Party switched on a searchlight that illuminated the tower to deter any last-minute saboteurs, climbed into their jeep and drove to the shelter 10,000 yards south at a cool-headed 35 mph. By the time they reached safety, the headcount inside each shelter had been checked and the well-rehearsed ten-minute countdown was already over halfway through.[26]

At the five-minute signals – rockets fired and a short siren blast – everyone in the open lay face down on the ground (in 'an earthen depression' if available), feet towards Point Zero. Those able to watch were told not to look until 'after the southern hills light up' – and then only through a square of heavily smoked welder's glass – and not to stand until after the blast wave had passed, which would take fifty seconds to hit the 10,000-yard shelters and nearly two minutes to reach Campania Hill.[27]

Two minutes – a long siren blast – and Oppenheimer was silent, staring straight ahead and hanging on to a post as if for strength.

One minute. Fifteen seconds later, humans were relieved of the responsibility for the split-second activation of the hundreds of instruments recording the birth of nuclear weapons. A 'robotic device' was switched on: a steadily rotating metal rod studded with steel pins which engaged with a trigger unit that travelled slowly along beside it. At two seconds to zero, a pin woke the camera that began tearing through 100,000 frames per second. The very last pin would unleash the 6,000-volt electronic pulse which Ernest Titterton had designed to fire simultaneously all the detonators embedded in Trinity, 5.7 miles away. With Oppenheimer 'a nervous wreck', watching over his shoulder, Titterton stared at the oscilloscope screen which monitored the firing circuit.

Zero. And almost before Titterton registered the upswing in the oscilloscope tracing, the world lit up.[28]

The moment had long been envisaged and rehearsed, but Trinity was a visceral assault on the senses and emotions that shook everyone to the core. Rudi Peierls later wrote that 'no amount of imagination would have given us a taste of the real thing'.[29]

Lying on his belly in the lee of a radio truck on Campania Hill, Otto Frisch had adjusted his eyes to the pre-dawn gloom when, soundlessly, the hills opposite were illuminated 'as if somebody had turned the sun on with a switch'. The glare was flat and colourless, like an overexposed flash photo. Frisch turned to peer behind him, but the sun-sized fireball 20 miles away was too bright to look at initially (Plate 25). It was round, red and expanded rapidly, climbing fast into the sky and flattening into a mushroom shape, outlined in a bright purple-blue glow and connected to the ground by a grey stem. Another mushroom sprouted out of its top, and the undersides of clouds high above turned white 'like a spreading pool of milk' as the shockwave slammed into them. When he guessed that the blast was about to hit, Frisch sat on the ground behind the truck and shoved his fingers into his ears. The bang was 'quite respectable' and was followed by loud rumbling 'like huge noisy wagons running around', which thundered on for over five minutes as the hills continued to throw back the echoes.[30]

Also on Campania Hill, James Chadwick squinted around his square of welder's glass and saw an intensely bright point of light swelling into a huge fireball 'of swirling debris' that bathed the desert in 'a great blinding light'. The cold-fish, agnostic Chadwick likened the bang to the skies cracking open and the fireball to 'a vision from the Book of Revelations ... as if God Himself had appeared among us.' Around him, hard men who hadn't seen the point of lying down at a range of 20 miles threw themselves to the ground in terror.[31]

Those closer to the awful majesty of Trinity saw extraordinary things. At 10,000 yards south, Groves's assistant General Thomas Farrell was awestruck by the pyrotechnics – 'golden, purple, violet, gray and blue' – which made 'every peak, crevasse and ridge' leap out of the darkness. Farrell thought the explosion 'magnificent, beautiful, stupendous, terrifying', while Kenneth Bainbridge settled for 'foul and awesome'. 'Very terrifying' was also the verdict of someone staring up at the cloud – 'ominous ... brilliant purple with all the radioactivity glowing'; it appeared to hang over them 'forever' and the thunderous echoes of the explosion 'never seemed to stop'. For Joan Hinton, 'It was like being at the bottom of an ocean of light ... The light withdrew into the bomb as if the bomb sucked it up.' The light and sound left some cold; Derrik Littler, a British physicist lying on open ground at Base Camp, was so disconcerted by the blaze of searing heat on his uncovered skin – 'at least as strong as the noonday sun' and lasting several seconds – that he forgot to look at the fireball.[32]

The emotional aftershocks of the explosion ranged from awe to primal fear, excitement to foreboding, and elation to guilt. Oppenheimer said later, 'A few people laughed, a few people cried, most people were silent.' Inside the Control Center, Oppenheimer's face immediately 'relaxed in tremendous relief'; George Kistiakowski staggered in from outside, where the blast had knocked him flat, put an arm around his shoulder and said, 'Oppie, you owe me two dollars.' Bainbridge told him, 'We're all sons of bitches now.' Others stated the blindingly obvious: 'Well, it worked.' At Base Camp, those lying on open ground broke out into loud cheering. Conant and Bush, still face down on either side of Groves, leaned across and shook him by the hand. For Enrico Fermi, the experiment was still running. With his body sheltered from the blast wave, he dropped small pieces of paper from a fixed height so that he could estimate the power of the explosion from how far they were displaced by the blast. High on

Campania Hill, Chadwick jumped in the air for joy. But James Tuck, without whose efforts the gadget might never have worked, asked a question which troubled many but was already too late: 'What have we done?'[33]

Abruptly, the show was over. Two short blasts on the sirens announced 'the passing of all hazard from light and blast', the signal that 'all personnel will carry on'. On Campania Hill, Peierls and Frisch were driven away as the sun was rising just before 6 a.m. – three and a half hours after arriving, and half an hour after detonation. Groves had ordered the bus to go straight to Los Alamos without stopping. It was a silent trip; Peierls thought that those still awake looked 'worn out and depressed' – and that 'an observer would have decided that the test had failed'.[34]

This was the belief of those back on the Hill who had waited up and gone to bed dispirited when nothing happened at 04.00. But in the hospital, a patient who couldn't sleep saw a brilliant flash on the southern horizon shortly before 5.30 a.m., ninety degrees to the right of where the sun would soon rise over the Sangre de Cristo mountains.[35]

When Robert Oppenheimer made his way to Base Camp from the Control Center, he found jubilation and a bottle of whiskey doing the rounds. He was walking on air – 'a kind of swagger . . . like High Noon' – and Groves told him, 'I'm proud of you.'[36]

Euphoria wasn't Oppenheimer's only emotion at that moment, because the detonation of Trinity had thrown another literary quotation into his mind. Not from a sonnet of John Donne's, but from the ancient Hindu scriptures the *Bhagavad Gita*, and the words uttered by the shape-shifting god Vishnu as he takes on his most terrifying persona:

I am become Death, the destroyer of worlds.[37]

And while those at Los Alamos came to terms with the success of Trinity, some of their colleagues were already 6,600 miles away on an island in the Western Pacific, preparing Little Boy and Fat Man to annihilate two Japanese cities in three weeks' time.

26

INSTRUMENT OF WAR

July–August 1945

The Potsdam Conference didn't start as planned on Monday, 16 July 1945, because Stalin arrived late. At 7.30 p.m. that evening, Stimson received a coded cable from George L. Harrison, his assistant in Washington. 'Dr Groves' was pleased with the 'operation' that morning; the diagnosis was 'not yet complete', but results 'had already exceeded expectations'.[1]

Next morning, the hard bargaining began over Japan and Russia's claims on post-war Europe. Truman was impressive, Stalin confident and bullish, and Churchill disorganised and lacklustre. They broke for lunch – Stalin with Truman at the yellow-painted mansion nicknamed the 'Little White House', and Stimson with Churchill and Clement Attlee in the Prime Minister's villa. Stimson showed Churchill the cable from Harrison and, while they walked down the garden, explained that Trinity had worked. Churchill was 'intensely interested' and 'greatly cheered up'. Then they went back to the afternoon session, and the bomb was forgotten.[2]

Meanwhile in Washington, Groves was assimilating the knowledge flooding in about Trinity – and determined to keep everyone in limbo until he'd finished writing his report for Stimson and Truman. The following morning, Wednesday, 18 July, Stimson was tantalised by a further bulletin from Harrison. The 'operation' had safely delivered 'a little boy ... as husky as his big brother'. The neonate possessed disturbing powers: 'his screams' could be heard at Harrison's farm

(50 miles from Washington), while 'the light in his eyes' had been visible at Highhold (250 miles away). Decoded: Trinity was more powerful than anticipated, at least matching the uranium bomb.[3]

The bomb was eclipsed by the business of the conference – deciding terms for a Japanese surrender and carving up the post-war world between Russia and the Western Allies. Truman joined Churchill for lunch and gave him the update from Washington; Churchill suggested telling Stalin that 'the United States and Britain have the weapon', without revealing specifics. At 3 p.m., they adjourned with their Secretaries of State to the Russian headquarters for 'a sumptuous buffet' with plentiful toasts in vodka. Thursday and Friday passed with no news from Groves to distract from the increasingly complex negotiations of the conference.[4] Finally, at 11.30 a.m. on Saturday, 21 July, a special courier delivered Groves's report to Stimson. This wasn't a no-frills military briefing but a rambling, verbose account peppered with biblical references. Nonetheless, the key facts blazed through: Trinity's yield was between 9,000 and 20,000 tons of TNT, greater than predicted. Plutonium was the nuclear explosive of choice.[5]

Stimson took Groves's report straight to Truman and then to Churchill, who had to break off reading to leave for the next Big Three session. The tempo picked up that evening with a cable from Harrison: the uranium bomb would be ready to use 'at the first favourable opportunity' in early August. At 9.20 a.m. the next day, Truman approved the accelerated schedule for the 'first operation' in Japan. Churchill finished reading Groves's report and declared the bomb 'a miracle of deliverance' which could solve all their problems – forcing surrender on Japan, blindsiding Stalin and tipping the post-war balance in favour of America and Britain. This was Sunday, 22 July, a week after the day earmarked for finding four-leaf clover at Los Alamos while Trinity waited on top of its steel tower.[6]

Monday, 23 July saw more arm-wrestling with Stalin over his intended encroachment into Europe – and dramatic progress behind the scenes in America and the Western Pacific. Harrison cabled the priority order of Japanese targets – Hiroshima, then Kokura, then Nagasaki – with indicative dates for their destruction. The first uranium bomb would be finished between 1 and 3 August, and could be dropped between 4 and 10 August. Stimson asked when the plutonium bomb would be ready. Harrison's reply: 6 August, just fourteen days hence.[7]

Tuesday, 24 July began with a difficult decision, taken in the hope that it would become redundant. Truman, Churchill and their Chiefs of Staff signed off the invasion of Japan on 1 November 1945. When that afternoon's session ended, Truman 'strolled around the conference table' to where Stalin was standing. It had been agreed that this would be done 'as casually and briefly as possible'; Truman didn't even take his interpreter, relying on Stalin's to get the message across. Truman told Stalin that America had 'a new weapon of unusual destructive force'. Stalin showed 'no special interest', almost as though this wasn't news at all. He said he was 'glad to hear of it' and hoped the Americans 'would make good use of it' and then walked away, leaving Truman 'lost for words'.[8]

That evening, plans to use the 'new weapon' moved closer to the brink. Harrison gave Stimson the news personally by radio rather than telegram: Groves had ordered the first 'special bomb' to be dropped as soon as possible on or after Friday, 3 August, on one of four targets – Hiroshima, Kokura, Niigata or Nagasaki. Additional bombs would follow as they became available. Truman approved, and the battle orders were wired to the Pentagon.[9]

The next day, Thursday, 26 July, was pivotal. Churchill had flown back to London to receive his nation's verdict through the General Election. By lunchtime, the Conservatives were routed and Churchill was yesterday's man, complaining that 'they did not want me for peace'. In Potsdam, Clement Attlee picked up the reins of power and slipped into the British hot seat for the bomb – which Churchill had hidden from him.[10] The Potsdam Declaration was issued that day, demanding that Japan surrender immediately and unconditionally or face 'utter devastation'.[11] Nine days earlier, Trinity had been detonated. In nine days' time, if Groves stuck to his timetable and the Japanese hadn't surrendered, the centre of a major Japanese city would cease to exist.

Green Glass

Trinity only just made the newspapers. The *Albuquerque Chronicle* reported 'an explosive blast' near the Alamagordo Bombing Range, before dawn on 16 July 1945.[12] The brief item, attributed to Associated Press, excited no interest.

In reality, Trinity had far exceeded all expectations. This was barn-door obvious to everyone who had been there, and confirmation came swiftly in the tsunami of data from the hundreds of instruments monitoring the test. Trinity's power was variously estimated at 9,000–20,000 tons of TNT; the consensus of 10,000 tons was the same result that Fermi had calculated from watching shreds of paper fall to the ground. The blast which knocked George Kistiakowsky over at a range of 7 miles had initially roared outwards faster than the speed of sound and could have flattened a city centre. The fireball had burned hotter than the sun, blazing out radiant heat for several seconds and instantly igniting samples of timber, cladding and roofing left up to 3,000 yards away. The inferno of radioactivity which created the 'spectral' purple-blue glow around the mushroom cloud had wrecked some sensitive equipment; it hadn't been properly explored because thunderstorms near the air base had grounded the instrument-packed B-29 from which William Penney had intended to drop photographic film canisters through the cloud.[13]

Trinity's most striking legacy, and a shocking prophecy of the fate awaiting Japanese cities, was the 100-acre vitrified wasteland around Point Zero. The steel tower, buildings, roads and all human artefacts had been vaporised, leaving a 'Lake of Green Glass' half a mile across and consisting of sand and debris fused into a man-made mineral that was later called 'trinitite'.[14]

William Penney had finished extrapolating the carnage of the British Blitz into the uncharted territory of an atomic explosion. Hiroshima, encircled by hills that would cup and concentrate the blast, would be an ideal target with the detonation point hoisted to 1,800 feet above the ground. Penney presented his findings in the cinema on the Hill on 21 July 1945, five days after Trinity. He worked methodically through the ways in which a nuclear explosion would kill, maim and incapacitate. Depending on distance, the flash would vaporise, carbonise or ignite the human body; severe burns would bring an agonising death in hours or days. The blast could dismember people, smash them against walls, crush them in collapsing buildings and rip off their burned skin. Acute radiation sickness would cause multi-organ failure, fatal in days or weeks. All injuries would run their natural course because medical services would be vestigial or non-existent.[15] Finally, the Manhattan scientists confronted the enormity of what their inventions would do. Physicist Philip Morrison later recalled Penney's prediction that a Hiroshima-sized city of 300,000–400,000 people would be reduced to 'nothing but a sink for disaster relief, bandages, and hospitals. He made it absolutely clear in numbers,' Morrison said. 'We knew it, but we didn't see it.'[16]

Leo Szilard mounted a last-ditch attempt to stop the bomb being used against people, with a seventy-signature petition from Met Lab scientists. They were outvoted, as most of their colleagues favoured targeting Japanese cities with no warning. On 1 August, to ensure that no disquieting murmurs could reach the President, George Harrison locked all the correspondence in his safe.[17]

The 'biggest and most expensive experiment in history' had already turned into an instrument of war, when the US Army symbolically handed the Met Lab a receipt for Trinity's plutonium hemispheres. After witnessing the Trinity explosion, General Thomas Farrell told Groves, 'This will end the War.' Groves replied, 'Yes, after we drop two bombs on Japan.'[18] That had been five days earlier. Now, the first of those two bombs could be dropped in two weeks' time.

Destinations

Project Alberta, to produce and deliver atomic 'combat bombs', was led by US Navy Captain William Parsons, who had been in charge of ordnance at Los Alamos. Alberta grew up far away from Trinity on a coral island in the Western Pacific. Tinian (codenamed 'Destination') is one of the Northern Mariana Islands, 6,000 miles from San Francisco and a quarter of that distance south of Tokyo. Within two months of its capture from the Japanese in July 1944, Tinian had been bulldozed to create the world's largest airport, with six 2-mile runways serving four separate airfields and 270 B-29 long-range bombers. Flying in from Los Alamos to join the team assembling Fat Man, Philip Morrison was transfixed by seeing 'long rows of great silvery airplanes' lined up as if on 'a giant aircraft carrier'. Tinian was fiendishly busy, echoing to the thunder of B-29s taking off at one-minute intervals on round-the-clock sorties to bomb Japan or the Philippines.[19]

The North Field, several miles from the other airfields, served the 509th Composite Group – 'composite' because it included massive C-54 transport aircraft as well as B-29s. The group was born in December 1944 at Wendover Air Base, Utah, and was commanded by Colonel Paul Tibbets, twenty-nine years old, who had left medical school to become 'the best damned pilot in the Air Force'. Only Tibbets knew that its purpose was to fly atomic bombing missions. Remote and top secret, Wendover was a termite-infested 'shanty town with plumbing', and the obvious place for tube insertion should 'North America ever need an enema'. But it was perfect for conducting 150 test drops of dummy Little Boy and Fat Man bombs, which were billed as conventional bombs of exceptional power and known as the 'Things'.[20]

In May 1945, the group relocated to a heavily guarded compound near North Field on Tinian. This housed workshops, labs, a theatre, a chapel and an air-conditioned, restricted-access factory for assembling the bombs. Two loading pits contained heavy hydraulic lifts to hoist the bombs, each weighing four tons, up into the belly of a B-29 strike aircraft. A third pit was dug at Iwo Jima, three hours' flying time north and halfway to 'Empire' (Japan), in case a strike aircraft had to be replaced in transit. Personnel had poured in since the run-up to Trinity, reaching full strength – 80 bomb specialists and aircrew, with 1,700 support staff – in late July.[21]

From Tinian, the group rehearsed precision bombing from 30,000 feet – beyond the range of anti-aircraft guns – initially using 6,000-pound 'pumpkin' bombs which fell similarly to Fat Man. These missions, targeting active Japanese airfields, ironed out potentially fatal flaws in the bombs and the aircraft that would drop them. In-flight adjustments included strengthening the bombs' square tail-units, which were revealed by high-speed photography to buckle while falling.[22]

The group's B-29s were special 'Silverplate' Superfortresses with fuel-injected engines and massive bomb-bay doors which could open and close in an instant so that the aircraft could be thrown immediately into a steep diving turn to flee the scene before the shockwave of the atomic explosion caught up with it. Even the Silverplates needed improvement. The twelve-hour round trip to Japan was at the limit of the B-29's endurance, and several aircraft overloaded with fuel had already crashed on takeoff for less-taxing missions. To help the aircraft into the sky with its atomic cargo, all armour plating and most guns were removed, and Little Boy was disarmed for takeoff to avoid the risk of 'accidental detonation of an active gadget'. Only when the aircraft was safely in level flight would the 'weaponeer' load the propellant charge into the gun's breech via a channel cut through the outer casing. This 'sufficiently simple' operation demanded skill, concentration and half an hour crouched in a freezing bomb bay, fiddling with a weapon that would soon annihilate a city.[23]

When Trinity exploded on 16 July, most of the components for several Little Boys and Fat Men had already reached Tinian or were on their way by sea or air. Project Alberta picked up seamlessly from where Trinity left off. The plutonium hemispheres for the combat Fat Man were ready on 23 July, a week after Trinity, followed the next day by the U-235 target for Little Boy. The U-235 target and its gun detonator reached Tinian on 26 July on the cruiser *Indianapolis*, which dashed across the Pacific from San Francisco in six days. The U-235 projectile to be fired onto the target consisted of nine fat washers that stacked to form a hollow cylinder; these had already been delivered at vast expense by three C-54 transporters, each capable of carrying 10 tons but now relegated to just three U-235 washers (about 8 kg) and an armed Alberta courier. While the combat Little Boy was being assembled, a dummy bomb – complete except for its U-235 nuclear explosive – was dropped and exploded bang on target above its ground zero.[24]

Delta City

The gods of weather were against them for the next four days, 1–4 August. On Sunday, 5 August, the forecast for the following morning was favourable over the primary target, Hiroshima. A B-29 (call sign 'Dimples 82') taxied to the loading pit where the combat Little Boy was hoisted into the bomb bay and anchored to a single fastening, which the RAF had invented to carry its 6,000-pound 'Tallboy' bombs (Plate 26). 'Enola Gay' was freshly painted on the aircraft's nose; Paul Tibbets, the pilot, had named it after his mother the previous day.[25]

During the afternoon, the crews of *Enola Gay*, its two companion strike aircraft carrying instruments and cameras, and three weather observation planes, were all sent to get some sleep. That night, the weather scouts went first, with the 'out of sacks' wake-up call at 22.30, briefing at 23.00, lunch at 23.30 and an optional Catholic or Protestant blessing. The three aircraft were fully fuelled – 7,400 gallons – and given 1,000 rounds per machine gun. They took off at 01.15 on Monday, 6 August, heading north over Iwo Jima and then diverging to reach the primary target, Hiroshima, and its two alternatives – Kokura and Niigata – soon after 07.00.[26]

The three strike aircraft followed an hour later. It was obvious that something extraordinary was afoot: it was 'just like Hollywood' on the apron, with crowds, floodlights, cameras, flash bulbs and instructions to pose beside the aircraft. But even after the final briefing at midnight, only Tibbets and William Parsons, the Commander of Alberta and the weaponeer for this sortie, knew that the Little Boy hanging in *Enola Gay*'s bomb bay was the world's first nuclear weapon. Also hidden from everyone else on board were twelve cyanide capsules, one for each man, given to Tibbets to ensure that that nobody could fall into enemy hands, and the pistol that he would use on anyone who refused a capsule. At 02.15, *Enola Gay* lumbered into the hot night air, 400 gallons of fuel lighter than the other aircraft but still 7.5 tons overweight. Eight minutes later, in level flight at 9,000 feet, Parsons lowered himself into the bomb bay and took twenty-five minutes to manouevre the propellant charge into the breech of Little Boy's detonator.[27]

The three strike aircraft passed over Iwo Jima on schedule just after 06.00. The flight was uneventful – 'a milk run' – and some of the crew relaxed by bowling oranges down the fuselage at the head of the radar operator, who was lying down to snatch some sleep. The horseplay stopped dead before Japan came into view. Tibbets informed the crew that their Little Boy was an atomic bomb which would destroy a city and end the war, and Parsons told them about Trinity.[28]

At 07.15, a coded message was relayed from the leading weather scout plane – visual bombing conditions over Hiroshima were ideal – and *Enola Gay* turned north-west towards its target. At 07.30, the plane began the long climb to the bombing altitude of 32,000 feet. Thirty miles out, the Archies were checked electronically from the cockpit and Little Boy was disconnected from the aircraft's power, leaving it running on its internal batteries. At 08.09, Tibbetts began his bombing run. Looking down through the plexiglass nose of the aircraft, the bombardier had a perfect view of the sea dotted with islands, and then the port and city of Hiroshima. The bomb-doors opened at 08.12 as Tibbets started the three-minute countdown.[29]

Thirty seconds after the scheduled time of 08.15, the aiming point – the T-shaped arch of the Aioi Bridge – glided into the cross-hairs of the bomb-sight and the bombardier pressed the button that released Little Boy. Through the plexiglass, he glimpsed the bomb beginning its descent as *Enola Gay* leaped upwards: it seemed to hover for a moment, then 'porpoised' and fell away. Tibbets threw the plane into the well-rehearsed escape dive. Little Boy took forty-five seconds to drop nearly 30,000 feet and exploded 1,850 feet above the ground, a few hundred feet off target.[30]

By then, *Enola Gay* was 7 miles away but the flash was so bright that Tibbets and his co-pilot were temporarily blinded and the rear gunner thought he would never see again. The blast caught up with them a minute later – a 'double slap' and a roar 'like snapping metal' or 'a giant hitting the aircraft with a telephone pole'. Back in stable flight, Tibbets became aware of his co-pilot thumping his shoulder and shouting, 'Look at that! Look at that!' The crew scrambled to the windows.[31]

Two minutes earlier, Hiroshima had been laid out in clear view below them. Now 'it had gone', submerged under a pool of grey-black smoke, shot with flames and heaving 'like bubbling tar'. A huge purple cloud with a fiery red core erupted from its centre, 'a mile or two wide ... sucking up pieces of houses ... rising towards us at unimaginable speed'. Someone shoved a portable disc recorder into the rear gunner's hand, and it captured a moment of sheer terror: 'Here it comes, the mushroom shape that Captain Parsons spoke about ... It's coming this way.' The mushroom cloud, 'blindingly bright' and blazing out 'more colours than I imagined existed', shot up to their altitude – 5 miles high – and continued 'boiling up like something terribly alive'. Down on the ground, the lake of smoke had spread to the encircling hills and fires were racing up their slopes.[32]

The mushroom cloud finally topped 40,000 feet. By then, its colours had burned out, leaving a massive pure-white plume. At 10.00, an hour and forty-five minutes after detonation, Tibbets wrote in his log, 'Still in sight of the cloud'; it was finally lost to view at 10.41, when *Enola Gay* was over 360 miles from Hiroshima.[33]

On the way home, the crew sat 'in stunned silence' and toyed with sandwiches. *Enola Gay* touched down at North Field at 14.58, twelve and a half hours after take-off, and received a heroes' welcome: over a thousand cheering servicemen lining the taxiways and 'more generals and admirals' than one crewman had ever seen. Back on earth, Tibbets had the Distinguished Service Cross pinned to his front by the chief of the Strategic Air Forces. The strike aircraft crews were debriefed in an atmosphere of elation, with just one jarring moment when a young scientist informed them that 'I'm not proud right now'. Then the party began: a game of softball, a jitterbug competition and a screening of *It's a Pleasure* starring Norwegian ice-skater Sonja Henie, with unlimited food and four bottles of beer for each man.[34]

Ground Zero

When a photoreconnaissance B-29 flew over Hiroshima four hours after detonation, the city was still hidden under its blanket of smoke.

That blanket was the lid on Hell. Beneath it was a twilight world, with the sunshine blotted out and the air thick with dust and smoke. The Delta City, famous for its beautiful rivers and bridges, was now 'a scorched wasteland' of flattened wooden houses, many still burning, with just three concrete skeletons of buildings standing to mark the city centre (Plate 27).[35]

Hell was painted in grey, black and red. Grey for the smoke and dust which coated everything, including the 'ghost people ... not of this world' who haunted the wasteland: stupefied, mute or begging for water, a doctor or their mother, some too dazed to pull shards of glass out of their bodies, others sitting down to die. Black for the charred wood and skin, which filled the air with the 'peculiar odour of burned human flesh'. Red, the fires which kept erupting, the blood that bubbled out of mouths, and the raw tissues exposed where the blast had ripped away burned skin.[36]

One eyewitness caught sight of Little Boy about forty seconds after the bombardier on *Enola Gay* lost it from view. She ran outside on hearing the unmistakable sound of a B-29 and was surprised to see no aircraft; instead, a tiny black dot was falling out of the clear morning sky. The flash burned permanent scars on her retinae but, incredibly, she survived. Of all those who didn't, the luckiest were those who left their pale shadows on scorched walls, like photographic negatives, as they burst into flames and had their ashes scattered by the blast seconds later. Everyday artefacts of a busy city told their own stories. Clocks and watches had stopped forever just after 08.15, with their mechanism seized or their hands fused to their face. A child's tricycle looked as though someone had taken a blowtorch to it. So did the three-year-old lad who'd been playing on it; they were buried together that evening, a few hours after the little boy died in agony. A wooden sandal with a scorched fabric strap was stamped with the shape of a left foot, outlined in charcoal. The foot belonged to a thirteen-year-old schoolgirl who had been helping to knock down wooden houses to create firebreaks in case of air raids. Her mother never found her but could easily have walked straight past her: just one of the thousands burned beyond recognition – young or old? male or female? front or back? Another mother was luckier: one of the nameless, faceless horrors opened its mouth and she heard her daughter's voice.[37]

The smoke had cleared when a Professor of Philosophy at the university climbed to a beauty spot on a hilltop that commanded a panoramic view over the Delta City. What he saw never left him. 'Hiroshima had disappeared. It was so shocking that I simply can't express what I felt.' Little Boy had ignited a firestorm that obliterated 6 square miles of the city centre. Most of the city's 76,000 buildings were wooden; 70,000 of them were completely destroyed. At 8 a.m., when the radio had warned of three high-flying B-29s, Hiroshima had sheltered 280,000 civilians, 43,000 soldiers and 10,000 prisoners of war. Twenty minutes later, 90,000 of them were dead or doomed to die within the day.[38]

On seeing the devastation, a Protestant minister foresaw the end of Japan and of humanity; this was 'God's judgment on man'. God was also invoked by the co-pilot in *Enola Gay*, who scrawled in his log: 'My God what have we done?'[39]

27

AFTERMATHS
August 1945

Fifteen minutes after the detonation of Little Boy, William Parsons dictated one of the most momentous radio messages of the twentieth century. Sent by the twenty-year-old radio operator on *Enola Gay* to Tinian, it read simply 'RESULT SUCCESSFUL'.[1] After debriefing the crews and reviewing the first photoreconnaissance images, a fuller account was radioed to the Pentagon, where Groves immediately wrote a five-page report for Oppenheimer. Transmission was entrusted to two young women operating telex machines in the Pentagon and 'Deep Creek' (Los Alamos). The latter kept breaking down, causing frustrated exchanges:

>> THE REPAIR MAN SAYS THERE ISNT ANYTHING WRONG WITH IT
HES BEEN HERE ALL DAY ... THIS IS A AWFUL MESS ISNT IT

>> SURE IS ... WELL JUST HAVE TO KEEP TRING AS THESE MESSAGES
AR IMP

After several attempts, the crux of the message reached Los Alamos more or less intact:

AFTERMATHS

CLEAJ CUT RESULTS COMMA IN ALL RESPECTS SUCCESS FUL[2]

'Triumph and relief' immediately erupted on the Hill where, fourteen hours behind Japan Standard Time, it was the evening of Sunday, 5 August. Exultant people cracked open bottles, piled into cars and headed for dinner or drinking in Santa Fe, or partied through the night at the Oppenheimers' and other houses. Others saw little cause for celebration. They included the two architects of the memorandum which, five and a half years earlier, had predicted that a U-235 'super-bomb' could annihilate a city centre. Otto Frisch was gripped by 'unease, indeed nausea', while Rudi Peierls felt elation at knowing that 'our work' should help to end the war, together with 'horror at the death and suffering'.[3]

Groves worked through Sunday night on his report for George Stimson, and also rewrote the statement which Truman – currently homeward bound from Potsdam on the cruiser *Augusta* – had prepared to announce the birth of nuclear warfare to the world.[4] At 7.45 next morning, with torrential rain hammering his Long Island retreat, Stimson took a phone call from George Harrison at the Pentagon. Harrison was just the mouthpiece, reading out Groves's script with Groves looking over his shoulder. Smoke still hid the full extent of the devastation, but Little Boy had flattened the centre of Hiroshima. Stimson listened carefully, then told Harrison to activate the plans for broadcasting the news, and to inform the President.[5]

At 11 a.m. Eastern War Time on Monday, 6 August – before the President himself received the news on board the *Augusta* – American radio stations began transmitting Groves's version of Truman's statement. Sixteen hours earlier, an American aircraft had dropped a single bomb on Hiroshima, an important Japanese military base. This new 'atomic bomb' had harnessed the energy of the sun and exploded with the force of 20,000 tons of TNT, over 2,000 times more powerfully than the biggest conventional bomb. Massive damage had been inflicted; Pearl Harbor had been 'repaid many fold'. Japan had chosen retribution by rejecting the Potsdam ultimatum, the only option to 'spare the Japanese people utter destruction'. If the Japanese did not now surrender on America's terms, 'they may expect a rain of ruin from the air, the like of which has never been seen on this earth'. More atomic bombs were in production, and 'even more powerful forms' were being developed: 'We have spent $2 billion on the greatest scientific gamble in history – and won.' America was now prepared to obliterate 'every productive enterprise the Japanese have above ground', to remove 'the power to make war'.[6]

Hiroshima came as a shocking bolt from the blue to the world at large, which included everyone in Manhattan who wasn't at top table or on Tinian. In the calutron lab at Berkeley, the pre-war pacifist Maurice Wilkins felt 'a joyful sense of achievement' on hearing the news on the radio, until a downcast friend admitted that he'd always hoped the bomb would fail and Wilkins realised that he agreed.[7] The *New York Times* proclaimed: 'First atomic bomb is dropped on Japan', and that Truman was threatening Tokyo with 'a rain of ruin'. In Hanford, a special edition of *The Villager* community newspaper finally came clean about what they'd all been doing for the last two years. Forget Kleenex, Pepsi-Cola or election-campaign badges: 'IT'S ATOMIC BOMBS'.[8]

Take Two

On 26 July 1945, a C-54 transporter flew into Hawaii to refuel. To a pushy US Army colonel, desperate to repatriate his battle-weary troops, the gigantic aircraft seemed like a gift from God: completely empty except for a lowly second lieutenant and a wooden chair, strapped to the floor. The colonel tried pulling rank to commandeer the plane, but not for long: the second lieutenant politely ordered him out, then pointed a pistol at him when he refused.[9]

Since leaving Albuquerque that morning, the wooden chair had mostly been occupied by a young physicist from Los Alamos and a small wire case containing 'Rufus', the two plutonium hemispheres destined to destroy a Japanese city. Rufus reached Tinian without further incident on 28 July and joined most of the other components of Fat Man. Still missing were the 'X-units', the improved detonation circuits which had been outsourced to an electronics company that didn't understand 'MOST URGENT'. The definitive Fat Man mock-up was tested at Wendover on 4 August, and the first and only live firing of the Rufus-free bomb followed on 8 August during a bombing sortie from Tinian. This dress rehearsal was dangerously rushed because the real thing had suddenly been brought forward two days, to early the following morning. An approaching monsoon, Truman and Groves were all responsible for the change in plan.[10]

The day after Hiroshima, 7 August, Truman ordered the blanket-bombing of forty-seven major Japanese cities – not with explosives, but with 'invitations to surrender'. The propaganda offensive now had terrifying credibility and the threat of more Hiroshimas to come. Over nine days, B-29s from Tinian and nearby Saipan dropped 16.5 million leaflets carrying graphic aerial photographs of the wasteland that had been Hiroshima. The message – surrender or suffer 'the complete destruction of your family, home, economy and nation' – was reinforced by radio broadcasts from Saipan, transmitted every fifteen minutes, day and night.[11]

The wishes of the Japanese people were never made known, but the response of their government was the same as it had been to the Potsdam ultimatum – silence. Behind this 'cold refusal', the six-man Cabinet which advised Emperor Hirohito was split and deadlocked. The Prime Minister, Foreign Secretary and one other minister wanted peace, as long as the Emperor could continue to rule his country; the other three, all military chiefs, refused to accept the occupation and disarming of Japan, and the prospect of being tried for war atrocities.

Little Boy had outperformed all expectations but had failed to bring Japan to her knees. The Japanese government's continuing silence bemused and angered Truman – and was exactly what Groves wanted to hear so that he could end the war by dropping his second bomb on Japan.[12]

Hiroshima had been an impeccably choreographed milk run. Dropping Fat Man was a string of blunders that could easily have made the bomb explode 1,000 miles off target or not at all. The first disaster was narrowly averted shortly before midnight on 8 August, just two hours before the strike aircraft were scheduled to take off. Fat Man, fully assembled and autographed by its attendants, was hanging in the bomb bay of the B-29 which its pilot, Frederick Bock, had christened '*Bockscar*'. The final check of the detonator circuit revealed two unconnected plugs, both female; the blueprints were urgently consulted and showed that some idiot had soldered two male plugs together. These were separated and reinserted correctly, and everything seemed fine.[13]

At the pre-flight briefing, General Farrell informed Bock that he was relegated to piloting the instrumentation escort plane and that Charles Sweeney – one of Tibbets's favourites – would have the honour of flying *Bockscar* to drop the world's first plutonium bomb. As if in protest, the pump that delivered fuel from a tank in *Bockscar*'s tail refused to work – a mandatory 'no-fly' fault that Sweeney, who had never flown that aircraft before, couldn't resolve. The two weather scouts had already left at 01.00 for Kokura, the primary target, and its alternate, Nagasaki. *Bockscar* and its companion photographic and instrumentation B-29s should have taken off at 02.00 but remained on the tarmac waiting for a decision about the defective fuel pump. Finally, the crew of *Bockscar* was ordered back on board to prepare for departure, even though the fuel locked away in the tail could make the difference between a safe return and ditching in the ocean.[14] The photographic aircraft (*Big Stink*) had its own problems. Oppenheimer wanted high-quality photos of Fat Man exploding to intimidate the Japanese, so *Big Stink* carried a state-of-the-art camera so complicated that only a skilled operator could use it. In the scramble to board the aircraft, the camera operator had grabbed a life-raft pack instead of the obligatory parachute and was left behind because it was too late to find him one. *Big Stink* took off with only a snapshot camera that a physicist shoved into the hands of a crewman.[15]

The three strike aircraft eventually left Tinian at 03.47 and headed separately to the rendezvous point over a sparsely populated island south of the Japanese mainland. *Bockscar* was several tons over its safe operating weight and a sitting duck because all its guns had been removed to help it fly. The weaponeer Frederick Ashworth and his assistant clambered down beside Fat Man and removed its outer casing to replace the array of thirty-six green dummy detonators with their bright-red live counterparts. Having armed the bomb, they refitted the casing and Ashworth settled down to sleep. The nightmare began three hours later, when the panic-stricken assistant shook him awake. A red warning light had started blinking, indicating that Fat Man was readying itself for detonation. With the light flashing faster, they frantically unbolted the casing and spread out the blueprints – and found that two small switches had been set in the wrong position. Ashworth took a deep breath and flipped the switches – and the red light went out.[16]

The nightmare continued at the rendezvous point, where there was no sign of the photographic escort, *Big Stink*. All radio communication was barred except *in extremis*, so *Bockscar* and the instrumentation aircraft circled for forty-five agonising minutes, unaware that *Big Stink* had already arrived and was circling at the wrong altitude, 9,000 feet above them. When they broke away and headed for Kokura, *Bockscar* had already used too much fuel to return to Tinian and would have to land on American-occupied Okinawa instead. Meanwhile, in desperation, *Big Stink* had broken radio silence and sent a coded message to Tinian: 'IS BOCKSCAR DOWN'. This was rushed to General Farrell in the canteen; he immediately bolted outside and vomited up his breakfast.[17]

Bockscar reached Kokura at 10.44 and found the city hidden under cloud and dense black smoke billowing from a steel factory, where quick-thinking workers had begun burning drums of coal tar on hearing radio warnings of high-flying B-29s. After forty-five minutes and three failed bombing runs, and with Japanese fighters being scrambled below, Sweeney gave up and turned towards Nagasaki, 100 miles to the south-west. When *Bockscar* arrived overhead shortly before 11.00, there was only enough fuel for one bombing run and landmarks were obscured beneath patchy cloud cover. At the last moment, a clearing opened up magically over the aiming point. Fat Man exploded only twenty seconds after release, at 11.02, 1,840 feet above the city centre (Plate 28). *Bockscar* re-enacted *Enola Gay*'s escape from Hiroshima: flash, blindness, the blast that seemed to tear the plane apart, and the petrifying terror of 'the mushroom cloud ... coming towards us'.[18]

The spectre of disaster sat with them every minute of the flight home; luckily, so did their guardian angel. Okinawa was 950 miles away, within range if the fuel in *Bockscar*'s tail tank had been available. As it wasn't, the crew pulled on their Mae West lifejackets and Sweeney sent Okinawa a Mayday cry for help. Nobody replied. When the fuel ran out, *Bockscar* was within gliding distance of Okinawa but the runway was busy with aircraft which all ignored the stricken B-29 making an emergency landing. The 25-foot bounce on touchdown could easily have thrown *Bockscar* into a line of B-29s, fully fuelled and loaded with bombs and incendiaries, but instead dumped the plane back safely on the runway. They were greeted with incredulity, because the first word of *Big Stink*'s message from the rendezvous point had been lost in transmission. 'BOCKSCAR DOWN', and the presumed loss of Fat Man, had pitched everyone into despair and thrown up General Farrell's breakfast. As Okinawa hadn't trusted the Mayday message from beyond the grave, there was no heroes' welcome, just chaos. One crewman on *Bockscar* later described the Nagasaki mission as 'the greatest foul-up that ever was', adding 'but we made it home and that's what matters'.[19] An official verdict was similarly terse: 'The bomb functioned well in all respects.' Translation: the equivalent of 21,000 tons of TNT; 2 square miles devastated; 40,000 dead; 40,000 injured.[20]

Nagasaki could not have come at a worse time for the Japanese Cabinet but had little immediate impact. Russia had declared war on Japan the previous day, 8 August, and was invading Manchuria. Prime Minister Suzuki requested an audience with Emperor Hirohito that morning; both agreed that peace was the only option, on condition that the role of Emperor was preserved. However, the three military chiefs still refused surrender, and the news soon after 11 a.m. that a second city had been destroyed by a single atomic bomb merely hardened their defiance. At a Cabinet meeting that night in the imperial bomb shelter, the Emperor's opinion was respectfully requested, and he favoured surrender. Deadlock in the Cabinet continued, even when Truman's threats escalated into 1,000-bomber attacks that turned several Japanese cities into firestorms, and a shot-down

American airman was tortured into claiming that America had 100 atomic bombs and that Kyoto and Tokyo were next.[21]

The shipment of the plutonium core of the second combat Fat Man was cancelled on 14 August, after the Japanese Cabinet finally began making overtures for peace and the three hard-line military chiefs committed suicide. At midday on 15 August 1945, Emperor Hirohito announced the surrender of Japan to his people on the state radio. He'd never broadcast before, and many were awed and shocked to hear the 'jewel voice' of a deity brought down to earth and forced to talk defeat into a microphone.[22]

28

WINNERS AND LOSERS
September 1945–January 1946

For almost everyone on the Hill, Nagasaki was another visceral thunderbolt. Many Americans took their cue from Truman, who had 'slept soundly' after approving the strike on Hiroshima. Having experienced aerial bombardment at first hand, the British were more hesitant. Ernest Titterton's wife, Peggy, was pragmatic: 'I'm just glad Hitler didn't get it first.' Some saw the bomb as 'a good thing', cutting short a horrifically destructive conflict and – perhaps – shocking humanity into abandoning the folly of war. Rudi Peierls couldn't identify 'any moral reason for dropping a second bomb', and both he and Otto Frisch still wished that the weapons' power had been demonstrated without the mass murder of civilians.[1] But it was too late. They'd all been swept along in an accelerating frenzy of scientific excitement, obsessed with making trailblazing experiments work.

The British Mission's sojourn on the Hill had been like an exhilarating, extended sabbatical: 'the finest scientific club you could ever belong to ... Unforgettable and enjoyable ... A most marvellous time ... Not to be missed for anything.'[2] However, all good things come to an end. American friends and colleagues were returning to their home laboratories; off scene, politicians and the military were haggling over the future of Los Alamos and America's nuclear assets. Groves was already clearing the stage for American supremacy by ridding the Hill of all foreigners. Pending their physical removal, the British were barred from any contact with their American colleagues in all 'restricted' domains – anything relevant to bombs or reactors – even if the work had been their own.[3]

Meanwhile, Britain was preparing to welcome home the saviours who would coax fire out of the cold embers of British atomic physics. From Washington, James Chadwick had been working with John Cockcroft, Mark Oliphant and others in Britain to begin developing the country's stand-alone nuclear weapons and reactors. During September 1945, a steady stream of British scientists made the long train journey from Santa Fe to Washington, to be interviewed for newly created jobs in Birmingham, Liverpool, Oxford, Cambridge and the planned nuclear research establishment at Harwell in Oxfordshire.[4]

To say thank you and goodbye to their American hosts, the British Mission threw a party which became part of the legend of Los Alamos. Smart printed invitations were sent out to celebrate 'The Birth of the Atomic Era' in Fuller Lodge – the Ranch School's pine-clad dining hall – on Saturday, 22 September 1945, with 'dancing, entertainment preceded by supper at 8 p.m.'. Funded by $500 from the British Embassy in Washington and the mission's ration vouchers, food shops in Santa Fe were plundered and Klaus Fuchs's Buick was loaded with drink. The entire mission turned out on the night, except for the Chadwicks who stayed in Washington; Rudi Peierls flew back from a planning meeting in England just in time. About seventy Americans came, mostly scientists. Groves, Bush and Conant weren't invited.[5]

Fuller Lodge was packed. Guests arrived as formally attired as they could muster and were announced by a uniformed 'footman'. The sit-down dinner began with pea soup (prepared in metal buckets by Genia Peierls), continued with an English roast (carved by Rudi) and ended with Winifred Moon's English trifle, a new and mystifying experience for many Americans. Toasts in decent port to the King, the President and the Grand Anglo-American Alliance were followed by the entertainment.

Babes in the Wood was a string of skits that lampooned the oddities of life on the Hill, mimed to a commentary rich in dry English wit and double entendre. Star turns included the wicked witch

Security (James Tuck, playing to the gallery as a red devil with a flashing light on the end of his tail), and an accident-prone, whiskey-swigging Native American maid (Otto Frisch in drag). The Trinity test was re-enacted with flashing lights and off-stage bangs, detonated by a cigarette end tossed into a bucket on top of a wooden stepladder (Plate 29). Dancing and inebriation continued long into the night. A US Army colonel was heard to remark that 'after all, the Brits have got a sense of humour', but even after eighteen months of living and working together, rapprochement was incomplete. Many of the jokes in *Babes in the Wood* fell flat and, long afterwards, untouched portions of trifle were discovered hidden in the drawers of the dining tables.[6]

Then it was time to go. As commanded by Groves, the mission handed in all Manhattan paperwork and equipment; as instructed by Chadwick ('After all our efforts, we cannot afford to leave all our information behind and go home with empty hands'), they wrote a comprehensive final report, kept a firm grip on their personal lab records, and made copies of any technical documents to which they'd contributed. This material was enough to reconstruct 'an almost complete knowledge of the bomb'. The few voids included the preparation and handling of plutonium, always barred to non-Americans.[7]

Parting was made easier by the grimness of Groves's clampdown and the deteriorating living conditions as support staff left the site. Otto Frisch made his way back to Britain via California and Canada, sending ahead a 2-foot-high pile of Native American woven blankets. Rudi Peierls used his remaining holiday allowance to take Genia and Klaus Fuchs on a motor-trip into Mexico, where the highlights included a bullfight in Mexico City.[8]

Few of the mission stayed on for the Hill's red-letter day on 16 October 1945, when Groves presented Oppenheimer with an 'Army-Navy Excellence' award honouring American companies for outstanding wartime achievement. The Los Alamos laboratories were recognised for 'valuable services rendered to the Nation in work essential to the production of the Atomic Bomb, thereby contributing materially to the successful conclusion of World War II'. This was

Oppenheimer's last day as Director of Los Alamos, and his response to Groves may have jarred: 'Our pride must be tempered with concern ... The time will come when mankind will curse the names of Los Alamos and Hiroshima.'[9]

The harsh winter of 1945–6 was made worse on the Hill by burst pipes and white worms (bleached by the heavily chlorinated water brought in by tanker) that wriggled out of taps. In January 1946, the Peierls family set off for Halifax, Nova Scotia, and the voyage home with two children, thirty suitcases and crates crammed with skis, Native American knick-knacks and cooking utensils that would be in short supply in England. Fuchs was persuaded to stay on for another few months and eventually returned to Britain in June 1946. Ernest and Peggy Titterton were the last of the mission to leave the Hill, in April 1947.[10]

Also Ran

As the war ended, the Americans rushed to seize German scientific assets before the Russians or French could grab them. Their ace acquisition was Wernher von Braun, mastermind of the V-1 and V-2 flying bombs that had terrorised London, who was spared embarrassing questions about his wartime activities and spirited away across the Atlantic to push America ahead in the space race.[11]

By contrast, the nuclear physicists captured by Alsos's Operation Big were a busted flush. The Nazis had scrapped their atomic bomb programme three years earlier and they hadn't managed to build a working reactor. What to do with them? Groves's suggestion – shoot them – may or may not have been a joke. They couldn't be released in case the Russians snatched them, so Eric Welsh, Michael Perrin and R.V. Jones conceived Operation Epsilon: isolating them for several months in a comfortable setting which might lull them into revealing secrets that gentlemanly interrogation had failed to extract.[12]

The eight members of the *Uranverein* were led by Werner Heisenberg and Carl von Weizsäcker, both theoreticians. Walther Gerlach from Hamburg had been the *Uranverein*'s administrative chief since late 1944. Paul Harteck and Karl Wirtz were experts in heavy water and reactor design. Kurt Diebner had headed atomic research in the army and invented the chandelier reactor in Haigerloch. Erich Bagge and Horst Korsching had worked on U-235 separation. Also detained by Alsos in the Black Forest were two innocent bystanders, both grand old men of the pre-war establishment. Otto Hahn had refused any military involvement in fission research, while Max von Laue, the father of X-ray crystallography, was bemused by 'the completely undeserved honour of being counted a nuclear physicist'. Both had been anti-Nazi, von Laue almost suicidally so by sending his gold Nobel medal to Bohr in Copenhagen.[13]

The detainees had a minder from MI6. Major Thomas Rittner was English born but his ancestry, university education and wife were all German. Calm and effortlessly bilingual, Rittner could talk man to man with his 'guests' – while relaying all information about them to Welsh, Perrin and Groves.[14] He managed to keep his charges 'in a good frame of mind' while they spent two months being shunted, incognito and incommunicado, around northern France and Belgium. They resented being treated as 'war criminals' rather than world-leading scientists, and became increasingly worried about their families in occupied Germany with whom they were denied contact. Rittner obtained permission for them to write to their loved ones, censoring all references that could reveal their whereabouts. He also sent brief character sketches to inform Groves and Welsh. Heisenberg – very friendly, genuinely anxious to cooperate. Von Weizsäcker – 'a diplomat ... but a good German'. Hahn – a man of the world, most helpful. Bagge – 'completely German', unlikely to cooperate. Diebner – unpleasant, untrustworthy.[15]

On 3 July 1945, the detainees were flown to England and confined in Farm Hall, a three-storey Georgian townhouse in Godmanchester, 15 miles north-west of Cambridge. Farm Hall had housed MI6 agents en route to occupied France and comprised several bedrooms, two sitting rooms, library, dining room, games room with a billiards table, and extensive walled gardens.[16] The guests were duly impressed by the grand English mansion with its oak panelling and oil paintings, the five German prisoners of war detailed to look after them, and good food. Their daily routine included English breakfast with English and American newspapers, two hours for personal study, an afternoon of games, reading and conversation, dinner, the nine o'clock news on the BBC, and listening to radio concerts or Heisenberg playing the piano. They continued the weekly scientific seminars which they'd begun in France 'to keep us mentally alert'. But they remained cut off from their families and colleagues, who didn't even know which country they were in.[17]

Soon after arriving, Diebner asked Heisenberg if he thought microphones had been installed. Heisenberg laughed: their captors weren't that smart. Wrong. A specialist team from Branch 19 of Military Intelligence had bugged the sitting and dining rooms and all bedrooms. The microphones fed into a secret listening room, where eight bilingual observers worked shifts to monitor the detainees' conversations. When something potentially important was heard, notes were taken and a recording machine switched on. German-language transcripts were reviewed by Rittner, who provided English translations of relevant extracts for weekly reports to Groves, Welsh and Perrin.[18]

The hidden microphones soon revealed the *Uranverein* to be hopelessly deluded and riven by jealousy and suspicion. They believed that they led the world in fission research; Bagge was convinced that the Allies had tried to 'imitate our experiments' and expected Russia or Argentina to snap him up after the war. Personal agendas and animosities were revealed. Bagge and Korsching disliked Heisenberg ('he was never the one for this job'), Hahn ('dictatorial') and von Laue. Gerlach wasn't trusted because his brother had connections with the SS. Rittner observed that 'the guests were at great pains to clear themselves of any connection with the Nazis' – especially Diebner and Bagge, who tried, unconvincingly, to argue away their Nazi Party membership.[19]

High drama came on the evening of 6 August, which the guests later christened 'Valhalla Day'. Just before dinner, Rittner went to Hahn's room with the news that the Allies had destroyed Hiroshima with an atomic bomb. Hahn was 'completely shattered' and told Rittner that he'd contemplated suicide if his discovery were ever turned into a weapon. Rittner calmed him with a drink, took him downstairs and left him to announce the news 'to the assembled guests'.[20]

They reacted with shock and incredulity. Could the bomb have contained uranium-235 or element 93, or was it just a trick to force the Japanese to surrender? Heisenberg: 'I don't believe a word of it ... Some dilettante in America has bluffed them.' Hahn: 'They are 50 years ahead of us ... If the Americans have a uranium bomb, then you're all second-raters ... Poor old Heisenberg.'[21] At 9 p.m., they clustered around the radio and were 'completely stunned' to hear the ghastly truth confirmed by the BBC news.[22] Von Weizsäcker: 'We were all convinced that the thing could not be completed during the War,' but they might have got somewhere if they'd started earlier. Korsching turned on Gerlach: 'real cooperation' had been impossible in Germany and 'colleagues on the bomb programme' had lacked courage. Gerlach exploded in fury and stormed out. Hahn demanded to know how much uranium-235 was needed for a bomb, because Heisenberg had mentioned weights between 30 kilos and 2 tons. Heisenberg confessed, 'I never worked it out as I never believed we could get enough pure 235'; he didn't know 'whether the bomb was made with 50, 500 or 5,000 kilos'.[23]

Hahn, obviously upset, went to his room, leaving the others playing cards until 1.30 a.m. During the night, the microphones picked up 'deep sighs and occasional shouts ... and considerable coming and going along the corridors'. Gerlach was suffering 'a real nervous breakdown' and threatening to kill himself 'like a defeated general', while Hahn, racked with guilt, cried alone in his room. Both were eventually calmed by the others, and von Laue stayed awake until the 'father of fission' was safely asleep.[24] Nagasaki came and went on 9 August without specific mention in the transcripts; perhaps the internees said nothing new or interesting. Incongruously, they 'listened intently to the King's broadcast on VJ Day and all stood rigidly to attention during the playing of the National Anthem'.[25]

The next days made it clear that the *Uranverein* could never have made an atomic bomb. After pressure from the others, Heisenberg gave them a seminar on the physics of the uranium bomb on 14 August. He correctly envisaged firing together two small masses of uranium and using a 'neutron reflector' (i.e. a tamper) to amplify the chain reaction, but, unlike the Frisch–Peierls memorandum, made no estimate of how much uranium was needed. Expert analysis of the twelve-page transcript of Heisenberg's talk delivers a harsh judgement. Heisenberg frequently missed 'essential points' and couldn't see beyond fission in his uranium reactor; his colleagues, supposedly Germany's nuclear elite, were 'even worse', talking 'desperation physics' with striking 'intellectual thinness'. In brief, the *Uranverein* never had any idea how to make a uranium bomb.[26]

They also believed that it was impossible to obtain enough plutonium for a bomb. When newspapers mentioned a new fissionable element discovered in 1941 and codenamed 'Pluto', they didn't think that this could refer to element 94, and instead wondered about protactinium because it also began with the letter P. They were thrown clean off the scent by a press report (planted by MI5) explaining that the new element now named 'plutonium' couldn't be used in an atomic bomb because it would take 70,000 years to amass enough of it. This strengthened their belief that the Americans had failed to build a uranium reactor.[27]

With no apparent end to their internment, the guests became angry and apathetic. The weekly colloquia petered out and Heisenberg threatened to break the gentlemen's agreement not to abscond. October brought partially good news. They would be returned to American- or British-occupied Germany, but not imminently; and Professor Otto Hahn had been elected in his absence to lead the new Kaiser Wilhelm Institute for Chemistry. Spirits were lifted by visits from British scientists, notably Patrick Blackett, who stayed for dinner and declared their incarceration 'just silly'. In early October, Heisenberg, Hahn and von Laue were driven to the Royal Institution in London to meet its President, Sir Henry Dale, and several Fellows of the Royal Society, all sworn to secrecy. After tea and a two-hour discussion, the three were returned to Farm Hall.[28]

Friday, 16 November was a momentous day for Hahn, who discovered from reading the *Daily Telegraph* that he'd been awarded the deferred 1944 Nobel Prize in Chemistry for discovering nuclear fission. The Royal Swedish Academy hadn't been able to notify him in advance because nobody knew where he was. This was a solo award, with no recognition for Lise Meitner or even Fritz Strassmann, who had alerted him to the crucial clue of the barium that appeared from nowhere when uranium was blasted with neutrons.[29] The internees celebrated with champagne and dinner; Rittner was invited and spent an 'awkward' evening sandwiched between Hahn and von Laue, both in tears. Afterwards, Diebner and Wirtz performed their 'Nobel Prize Song', which ran to fourteen stanzas with the refrain 'Otto Hahn's the culprit's name'.

Even though the Farm Hall recordings confirmed that Hahn had refused all military work, the British authorities barred him from attending the prize ceremony in Stockholm on 10 December 1945. And they refused to disclose that he was being held incommunicado in England.[30]

On 3 January 1946, the guests' internment in Farm Hall ended after exactly six months, the maximum period in Britain for detention without charge. They returned to Germany as heroic nuclear pioneers with clear consciences because, in contrast to the Americans and Anglo-Saxons, they had never set out to make an atomic bomb.

The eight members of the *Uranverein* had undergone a remarkable metamorphosis at Farm Hall. On the morning after Valhalla, von Laue witnessed the birth of '*eine abwechselnde Lesart*' ('an alternative version') of how they had spent the war. It began when von Weizsäcker remarked, 'History will record that the Americans and English made a bomb, a ghastly weapon of war,' whereas 'Germany under the Hitler regime had produced the peaceful development of the uranium engine.' That evening, von Weizsäcker developed the theme: Germany hadn't made a bomb because 'all the physicists didn't want to do it ... If we had had all wanted Germany to win the War, we would have succeeded.' Bagge objected – 'It is absurd for us to say we didn't want the thing to succeed' – but von Weizsäcker's 'alternative version' steadily gained traction.[31]

In September, the internees wrote a memorandum to counter newspaper rumours about the so-called German atomic bomb. After a preliminary feasibility study, they claimed, they'd abandoned all work on a weapon. The *Uranverein* hadn't made a bomb – not because they couldn't, but because they were men of virtue who made a principled decision to turn their backs on nuclear weapons and dedicated themselves to the 'peaceful' reactor. It was solely thanks to them that Hitler never had the atomic bomb.[32]

Although this sounded plausible, the microphones had picked up evidence to the contrary – including the secret patent for an atomic bomb which the KWIs had filed in 1941. But with luck, and assuming that the walls of Farm Hall didn't have ears, nothing could undermine the nobility of their stance.

Epitaph for a Bomb That Never Was

The Nazis' atomic bomb never stood a chance. It died of neglect soon after conception and long before any hope of a live birth. Its death was due partly to starvation by its parasitic twin, the *Uranbrenner* reactor, but mainly because nobody in Germany cared enough about it.

The Reich's top brass saw no need for the bomb and refused to invest in it. Compare with the V-1 rocket: also hugely expensive but absolutely credible, because Wernher von Braun had presented the military chiefs with blueprints, working prototypes and grandstand seats at test-firings on Peenemünde. By contrast, Heisenberg's U-235 and plutonium bombs existed only as equations – a pineapple-sized piece of shallow thinking that hadn't even crystallised into a rough sketch indicating the weapon's size and how it would be detonated.[33]

Rudi Peierls had fretted that the notion of a pure U-235 bomb could strike 'any competent nuclear physicist'. He needn't have worried about Heisenberg, whom he knew personally as 'a brilliant theoretician but very casual about numbers'.[34] Heisenberg never attempted to calculate how much U-235 was needed in a bomb and the seminar he gave to his peers at Farm Hall eight days after Hiroshima demonstrated conclusively how little he understood about atomic bomb physics. He was also the weak link in the chain of command: selfish, divisive and the antithesis of the leader that the *Uranverein* needed.[35]

Back in 1940, Sir John Anderson had dismissed the Nazi bomb with 'I think we can all sleep soundly in our beds'.[36] In truth, the threat never advanced beyond that stage.

29

WHODUNNIT

Both the atomic bombs dropped on Japan in August 1945 were created by a vast, top-secret collaboration that brought together the best scientific brains, industrial resources and big-project expertise in America and Britain. The crucial importance of that collaboration was underlined by President Truman in his address to the nation on 6 August 1945, hours after the bombing of Hiroshima:

> Beginning in 1940, before Pearl Harbor, scientific knowledge useful in war was pooled between the United States and Great Britain ... Under that general policy, the research on the atomic bomb was begun. With American and British scientists working together we entered the race of discovery against the Germans.[1]

Yet just six days later, the British found themselves written out of the official US government history of the Manhattan Project. On 12 August, the report by Professor Henry Smyth of Princeton, commissioned sixteen months earlier by Groves, was splashed across American newspapers and radio.[2] Smyth described in detail the contributions of the Americans credited with making the bomb, but dismissed the British scientists in a few lines of faint praise for their 'valuable advice'. He didn't specify what they'd done or explain its significance. Frisch's landmark dragon experiment was anonymised as 'a method which permitted calculation of the critical size', while the revolutionary nickel diffusion membranes which enabled gaseous diffusion to concentrate U-235 were 'the ideas of several men' in American companies rather than a gift from their British inventors.[3] In the gospel according to Smyth, both bombs were exclusively American inventions, and the British merely camp-followers.

Unexpectedly, the Smyth Report was a bestseller and eventually sold over 150,000 copies, including 30,000 of a bootleg Russian version. After British lobbying, a twenty-page appendix entitled *Statements Relating to the Atomic Bomb* was added to a revised edition published in October 1945.[4] *Statements* was introduced by Churchill, who concluded that 'by God's mercy, British and American scientists outpaced all German efforts'.[5] Then, on behalf of the Tube Alloys Directorate, Michael Perrin summarised 'the work in the United Kingdom and the share taken by British scientists in the Project'. In places, Perrin's analysis read as though it had been thrown together in haste, as indeed it had: he wrote it non-stop through the night of 12 August.[6] Chadwick was 'most disappointed' and later complained that Perrin had 'failed, almost completely, to emphasise the value of the British contribution'.[7] Tacked on the end of 260 pages of American self-praise, Perrin's *Statements* presented little challenge to Smyth's claim that 'no other country in the world had been capable of such an outlay in brains and technical effort'.[8]

Having excised the British from the narrative before the war ended, the all-American version of the Manhattan story steadily tightened its grip on public perception. In *Scientists Against Time* (1947), 'a brief official history' of how American scientists won the war, John Baxter praised two British inventions – the cavity magnetron and the Rolls-Royce Merlin engine – but identified the 'Triumph of the Manhattan District' as purely American.[9] Three hefty American histories reached the same conclusion. The anonymous author of *Manhattan Project: Official History and Documents* (1947) stated that British input was 'in no sense vital and actually not even important. To evaluate it quantitatively at 1% of the total would be to overestimate it.'[10] In 1961, David Hawkins's 550-page *Project Y: The Los Alamos Story* dismissed the British Mission in three pages; Frisch's dragon experiment was accorded 'historic importance' but without naming him.[11] *The New World 1939/1946* (1962) by Richard Hewlett and Oscar Anderson mentioned the Frisch–Peierls memorandum and M.A.U.D. and credited Frisch and Peierls for the dragon and innovations in U-235 separation, but didn't acknowledge their vital significance to Little Boy.[12]

In *Now It Can Be Told* (1962), General Leslie R. Groves promised 'first-hand knowledge of the facts and first-hand opinions' rather than 'historians' conjectures'.[13] He explained that the atomic bomb was the pure-bred offspring of America's scientific, engineering and military brilliance. The British had begun 'first work on U-235 separation in the UK' but fell behind and sent 'a strong delegation' to America to 'learn of our plans'. Groves acknowledged that Peierls, Skyrne [sic] and Fuchs had been valuable 'in solving some of our theoretical problems', but it was he, not Chadwick, who 'assigned some 28 British scientists to the work that was under my control'. His conclusion: 'the contribution of the British was helpful but not vital. Their work at Los Alamos was of high quality but their numbers were too small to enable them to play a major role.'[14]

And so to *American Prometheus: The Triumph and Tragedy of J. Robert Oppenheimer* (2005) by Kai Bird and Martin Sherwin. The atomic bombs which turned Oppenheimer into an icon couldn't have been made without British help, yet the British Mission barely featured; Chadwick appeared only once at Los Alamos, hosting the dinner party where Groves bragged that the bomb would be used against the Russians.[15]

Cinema and television have also retold the story of the birth of the bomb. *The Beginning or the End* (1947), a 'hastily shot' feature film celebrating 'America's greatest men', was criticised in *Nature* for 'leading the non-scientist into believing that most atomic energy research has been American'.[16] Chadwick complained that 'it leaves the British contribution in the shadow', while *Atomic Scientists' News* (London) dismissed it as 'sentimental glorification of American science and Armed Forces'. The 'Americanisation of the making of the atomic bomb' was particularly blatant in *Fat Man and Little Boy* (1989), which completely omitted the British from Los Alamos;

Frisch's dragon experiment was conducted by the film's central character, a fictitious American physicist.[17] And the blockbuster *Oppenheimer* (2023) recycled the same biased narrative as *American Prometheus*, on which it is based.[18] The British Mission was represented only by Klaus Fuchs, included for his perfidiousness and not his scientific prowess.

The campaign to expunge the British from the history of the atomic bomb was conceived, orchestrated and executed by one man: Groves. In a confidential memorandum in 1947, he stated: 'I cannot recall any direct British contribution to our success in achieving the bomb.'[19] He ensured that the Smyth Report could only portray Manhattan as a purely American triumph, by granting Smyth access to 'every element' of Manhattan[20] – but not the British Mission. Smyth's report therefore took the British by surprise. Chadwick knew nothing about it until he was summoned to Washington in early August 1945 to approve the final draft, to be published soon after the dropping of Fat Man.[21]

Why did Groves exclude the British? First and foremost, he was an unquestioning American patriot, programmed to assume that 'no other country' could have made the bombs and that the story was so big it could only be American. He was an Anglophobe who often 'made plain his dislike of the British' and 'British rascality'. After the war, Groves confessed to 'doing everything to hold back' cooperation with Britain. He admired the 'masterful way' in which Vannevar Bush 'conducted the ticklish discussions with the British', and agreed 'completely' with him that 'a complete interchange of atomic information was a mistake [and] a pure waste of time'; he felt 'no pangs of conscience' about shutting out the British in July 1943 and then barring them from Smyth's report. Groves was also an egotist who claimed personal credit 'for all past events' (such as building the Pentagon) and resented having to share glory or limelight. Finally, he was driven by self-preservation. Having siphoned off $2 billion from Congress into the project 'for which I became responsible', his reputation and career hung on demonstrating 'the success of the use of the bombs on Japan'[22] – and their all-American origins.

Giving All

Millions of people around the world have been conditioned to believe that the atomic bombs dropped on Japan were entirely American inventions, but we should revisit the question which prompted me to write this book. What, if anything, did British scientists actually contribute to the creation of the world's first nuclear weapons?

Groves wrote off the British as 'helpful but not vital' because they were too few to play a major role in the Manhattan Project.[23] They were certainly thinly scattered – twenty-four at Los Alamos and another thirty-five at Berkeley and Oak Ridge (Appendix II). However, their impact has been judged 'out of all proportion' to their numbers,[24] and the evidence is incontrovertible that several were crucial to deploying the uranium and the plutonium bombs in August 1945.

James Chadwick was the essential glue which bonded together the British Mission and Manhattan and enabled the British to 'give the Americans all help'. He identified critical gaps across the Manhattan sites – experimental physicists at Berkeley, industrial chemists at Oak Ridge, experts in bomb damage at Los Alamos – and plugged these with top-rank scientists from Britain, extracted at great cost to their home institutions. Everything hinged on his 'extraordinarily good' working relationship with the 'exceedingly tough' Groves, who respected Chadwick's 'frankness and essential fair-mindedness'. This demanded the 'highest qualities of diplomacy' from Chadwick, a shy man whose comfort zone ended at the door out of his laboratory.[25] Without Chadwick, the Manhattan Project could not have fully exploited the contributions of the British Mission.

Mark Oliphant was the leading British light at Ernest Lawrence's calutron lab in Berkeley. In late 1943, he answered Lawrence's plea for a 'joint intensive attack' to perfect the calutron, by bringing in his entire research group from Birmingham. Oliphant did 'magnificent work' as Lawrence's right-hand man and deputy, while Harrie Massey, a brilliant physicist who had been working on mines for the British Admiralty, headed the Theoretical Group at Berkeley. The British contingent included experimental physicists, chemists and J.P. Baxter,

Research Manager at ICI, who had galvanised the ICI–Tube Alloys collaboration in 1940 by giving Chadwick hex and uranium. Oliphant was awed by Oak Ridge – 'it surpasses all imagination' – yet his team played 'a real part in removing the difficulties' in recovering U-235 from the calutrons.[26]

When Chadwick first heard of Los Alamos in September 1943, 300 American scientists were already on the Hill but Oppenheimer was in despair because they were so far behind schedule. The twenty-four resident members of the British Mission – 0.7 per cent of the total workforce at Los Alamos – were disproportionately effective. They eventually included six group leaders for both Little Boy and Fat Man, and neither weapon could have been battle-ready by August 1945 without the work of specific members of the mission, notably Otto Frisch and Rudi Peierls.[27]

Frisch's seminal contribution was the dragon experiment, which proved conclusively that the gun detonation method would work with U-235 and also determined the critical mass of U-235 for Little Boy, thus setting the timetable for completing the bomb.[28] Rudi Peierls and his close collaborators, especially Klaus Fuchs, were essential in bringing both bombs to fruition. Without their input, the K-25 gaseous diffusion plants at Oak Ridge, which did the heavy lifting in enriching U-235, might well not have delivered Little Boy's nuclear explosive in time. In late autumn 1943, Peierls went to New York as 'consultant' to the Kellex company, which was building the separation plant. Groves had recently frozen the plant's design, ignoring the fact that it wasn't working and the 'gloomy prognostications' that it never could. Peierls brought experience from Francis Simon's plants ('the pride and joy of Tube Alloys'), and his formidable mathematical firepower was soon enhanced by the arrival of Fuchs and Tony Skyrme, another gifted theoretical physicist from Birmingham. The three identified the pinch-points which made the flow of hex through the plant unstable and unpredictable.[29] Also transformational was the sintered nickel diffusion membrane produced in South Wales, which finally enabled an American prototype separation unit to enrich U-235. The 3,000 tons of sintered nickel that made K-25 functional were supplied free by the British government.[30]

Peierls and Fuchs were later transferred to Los Alamos, to model the implosion that would crush the plutonium ball at the heart of Fat Man. Their work interfaced with the implosion experiments using shaped composite charges conducted by James Tuck and George Kistiakowsky. Peierls said modestly that he arrived 'at the right time' to crack the problem, while Hans Bethe viewed Fuchs as 'one of the most valuable men' in the Theoretical Division for modelling blast waves. Their work was essential in determining the number and geometry of the thirty-two detonation points on the high-explosive sphere surrounding Fat Man's plutonium core. With Robert Christy, Peierls was co-inventor of the implosion gadget for Fat Man. Without this, the plutonium bomb could not have been ready to use in August 1945.[31]

Fat Man owed its existence to two other members of the British Mission. Seth Neddermeyer first suggested that implosion could create a supercritical mass of plutonium, but two technological breakthroughs were needed to create the spherical implosion shockwave modelled by Peierls and Fuchs: each explosion had to be focused precisely onto the target plutonium ball; and all thirty-two charges had to be detonated within a microsecond. Focusing was achieved with 'composite lenses' of fast- and slow-detonating explosives, suggested by James Tuck from his extensive experience with armour-piercing shells and bombs. Tuck also helped to design the urchin initiator; according to American physicist Bruno Rossi, neither Trinity nor Fat Man could have been detonated on schedule without Tuck's input.[32]

The microsecond-precise choreography of detonation was the brainchild of Ernest Titterton, a world expert in 'fast timing' whose ultra-rapid signal trains had enabled the cavity magnetron to outperform all competitors. At Los Alamos, Titterton's skills were exploited for 'flash' X-ray imaging of imploding metal targets, and ultimately in the 6,000-volt signal that fired all thirty-two of Trinity's explosive lenses simultaneously. On 16 July 1945, Titterton took his place in world history as the first person to detonate an atomic bomb.[33]

This evidence indicates that the British contributions were crucially important in their own right and so tightly interwoven with the work of their American colleagues that without them, neither the U-235 nor the plutonium bomb could have been ready to detonate in August 1945. One of the British Mission interviewed by Margaret Gowing, the official historian of the British nuclear programme, believed that the British had 'hastened by two or three months the time when the first bombs were ready'. An anonymous American – 'one of the foremost in the Manhattan Project' – was more generous, opting for 'at least a year'.[34] Even three months could have made the difference between life and death for the hundreds of thousands of troops destined to invade Japan in early November 1945, just thirteen weeks after Nagasaki – a campaign that budgeted for 250,000 American deaths.

Ferenc Szasz, Professor of History at Albuquerque, also recognised the British as crucial to Manhattan. In 1992, he wrote: 'A historian may single out three factors that were absolutely critical to its success: the leadership of Oppenheimer, the supervision of Groves, and the contributions of the British to Los Alamos.'[35]

Like a three-legged stool, Manhattan would have fallen over if the British had been pulled away. This begs an even more fundamental question. Without the British, would the Manhattan Project ever have been born?

Sleeping Giant

The Nazis' atomic bomb died in summer 1942, before it could begin embryonic development. The American atomic bomb had been snuffed out almost three years earlier, even before conception.

It was killed during the first meeting of the Advisory Committee on Uranium on 21 October 1939, ten days after Roosevelt demanded 'action' on hearing Alexander Sachs's warning that 'the Nazis are going to blow us all up'. The committee set the dangerous precedent for entrusting existential decisions about atomic weapons to people who knew nothing about nuclear physics. None of its members – chairman and ex-soil scientist Lyman Briggs, and two military armaments experts – had any understanding of what was on the table. They supported Fermi's electricity-generating nuclear pile but advised Roosevelt against funding the bomb because they thought it too uncertain, expensive and slow to deploy.

Successive iterations of Briggs's Uranium Committee continued to veto the bomb, even when heavyweight scientists were brought in. Their negativity reflected the general conviction of American nuclear physicists that pursuing an atomic bomb was 'a sheer waste of time'. Much effort was invested in concentrating U-235, but at that stage only to produce more efficient fuel for Fermi's pile.

In October 1940, to kick-start the Anglo-American exchange of scientific intelligence after the Tizard Mission, the Frisch–Peierls memorandum and all the M.A.U.D. Committee's documents about the U-235 super-bomb were sent to Briggs and to Vannevar Bush, America's scientific supremo. In Britain, the memorandum's importance had immediately been recognised, triggering a chain reaction through Mark Oliphant, George Thomson and James Chadwick which culminated in the formation of the M.A.U.D. Committee. By contrast, America showed no interest. Why not? Because all this information fell on deaf ears and closed minds.

Like Briggs, Bush was hopelessly out of his depth. He confessed that 'I am no atomic scientist ... most of this is over my head' – but instead of standing aside for someone competent, he clung to power and threw his weight around. When Ernest Lawrence advised James Conant to 'light a fire under the Briggs Committee', Bush (whose 'ego was commensurate with his responsibilities') took this as a challenge to his authority and 'flatly' told Lawrence to fall in line or quit the committee.[36] Bush's only commendable action, in April 1941, was to commission the National Academy of Science to advise him on the viability of the atomic bomb. The first NAS report confirmed Bush's prejudices: the bomb was too speculative and 'not likely to be of decisive importance in the current war'.

The next decision point should have occurred in July 1941, when the final M.A.U.D. report and Rudi Peierls's detailed technical memoranda on bomb design and U-235 separation were sent to Briggs, Bush and Conant. The data had been sifted by five British Nobel Prize winners who concluded that the U-235 bomb would work and must be built, but the Americans didn't respond – even after 'the

British had been trying all winter and spring to pass the word'.[37] Bush requested a second NAS review, which again advised against developing an American atomic bomb. America's top scientists, including Nobel Prize winners Ernest Lawrence and Arthur Compton, were still mired in the assumption that the weapon was far beyond the reach of current technology.

Astonishingly, that stance was reversed in just three months. In late October 1941, Bush directed the NAS to produce a third report on the atomic bomb. This time, the NAS performed a complete U-turn and concluded that a British U-235 super-bomb of 'superlative destructive power' was 'entirely plausible' and could determine military superiority within a few years. This view was endorsed by George Pegram and Harold Urey when they returned from their fact-finding tour of British nuclear research. Suddenly, Bush became the bearer of urgent news to the White House. On 6 December 1941 – the night before Pearl Harbor – Roosevelt ratified full Anglo-American cooperation to build the U-235 bomb. The Manhattan Project was established in June 1942, and Groves, Oppenheimer, Los Alamos, Little Boy and Fat Man duly followed.

The bomb's dramatic change in fortune didn't happen because an American savant finally saw the light. It was triggered by Mark Oliphant blowing his top on discovering that the nuclear physicists on the Uranium Committee hadn't seen the decisive M.A.U.D. documents because the uncomprehending Briggs had locked them unread in his safe. Oliphant's furious protest swept up Ernest Lawrence, then Arthur Compton and finally Bush and Conant – into whose hands George Thomson personally pressed copies of the M.A.U.D. final report. This was a wise precaution, as both those men remained unconvinced by the bomb. As late as January 1943, Wallace Akers wrote scathingly: 'Bush knows nothing whatever about the scientific side of this project [the bomb] ... Unfortunately, Conant also knows and understands practically nothing about the science ... Neither will delegate any authority and neither has time or knowledge to deal with problems.'[38]

In *The Making of the Atomic Bomb* (1986), Richard Rhodes credited Oliphant with helping to 'goad the American program over the top' by confronting Briggs and igniting the chain of disturbance which, hours before America entered the war, persuaded Roosevelt to cooperate with Tube Alloys. Rhodes quoted Leo Szilard: 'If Congress knew the true history of the atomic energy project, I have no doubt that it would create a special medal to be given to meddling foreigners for distinguished services ... Dr Oliphant would be the first to receive one.'[39]

Before Oliphant, there were no prepared minds in America, only the institutional assumptions that the USA could win any imminent war with its conventional might, and that the atomic bomb was neither feasible nor affordable. Left alone, the Americans would undoubtedly have built their own nuclear weapons – but at their own tempo after the war. The inescapable truth is that without the scientific prowess and tenacity of British physicists, the Manhattan Project would never have been set up during the Second World War.

Stiff Upper Lip

As well as throwing away its ownership of key ideas and inventions, Britain made huge sacrifices to ensure that the Americans made 'their' bombs. With the country already on its knees after five years at war, Chadwick and Oliphant gutted British universities and industry of their best nuclear physicists, electronics experts and chemists, and sent them to the USA. So why didn't the British protest about being stabbed in the back and erased from the historical record by the Americans?

First, they had bigger fish to fry. Their top priority was to resuscitate British nuclear physics and create Britain's own nuclear reactors and weapons. Second, none of the key players would have wasted time trying to rewrite history. They all saw science as a procession of projects; having put Manhattan to bed, they'd moved on to the next challenge. Third, they weren't glory-grabbers. Even outside any constraints of the Official Secrets Act, Chadwick and Cockroft were 'men of few words', totally focused on hard facts and what had to be done next, while Frisch and Peierls were 'basically shy men' who clearly felt guilt and remorse after the bombs were dropped on Japanese cities.[40] For once, the volatile Mark Oliphant also said nothing, perhaps because he was frantically busy building up Britain's post-war nuclear network.

Years later, Chadwick decided not to write his own account of the birth of the bombs, and no other British atomic scientists publicly challenged the American version of history. That task fell to Ronald Clark, a former war correspondent who heard about Hiroshima while standing beside the ruins of the KWI for Chemistry in Berlin. In 1961, Clark published *The Birth of the Bomb: The Untold Story of Britain's Part in the Weapon That Changed the World*, which argued that the Hiroshima bomb[41] was not 'due entirely to the USA'. Britain had first 'seriously set about making a specific nuclear weapon' but this work had been 'overshadowed ... by American propaganda [which] staked America's claim on nuclear history'. Clark described vividly the initial British research on concentrating U-235 for the Frisch–Peierls super-bomb. He viewed the M.A.U.D. Committee's final report as 'a watershed in human affairs', while Pegram and Urey's visit to Britain in late 1941 was 'decisive' in persuading the Americans (with great difficulty) to 'grapple with the problems of using fission as a weapon of war'. Subsequently, British scientists contributed materially to building the uranium bomb. Clark concluded that the Americans would eventually have made the uranium bomb but, without the efforts of Chadwick, Frisch, Peierls and Thomson, 'probably too late for this particular war'. In Britain, *The Birth of the Bomb* was enthusiastically reviewed, provoked 'surprise and incredulity' and sold briskly.[42] In the USA, it didn't stand a chance against the all-American *Project Y*, *The New World* and *Now It Can Be Told*, which all came out within a few months of its publication.

The next attempt to shoehorn the British into the story of Manhattan was made by Margaret Gowing, later the founding Professor of the History of Science at Oxford. In 1962, newly appointed as historian to the UK Atomic Energy Authority and only recently able to 'tell an atom from a molecule', Gowing began writing her account of the UK's atomic energy programme. She was granted exceptional access to top-secret documents that wouldn't be declassified and available to ordinary historians for another thirty years; everything she wrote was vetted to avoid compromising national security. In addition to the British protagonists, she interviewed Bohr, Oppenheimer and Groves. All respected her, and Rudi Peierls became a close friend. In *Britain and Atomic Energy 1939–1945* (1964), Gowing detailed the 'heroic contributions' of scientists at Cambridge, Manchester and Birmingham which paved the way to Manhattan, and reproduced in full the Frisch–Peierls memorandum to demonstrate the origin of the first feasible atomic bomb. In Gowing's opinion, the M.A.U.D. Committee's final report was a pivotal event in the saga of the bomb. Without its 'clarity of analysis, its synthesis of theory and practical programming ... the Second World War might well have ended before an atomic bomb was dropped'.[43]

Britain and Atomic Energy came out three years after *The Birth of the Bomb*, but the American narrative was so deeply embedded that Clark's book had already been forgotten. Gowing's story – written because 'American accounts of the nuclear story needed a British companion' – was 'unfamiliar to the British public and little known even to many in senior government circles'. In Britain, her book was hailed as 'a triumph', received 'rave reviews' and propelled her 'into the limelight as a national treasure'. However, it was dense and academic and, even in Britain, failed to break the grip of American indoctrination. In the USA, Gowing barely broke the surface. When she visited the museum at Los Alamos in 1983, she was left 'deeply dismayed by the absence or distortion of the British contributions'.[44]

In 1992, Ferenc Szasz published *British Scientists and the Manhattan Project: The Los Alamos Years*. Szasz focused on the twenty-four members of the British Mission who lived on the Hill between December 1943 and September 1945.[45] During his decade of research, Szasz interviewed many of the mission and reconstructed the 'social and scientific dimensions' of their sojourn at Los Alamos. His documentation of how the British were instrumental in completing both Little Boy and Fat Man is painstaking, balanced and authoritative. For these reasons, the last words must go to him:

WHODUNNIT

The British contributions to the development of the atomic bomb were crucial. Had the members of the British Mission at Los Alamos not done what they did when they did it, the atomic bomb would not have been available to end the war in August of 1945.[46]

30

FOLLOW-UP

This last chapter returns to autumn 1945 to pick up the stories of the British Mission to the Manhattan Project. At its heart are six men without whom history would have to be rewritten: Oppenheimer would be remembered as a brilliant, pro-Communist physicist at Berkeley and Groves as the architect of the Pentagon, while Manhattan, Trinity, Little Boy and Fat Man would be a random cluster of names without any historical significance. And one of these six men is the reluctant recipient of a title that others have fought over: the 'father of the atomic bomb'.

Demob Happy

The mammoth task of creating Britain's post-war nuclear programme was made even harder by four years of asset-stripping to give the Americans 'all help'. Ten of the British Mission went to the nascent Atomic Energy Research Establishment (AERE) on a decommissioned airfield at RAF Harwell, near Oxford. The AERE aimed to generate electricity from nuclear fission and to produce plutonium for bombs. Its first director was John Cockcroft, who gave up his Jacksonian Chair of Physics at Cambridge and moved his family into the RAF commander's house while the AERE was still a building site. Cockcroft had transformed Halban's demoralised no-hopers in Montreal into a 'happy and efficient group' led by Kowarski; their reactor went critical in September 1945, a month after Hiroshima. The 'wise and modest' Cockcroft similarly guided the AERE to its inaugural triumph in 1956, when the Calder Hall reactor was the world's first to feed electricity into a national grid.[1]

Cockcroft's own career flourished in tandem, with a knighthood (1948) and – better late than never – the Nobel Prize in Physics (1951), which he shared with Ernest Walton for having split the atom twenty years earlier. Having 'no wish to spend the rest of my working life perfecting means of blowing the world to pieces', he argued with 'authority and dignity' for international control of atomic weapons testing (he did, however, ask his daughter not to march with the Campaign for Nuclear Disarmament). During the early 1960s, he was a fixture at the Pugwash Conferences, answering the call by Einstein and philosopher Bertrand Russell to unite the world's scientists in preventing nuclear warfare.[2]

In 1959, Sir John Cockcroft was appointed the founding Master of the new Churchill College in Cambridge. Some saw Cockcroft's traits mirrored in the 'four-squaredness and solidity' of the college's architecture; his ground-floor office gave undergraduates a clear view of their Master sitting 'imperturbable and sphinx-like' at his desk. Cockcroft's 'full life' ended on 18 September 1967 on the bathroom floor of the Master's Lodge. He'd suffered chest pains the previous day but, typically, was either too busy or too diffident to make a fuss.[3]

Harwell's military counterpart was the Atomic Weapons Research Establishment at Aldermaston, some 15 miles away. Its first director was William (Bill) Penney, who turned down a chair at Oxford to take the job and went on to chair the UK Atomic Energy Authority. In diametric opposition to Cockcroft, Penney was convinced that 'the possession of atomic weapons was the safest way to keep the peace'. On his watch at Aldermaston, Britain built its own fission bomb (1952) and a thermonuclear hydrogen (H-)bomb (1957). Along the way, Penney picked up a knighthood, a peerage, the Order of Merit and great happiness: after returning to Britain in September 1945, he fell for and married the nurse who had looked after his sons while his wife faded away in the psychiatric hospital. Penney was variously described as a man of 'obvious sincerity and honesty', 'a vigorous and aggressive gardener who concentrated on vegetables', and a 'teddy bear'. Despite his benign smile, he made life difficult for the Royal Society's biographer: on learning that he had aggressive cancer and little time left, he burned all his papers. Bill Penney died in March 1991.[4]

After the War, 'Messrs ICI' – who had been so prominent in the M.A.U.D. report – backed away from everything nuclear, but Wallace Akers continued advising the government about 'the whole problem of atomic energy'. He was knighted in 1946 for his services to Tube Alloys and, as ever a 'very clubbable man', pursued his passions in art (Trustee, National Portrait Gallery) and music. Nineteen fifty-three was a busy year: he retired from ICI, was delighted to be appointed a Fellow of the Royal Society, and – unexpectedly for a 'confirmed bachelor' whose pad was in the Royal Thames Yacht Club – married a French lady. Sir Wallace Akers died at home the following year.[5]

Michael Perrin returned to ICI but was increasingly consulted as a top-level nuclear adviser alongside Cockcroft and Penney. In 1950, Perrin's legendary people skills and the fact that 'no one knew the work or the men so well' lined him up for a hard day's work that left him feeling 'years older' – hearing the confession of a trusted Tube Alloys scientist who had also been a Russian spy (see below). Perrin changed direction in 1953 and became Chairman of the Wellcome Foundation, which he transformed into the Wellcome Trust, today a leading funder of biomedical research. He was knighted in 1967, retired in 1970 and died in 1988.[6]

Sir Michael Perrin's papers live on in the Archives at Churchill College, a few sculptures away from the Master's Lodge where Cockcroft lived and died. Perrin's diary provides many sideways glances at the story of the bomb. His diagram of the US Uranium Committee (28 October 1941) includes 'Briggs's Safe', while the strip of grey metal sellotaped onto the entry for 21 October 1943 is the game-changing nickel membrane which was crucial for concentrating U-235 at Oak Ridge. The many blank days after July 1944 reflect his increasing involvement with MI6 and Alsos, including the dramatic culmination of Operation Big at Haigerloch.[7]

After the war, the Americans drafted in several of the British Mission to fill critical gaps, notably in the atmospheric and underwater tests of Fat Man-type bombs at Bikini Atoll in July 1946. Groves insisted that 'our people' made and detonated the weapons, but this wasn't true.[8] 'On the grounds of irreplaceability', Ernest Titterton again took charge of 'fast timing' and detonation, and was also the anonymous 'Voice of Abraham' who announced the countdown over loudspeakers. In 1950, Titterton left Harwell with Mark Oliphant for Australia, where he headed Nuclear Physics at the University of Canberra and acquired a knighthood. He'd planned to write his memoirs in retirement, but was left quadriplegic after a driving accident in 1987. He tried dictating into a speech-activated recorder but the Voice of Abraham wasn't up to the task. Sir Ernest Titterton, the first man to detonate a nuclear weapon, died in 1990 and was returned to his roots; his ashes were scattered from a cliff overlooking the English Channel.[9]

James Tuck also ended up in a different continent, which he knew and loved. In 1950, after five restless years in the Clarendon Laboratory at Oxford, he returned to Los Alamos, 'less glamorous but with the old charm still operative' (his words), and 'the only town in the USA in which I'll live' (his wife's). They never left. Tuck retired in 1973 after a 'brilliant and eccentric' spell as Associate Leader of the Physics Division, and died 'a legend' at Los Alamos in 1980.[10]

Inner Circle

Next come the six men who carried the torch from Birmingham in March 1940 to Los Alamos over five years later. Ironically, only James Chadwick and George Thomson were British born. Mark Oliphant had spent eighteen years in England but was Australian. The other three – Otto Frisch, Rudi Peierls and Francis Simon – were Jewish refugees and British by recent adoption. Klaus Fuchs, a Communist, also had to flee Germany.

George Thomson masterminded the M.A.U.D. Committee ('one of the most effective committees ever set in motion') which concluded that the U-235 bomb could and must be built. He hammered the message home in October 1941 by personally handing Vannevar Bush and James Conant copies of the final M.A.U.D. report to prevent them from claiming they hadn't received it. Thereafter, he cut himself off completely from the bomb. Thomson had moved to Ottawa as Britain's scientific liaison officer in order to be near his family, who had been evacuated to America. After his wife died in late 1941, Thomson took his children back to Britain; he became scientific adviser to the Air Ministry and was knighted in 1943. He heard about Hiroshima while sailing with his son in Christchurch Harbour. After the war, Sir George Thomson switched his research to an even more elusive Holy Grail – generating power by nuclear fusion – and in 1952 became Master of his old Cambridge college, Corpus Christi. His popular science book, *The Foreseeable Future*, concluded that the future couldn't be foreseen, but Thomson told a friend that he believed firmly in the afterlife. His opportunity to discover the truth came on 10 September 1975.[11]

Francis Simon also turned his back on the bomb, in favour of thermodynamics. His lab – 'a happy place and never dull' – once again led the field in the strange world just above absolute zero. Simon was awarded a CBE in 1946 for his work with Tube Alloys, upgraded to a knighthood in 1955; he joked that he was 'probably the only man' thus honoured who also had the German Iron Cross First Class, and used to re-enact his ennoblement using fire tongs in lieu of the Queen's sword. And finally, the self-styled 'Vice-President of the Broken-English Speaking Union' was taken for 'a typical Britisher', albeit by an American. In October 1956, Sir Francis Simon succeeded his old friend Lord Cherwell as head of the Clarendon Laboratory in Oxford. But it was not to be. Simon had suffered worsening cardiac symptoms and had barely set foot in his new empire when he suffered a fatal heart attack.[12]

Everybody knows the man who comes next, thanks to his cameo appearances in all renditions of the Manhattan story: the treacherous

'Atom Spy' who perpetrated 'the crime of the century' by handing the bombs' secrets to the Soviets. Klaus Fuchs left Los Alamos in June 1946 and became a much-admired head of Theoretical Physics at Harwell. Everything fell apart in January 1950, when the FBI discovered that an unknown member of the British Mission had fed top-secret intelligence to the Soviet Vice-Consul in New York. On being accused, Fuchs simply asked for Michael Perrin to take his confession. Between 1940 and 1947, Fuchs had passed detailed information about gaseous diffusion and the Trinity bomb to Russian agents in Birmingham, New York, Santa Fe and Berkshire. His espionage had a massive impact. In Potsdam, Stalin wasn't surprised by Truman's news about America's super-weapon, because he already knew. And in August 1949, Russia became the world's second nuclear power by exploding a Fat Man-type plutonium bomb – probably two years earlier than if Fuchs and other spies hadn't accelerated the Soviet nuclear weapons programme.[13]

At his trial in London in March 1950, Fuchs pleaded guilty, thanked everyone for their fairness and accepted his fourteen years' imprisonment for violating the Official Secrets Act. He was a model prisoner, reading physics journals, organising teaching for his fellow detainees and sewing mailbags. Fuchs was terrified of being extradited to the USA, with good reason. Two other Soviet agents who infiltrated Manhattan, Julius and Ethel Rosenberg, went to the electric chair in June 1953; Mrs Rosenberg's heart was still beating after three supposedly lethal shocks, so she was put back in the chair to finish the job. After nine years of good behaviour, Fuchs was freed. Wearing a cheap brown suit, 'Mr Strauss' dodged the press at Heathrow and flew to East Berlin, never to return to the West. He was welcomed with a top nuclear job and the Soviet equivalents of the gongs and prizes which Britain had showered on his fellow mission members.

Fuchs wasn't driven by money, sex or revenge. He refused to accept $1,500 for his diagram of Trinity, and despite being interned and imprisoned, never held any grudge against Britain. Rather, he was a devout Communist with undying allegiance to Russia. They both suffered greatly during the war; Fuchs's hammer-and-sickle lapel badge nearly had him killed in Kiel, and he believed that 'the Western

Allies deliberately allowed Russia and Germany to fight each other to the death'. Fuchs saw only one solution – 'The American monopoly on the bomb had to be broken' – and that's what he did. He hoped to 'repair the damage ... to my friends' but never expressed remorse; if re-run, history would simply have repeated itself.[14] Klaus Fuchs died peacefully in East Germany in 1988.[15]

Gang of Four

This leaves just four men, who between them conceived a workable atomic bomb, guided it through five arduous years of threats and pitfalls, and ensured its birth. You can judge their importance by playing 'what if?' What if the whistling bomb that terrified Otto Frisch during the Liverpool Blitz really had his name on it? Or if Rudi Peierls had been one of the firefighters killed during the bombing of Birmingham? Or if the plane carrying Mark Oliphant to America in August 1941 had crashed in the Atlantic? Or if James Chadwick had decided against sacrificing British nuclear physics to help the Americans? If fate had picked off any of the four before their decisive interventions, atomic bombs wouldn't have been born during the war.

William Laurence, a veteran *New York Times* journalist, was among the VIPs who watched the Trinity test from Campania Hill. After Hiroshima, Laurence wrote of Chadwick, 'Never before in history had any man lived to see his own discovery materialise itself with such telling effect on the destiny of man.' Those words wouldn't have inspired pride or elation in the discoverer of the neutron. Chadwick had recently written to Rotblat in Liverpool, 'I am living in a different world from which I cannot easily escape.' After realising that the bomb would be dropped, he'd started taking sleeping tablets every night and was exhausted and depressed by the 'gruelling labour, endless paperwork, conferences and much travel'. When he and Aileen finally returned to Liverpool in mid-summer 1946, Chadwick looked 'physically, mentally and spiritually drained' and was wrestling with his conscience over the bomb.[16] His team and the biggest cyclotron outside America no longer provided the tonic he needed, and he jumped sideways in 1949 to become Master of Gonville and Caius, his old Cambridge college. Under Chadwick's leadership, Caius thrived but he was victimised by a 'Peasants' Revolt' of back-stabbing dons and had to decline the Vice-Chancellorship of Cambridge on medical grounds. After months of despondency and bed rest, he retired in 1958 and took Aileen to the Vale of Clwyd in North Wales. They enjoyed a decade of quiet life and entertaining visitors, including the Bohrs, and in 1968 returned to a modest house in Cambridge to be near their daughters. Chadwick was delighted when the Queen made him a Companion of Honour in 1970 and he put on a cheerful show for his eightieth-birthday party in Liverpool in 1971, despite depression and failing eyesight. Robbed of the energy to write his memoirs, he spent his last months listening to gramophone records.[17]

James Chadwick was a modest, self-effacing man who hid genuine warmth and humour (his own brand) behind a carapace of inscrutability. He never made a fuss about what he did during the war; his daughters only found out from others long afterwards. And he would have been entirely happy to see no mention of himself in the film *Oppenheimer* (2023). As one of his daughters said, 'Who would play him?' On 23 July 1974, Sir James Chadwick was put to bed by Lady Aileen, his wife of forty-nine years and his carer for the last two. When she woke beside him next morning, she found that he'd died in his sleep.[18]

After a couple of years in Birmingham, Mark Oliphant took his biggest ever project – the world's most powerful proton accelerator – back to his native Australia. He co-founded the Australian National University in Canberra and became the first Director of its Research School of Physical Science – with an administrative burden which made his super-accelerator fail so dismally that it was nicknamed the 'White Oliphant'. He enjoyed greater success as the creator and first President of the Australian National Academy and developed keen interests in ecology and solar energy, decades before these became fashionable. Oliphant was knighted in 1959.[19]

News of Hiroshima broke while he was on a family holiday in Wales. Horrified by the carnage, he became a 'belligerent pacifist', campaigning against nuclear weapons, Americans in Vietnam, and French nuclear tests in the Pacific. In America, his 'outspokenness' raised McCarthyite hackles and had him branded a Communist sympathiser. In 1971, aged seventy and 'still vigorous', he moved into politics and served as Governor of South Australia for five years. He lived long in retirement, but spent his late eighties exhausted emotionally and financially, living in a bungalow and sitting for hours each day with his wife, bed-bound in an expensive nursing home.[20]

Sir Marcus Oliphant died in July 2000, aged ninety-eight. The plaque on his statue in Canberra lists 'Leader in the development of microwave radar and atomic energy in Britain and USA' among his achievements. The US Congress nominated him for its prestigious Congressional Medal of Freedom with Gold Palm, but this was blocked (without Oliphant's knowledge) by the Australian government. Sadly, the US Congress never followed up Leo Szilard's suggestion of awarding him a medal for 'meddling foreigners', to honour the man whose outspokenness ultimately spawned the Manhattan Project.[21]

And so to the last two, who initiated the vast five-year enterprise that gave birth to the atomic bombs dropped on Japan (Plate 31).

In spring 1946, Rudi Peierls turned down a chair in Cambridge and returned to Birmingham. He became a consultant at Harwell and designed the British version of the U-235 separation plant at Oak Ridge. The revelation in February 1950 that Klaus Fuchs had spied for the Russians was 'a shattering blow' for Rudi and Genia. Rudi visited Fuchs in Brixton Prison and was astonished by his 'arrogance and naiveté'. Genia sent Fuchs a tear-stained letter, which Rudi had to rewrite: 'I considered you the most decent man I know ... I was so very fond of you ... You have burned your god.' Even worse, a shoddily researched book accused Peierls of being a Soviet agent; retribution was swift and expensive, with the book pulped and hefty damages from the publisher, which Peierls donated to the Pugwash Conferences. It transpired that MI5 had opened a file on Peierls after he attended a conference in Russia in 1938 and kept it live until 1953. Ripples also crossed the Atlantic, culminating in 1957 with an

American demand for Harwell to revoke Peierls's security pass. When they did, he resigned his consultancy.[22]

In 1963, after twenty-six years in Birmingham, Peierls moved to Oxford and the Wykeham Chair of Physics. His group continued to attract the brightest and best across a broad spectrum of theoretical physics, and his knighthood in 1968 reassured him that his adoptive country had finally accepted him. Peierls became increasingly preoccupied with the proliferation of atomic weapons and the 'unthinkable disaster' of a nuclear holocaust. It was too late to feel guilty about his own roles in creating the weapons, but he strongly supported the Pugwash Conferences and served as secretary from 1969 to 1974. He retired in 1977, aged sixty-seven.[23]

Since 1931, Genia had been Rudi's soulmate, pillar and intellectual fencing partner – and the 'Genia factor' was a powerful magnet for potential recruits to his group. For her seventieth birthday in 1978, friends invented the 'genia', defined as a unit of 'loudness, bigheartedness, self-confidence, loving concern and irresistible kindness'; as she was much larger than life, the one-thousandth quantity or 'milligenia' was judged 'perfectly adequate for normal purposes'. In 1985, Rudi and Genia cancelled a long-anticipated trip to Russia because she needed surgery for a benign brain tumour. While she was recovering, Rudi finished typing up his autobiography on the word processor that she'd given him as a seventy-fifth-birthday present. In deference to the nomadic lifestyle which they both loved, the book was called *A Bird of Passage*. It was published in 1986 and ended with them settled in Oxford, 'within walking distance of the laboratories and the shops'. In June that year, they enjoyed a 'glorious' holiday in Greece; four months later, Genia was dead. She missed by months Rudi's award of the Copley Medal – the Royal Society's greatest – and the big symposium in Oxford that celebrated his eightieth birthday. Sir Rudolf Peierls, the wandering bird of passage, finally came home to roost seven years later. Still mentally alert and forever curious, he died in his retirement home in September 1995.[24]

While travelling home from Los Alamos in autumn 1945, Otto Frisch took a phone call from John Cockcroft and was offered a top job at

THE IMPOSSIBLE BOMB

Harwell, still under construction. That winter was harsh, and water froze inside Frisch's prefabricated house. He concentrated on smoothing out fluctuations in the chain reaction inside reactors – valuable work, although he later said that the paper he published contained 'so many mistakes that I don't dare look at it'. In spring 1947, he was 'quite unexpectedly' offered the Jacksonian Chair of Physics at Cambridge, recently vacated by Cockcroft. Chadwick told him that 'Only a fool would refuse it', so Frisch accepted. In 1954, he became head of the Cavendish Laboratory and was instrumental in making it a major research centre in particle physics.[25]

In 1948, Frisch's parents came to live in Cambridge. His father, a concentration-camp survivor, soon died of lung cancer. Frisch built a wheelchair for his mother, frail and immobile after fracturing her hip, in which four burly students carried her upstairs to his 'palatial rooms' in Trinity College to hear him playing his Bechstein piano. Meanwhile, he'd become engaged to Ulla Blau – Viennese and a lover of music and culture who opened Otto's eyes to 'a world of visual art to which I had paid little attention'. Their engagement ring was the one which Otto Hahn had given Lise Meitner when she fled Berlin in July 1938, in case she needed to bribe her way to safety. Otto and Ulla were married in spring 1951, and a daughter and a son followed 'in rapid succession'. After Frisch's mother died, his aunt Lise, still robust at eighty-one years of age, left Stockholm and moved into a nearby flat. In 1963, science historian Thomas Kuhn and Frisch interviewed Meitner for a series of 'atomic histories'. The transcript (in German) shows Kuhn addressing her respectfully with the formal *Sie*, while her nephew calls her *du*, as would befit a family member, a dear friend, a close colleague or all of the above. Meitner died in 1968 and was buried in Hampshire, near her youngest brother.[26]

Frisch notionally retired in 1974 but remained busy. He led a team that built a computer-assisted laser-scanning device to interpret the trails of subatomic particles in the recently invented bubble chamber, and as a result became chairman of the first company in the Cambridge Science Park. In 1979, he published his autobiography, *What Little I Remember*, dedicated to his daughter, Monica, 'who made me write this'. This is a fistful of gems spanning fifty momentous years, full of insight, wisdom and self-deprecating wit, and enlivened by Frisch's cartoons of key characters. After Nagasaki ('few of us could see any reason for dropping a second bomb'), he mentions nuclear weapons only once – explaining that his cataracts were 'probably due to exposure to neutrons during my A-bomb work ... Some may think I deserve more punishment'. On leaving Los Alamos, Frisch severed all links with nuclear weapons, like a parent disowning a child whose birth was regretted. And being the first person to realise how neutrons would behave in a mass of pure U-235, he was ultimately the father of the atomic bomb.

Otto Frisch ended *What Little I Remember* as a 'lucky man' content with his lot, working at home and breaking off 'to play the piano whenever I like'. A few months later, he suffered an unlucky fall and a freak complication. He died on 22 September 1979, a week short of his seventy-fifth birthday.[27]

Farewell Tour

During the 1990s, some bright spark thought of luring tourists to Santa Fe with a 'Klaus Fuchs Tour'. The idea sank without trace, but it chimes with the ending I'd envisaged for this book – a pilgrimage that takes in twelve places of significance and ties up some loose ends. It starts in the country codenamed 'Empire', then crosses the Pacific to America before heading to England and finally mainland Europe.

Japan

First must come two cities which will forever be twinned in catastrophe, and which we desperately hope will be the last of their kind. When the American journalist John Hersey visited Hiroshima soon after the bombing, he was horrified to see that American troops had cleared rubble and corpses to create an 'Atomic Bowl' for playing football. Returning in 1985, Hersey found that 'a gaudy phoenix had risen from the ruinous desert'. Forty years later, the Delta City is beautiful and prosperous, and Nagasaki 'a vibrant port city and a hub for fishing, tourism and manufacture'.[28]

There are still relics of what turned these cities into household names. Hiroshima has a World Heritage Site, the skeletal steel cupola of the Genbaku Dome which topped its Industrial Promotion Centre; in Nagasaki, the shattered bell tower of the Urakami Catholic Cathedral lies where it fell. The bombs killed at least 90,000 at Hiroshima and 40,000 at Nagasaki, with a vastly longer legacy than high explosives or napalm. Many of the 180,000 *hibakusha* (survivors) died years later from radiation-induced organ failure or cancers, or suicide. The *hibakusha* were feared and shunned by their countrymen, unemployable and unmarriageable; some of those who had a family transmitted the curse of untouchability to their children and even their children's children.[29]

Both cities now have energy and vitality, but the dead are never far away. In parts of Hiroshima, you will hit human bones if you dig down a couple of feet. The Peace Memorial Museum at Hiroshima displays watches that stopped at 8.15 on the morning their owners were incinerated. Here too is the blowtorched tricycle belonging to three-year-old Shinichi Tetsutani, dug up from their shared grave when the little boy's remains were reburied in his family's tomb. Ground zero for a different Little Boy is marked by a plaque on a nondescript grey building where Hiroshima's Medical College stood at 8.14 a.m. on 6 August 1945.[30] The plaque is a terse epitaph to the 900 doctors, nurses and students who died there – and to the pre-nuclear age of the whole of humanity. Just two words would have sufficed: *Never again*. But three days later came Nagasaki.

America

I've selected two American sites, which symbolise success and failure.

'Success' is a 15-foot obelisk faced with black basalt. It's a National Historical Monument but you can only visit twice a year, when this corner of the White Sands Missile Range at Alamagordo is open to the public. On 15 July 1945, the spot was occupied by a 100-foot steel tower crowned by Trinity, awaiting its destiny. The crater is still there, but you won't see the fabled Lake of Green Glass. In 1953, as if to cover up a dreadful mistake, bulldozers cleared the acres of fused sand and buried the evidence at a secret location. If you're lucky enough to spot a shard of the man-made mineral trinitite, glinting green like a peridot in the rubble, don't pick it up, because stealing trinitite is punished by a hefty fine. Instead, picture the moment, perhaps sitting beside Otto Frisch 20 miles away on Campania Hill, or in the Command Center with Oppenheimer and Ernest Titterton, waiting for the blip on the oscilloscope screen to be convulsed by the 6,000-volt firing signal.[31]

'Failure' takes us back to late 1946 and Einstein's refuge at Peconic on Long Island. Einstein is sucking on his china-clay pipe, perusing a sheet of paper under the attentive gaze of Leo Szilard, surprisingly dapper for a man who lives out of a suitcase. They strike several poses, because they are re-enacting the signing of the letter, drafted by Szilard, in which Einstein urged Roosevelt to make an American uranium bomb before the Nazis (Plate 32). Einstein later said that his greatest regret was signing that letter, and that he 'wouldn't have lifted a finger' if he'd known there was no Nazi bomb. His conscience should have been clear because he played no role in the bombs dropped on Japan – but he died without knowing that, between them, Alexander Sachs and Lyman Briggs had ensured that his letter to the President went nowhere useful.[32]

Britain

An entire pilgrimage could be concocted within Britain's borders, but I've restricted myself to four sites. The first is James Chadwick's lab, with his hard-won cyclotron, the windows blown out in the Liverpool Blitz, the crates of whale blubber from Port Sunlight and a fatally forgettable high-voltage switch. A confession: this choice is partly motivated by guilt. During the fifteen years I worked in Liverpool, I walked past the Chadwick Building hundreds of times but never once wondered who he'd been or what he'd done.

From Liverpool, it's a 20-mile hop across the rivers Mersey and Dee into Wales, to the 'gaunt and majestic' remains of the poison-gas factory at Rhydymwyn, alias 'Valley'. The only structure in Valley which isn't Grade II listed for historic interest is the building designated 'P6.' Unlike the others, P6 didn't make mustard gas but instead housed the prototypes of Francis Simon's U-235 separation units. These never worked but drove the evolution of Oak Ridge's K-25 plant which fuelled Little Boy. Like Chadwick's lab, Valley was sacrificed for the greater good of Manhattan.[33]

By contrast, the Georgian townhouse at 24 West Street in Godmanchester, Cambridgeshire, is listed and well preserved. Farm Hall looks much as it did when Major Tom Rittner's 'guests' spent the second half of 1945 here, comfortable but heavily bugged. This was where Carl von Weizsäcker began spinning his deceitful 'alternative version' of the *Uranverein*: these were honourable men who actually prevented the Nazis from making atomic bombs. Von Weizsäcker fooled Robert Jungk, whose *Brighter Than a Thousand Suns* (1956) praised Heisenberg as the humanitarian genius who knew how to make the bomb but hid his secrets from the Nazis (von Laue said of this book, 'I had to stop reading'). The 'alternative version' died immediately when the transcripts of the Farm Hall recordings were released in 1992. These should have been made public after thirty years in 1975 – but for undisclosed reasons were kept secret by the British Foreign Office for another seventeen years, despite protests from historians and the Royal Society.[34]

The final stop in England is the church of St John the Baptist at Charlton, near Malmesbury. A stained-glass window commemorates a life that ended abruptly and painlessly on 12 May 1941, taking seven other souls with it. The place was the 'bomb cemetery' on a marsh in south-east London, where someone had chalked 'OLD FAITHFUL' on a 250-kg, double-fused bomb because it hadn't done anything since falling on a factory six months earlier. It was UXB No. 35 for the still-lucky 'Holy Trinity'; Jack Howard, Earl of Suffolk and Berkshire, was demonstrating how to remove its main fuse, assisted by Fred Hard and with Beryl Marsden taking notes in a nearby lorry. When the bomb began ticking, Howard took his time because he believed that he had thirty minutes to deactivate this particular type. Old

FOLLOW-UP

Faithful left a crater 50 feet across. Beryl Marsden was the only one left recognisable; she died on the way to hospital. None of the body parts was confidently matched to Fred Hard. Jack Howard, who'd been sitting astride the bomb, was identified from shreds of his orange-spotted scarf and his engraved silver cigarette case. The stained-glass window in Charlton carries a verse by John Masefield:

Probing and playing with lightning thus
He and his faithful friends met death for us.[35]

France

Jack Howard could have led us to the docks at Bordeaux, ushering French scientists and *matériel* on board the *Broompark* as bombs fell all around. Instead, we'll rewind to the day before and an emotional moment in Frédéric Joliot's temporary lab in Clermont-Ferrand. Joliot is staying behind to look after his sick wife, Irène, while Hans Halban and Lew Kowarski are about to leave for Bordeaux and England. The three musketeers solemnly shake hands; will this be *au revoir* or *adieu*?

It is 1948, and they are playing themselves in the French–Norwegian film *La Bataille de l'Eau Lourde* (*The Battle of the Heavy Water*). Also playing himself in the film is the spectacled, bland-faced man who appears at Oslo airport, furtively transferring heavy grey canisters from the Amsterdam flight to the plane bound for Dundee. Jacques Allier is moonlighting from his post-war day jobs as Deputy Director of the Banque de Paris and Vice-Chairman of Norsk Hydro. And as it was Allier who sent the telegram ordering that 'RJUKAN MUST BE DEFENDED AT ALL COSTS', he is the perfect link to the next destination.[36]

Norway

Nowadays, there is much to delight the visitor to Telemark: winter sports, bungee jumping and, since 2015, a World Heritage Site featuring 'an exceptional combination of industrial assets and themes associated with the natural landscape'. Centred on the vast hydroelectric plant at Rjukan, its ninety-seven 'culturally significant structures' include the railway, ferries and the former ammonia/heavy-water factory at Vemork, now an industrial museum.

During the summer, you can take a bus from the car park at Vemork up to an isolated hut high in the alpine meadows of the Hardanger Plateau, and walk back down along a well-marked route. The scenery and wildflowers are magnificent, and the going is 'moderate', except for a steep descent at the end. This is the 'Saboteurs' Trail' taken by Joachim Rønneberg's commandos on their successful demolition mission in February 1943. Every year, locals retrace their path – but by day in July, because only fools with a death wish or very brave men would attempt this by night in winter and on skis.[37]

Germany

Our penultimate stop is another tourist destination: Haigerloch, near the Black Forest. The baroque church where Heisenberg played the organ looks out from its clifftop across the river, but the real attraction lies in the cave below. The *Atomkeller* ('Atomic Cellar') Museum houses a reconstruction of the B-VIII *Uranbrenner* reactor, with wooden blocks strung on wires to simulate the chains that carried 664 2-inch uranium cubes suspended in heavy water. Most of the cubes were sacrificed to produce a kilo of U-235 for Little Boy; others were black-marketed to Russia or found use as paperweights or doorstops. Some have been dissected by 'nuclear forensics', which proved that the reactor came 'nowhere near' sustaining a chain reaction.[38] Even if it had, and even if that had been achieved three years earlier, there would never have been a Nazi atomic bomb.

Bohemia

Our final destination is where this story began almost 500 years ago, in the silver rush that made St Joachimsthal the wealthiest town in central Europe. Here surely is the place to wish that things could have been different, even though this would create a parallel world so far-fetched that it's futile even to dream about it.

A fairy-tale world where pitchblende was left in the ground because it was worthless, and where 'a weapon beyond the wildest nightmares' was never invented. And where the worst thing that could ever happen is only another world war – like the one which was brought to its end many months after Victory in Europe, by the American-led invasion of Japan that began in November 1945.

POSTSCRIPT
Full Circle

Ignorance made me write this book. I knew nothing about Frisch, Peierls, Oliphant, the M.A.U.D. Committee, Tube Alloys or the British Mission to Los Alamos. Those names hadn't appeared in anything I'd read or heard about the birth of the atomic bomb, and I wanted to find out why. I began writing with no clear idea of where it would take me. Over time, three intertwining themes crystallised: telling the story of the British scientists who attended the birth of the atomic bomb; discovering what they actually did; and exploring why, in the words of Ferenc Szasz, 'the average person will not recognise these names'.

I found the British co-creators of the atomic bomb every bit as captivating and thought provoking as any of the characters usually given centre stage in the saga of the Manhattan Project. With all the information available today, there can be no doubt that without the British, the race to develop nuclear weapons would have ended years later and not at Los Alamos. Untangling the third strand of the saga – why the British contribution has been forgotten – revealed that this story wasn't simply 'untold', as Ronald Clark put it in 1961; it had been strangled at birth by General Groves, and the few British scientists who could have contradicted him chose not to do so.

Writing this book has been a journey of discovery and surprises. It's been a thriller, a horror story and an eye-opener – and, despite myself, there have been many moments when I wished that I could have been there. I still loathe nuclear weapons because they terrify

POSTSCRIPT

me and always will, but my hatred is no longer channelled at those who brought them into the world. Context is everything; I've never had to face an existential threat, so I can't know how I would have reacted if I'd been standing there as the Luftwaffe's bombs fell on Liverpool, Birmingham, London or Cambridge.

I hope that, like me, you've been moved and excited by the stories in this book and can now appreciate why 'the man from Aldermaston', who featured in the Preface, was so frustrated by the obliteration of the British contribution from popular consciousness. If those forgotten names are now ones you'll remember for who they were and what they did, then I'll be content. And who knows? Maybe the Manhattan Project will come to be seen for what it was: in terms of scientific achievement and impact on humanity, the most significant international collaboration of the twentieth century.

This brings me back to the man who prefabricated the history of the birth of the bomb: General Leslie Groves. When I first tackled his bombastic *Now It Can Be Told*, I skipped the penultimate chapter on 'Postwar Developments' because it seemed beyond my brief and he'd already lectured me enough about American supremacy. When I eventually read it, I found an uncharacteristically sober review of the British contributions and a throwaway line with an extraordinary twist. On page 408, Groves wrote: 'I cannot escape the feeling that without active and continuing British interest there probably would have been no atomic bomb to drop on Hiroshima.'

He didn't explain this shuddering U-turn, which contradicts everything he'd ever said before. Perhaps he looked back, seventeen years after the event, and at last saw things that hadn't been refracted through a lens distorted by patriotism, Anglophobia and the crushing responsibility of pushing Manhattan to a successful conclusion. Maybe he'd relaxed, because the world had accepted the armour-plated version of history which he'd created. And alien though such emotions might appear to his nature, it's even possible that he'd fallen victim to gratitude, regret or remorse.

What if Groves had inserted those two lines in the Smyth Report back in 1945? The myth of the all-American atomic bomb would never have been born, and the world would have accepted that British scientists had been a small but essential cog in the massive American war machine at Los Alamos. But then Ronald Clark wouldn't have had an 'untold' story to recount in 1961 – and this book would never have been written.

APPENDIX I

FRISCH-PEIERLS MEMORANDUM

Strictly Confidential
Memorandum on the properties of a
radioactive "super-bomb".

The attached detailed report concerns the possibility of constructing a "super-bomb" which utilizes the energy stored in atomic nuclei as a source of energy. The energy liberated in the explosion of such a super-bomb is about the same as that produced by the explosion of 1000 tons of dynamite. This energy is liberated in a small volume, in which it will, for an instant, produce a temperature comparable to that in the interior of the sun. The blast from such an explosion would destroy life in a wide area. The size of this area is difficult to estimate, but it will probably cover the centre of a big city.

In addition, some part of the energy set free by the bomb goes to produce radioactive substances, and these will emit very powerful

APPENDIX I

and dangerous radiations. The effect of these radiations is greatest immediately after the explosion, but it decays only gradually and even for days after the explosion any person entering the affected area will be killed.

Some of this radioactivity will be carried along with the wind and will spread the contamination; several miles downwind this may kill people.

In order to produce such a bomb it is necessary to treat a few cwt. of uranium by a process which will separate from the uranium its light isotope (U235) of which it contains about 0.7%. Methods for the separation of isotopes have recently been developed. They are slow and they have not until now been applied to uranium, whose chemical properties give rise to technical difficulties. But these difficulties are by no means insuperable. We have not sufficient experience with large-scale chemical plant to give a reliable estimate of the cost, but it is certainly not prohibitive.

It is a property of these super-bombs that there exists a "critical size" of about one pound. A quantity of the separated uranium isotope that exceeds the critical amount is explosive;[+++] The bomb would therefore be manufactured in two (or more) parts, each being less than the critical size, and in transport all danger of a premature explosion would be avoided if these parts were kept at a distance of few inches from each other. The bomb would

---

[+++] yet a quantity less than the critical amount is absolutely safe.

be provided with a mechanism that brings the two parts together when the bomb is intended to go off. Once the parts are joined to form a block which exceeds the critical amount, the effect of the penetrating radiation always present in the atmosphere will initiate the explosion within a second or so.

The mechanism which brings the parts of the bomb together must be arranged to work fairly rapidly because of the possibility of the bomb exploding when the critical conditions have just only been reached. In this case the explosion will be far less powerful. It is never possible to exclude this altogether, but one can easily ensure that only, say, one bomb out of 100 will fail in this way, and since in any case the explosion is strong enough to destroy the bomb itself, this point is not serious.

We do not feel competent to discuss the strategic value of such a bomb, but the following conclusions seem certain:

1. As a weapon, the super-bomb would be practically irresistible. There is no material or structure that could be expected to resist the force of the explosion. Using the bomb for breaking through a line of fortifications, If one thinks of using the bomb for breaking through a line of fortifications, it should be kept in mind that the radioactive radiations will prevent anyone from approaching the affected territory for several days; they will equally prevent defenders from reoccupying the affected posi-

tions. The advantage would lie with the side which can determine most accurately just when it is safe to re-enter the area; this is likely to be the aggressor, who knows the location of the bomb in advance.

2. Owing to the spreading of radioactive substances with the wind, the bomb could probably not be used without killing large numbers of civilians, and this may make it unsuitable for use by this country. (Use as a depth charge near a naval base suggests itself, but even there it is likely that it would cause great loss of civilian life by flooding and by the radioactive radiations.)

3. We have no information that the same idea has also occurred to other scientists, but since all the theoretical data bearing on this problem are published, it is quite conceivable that Germany is, in fact, developing this weapon. Whether this is the case is difficult to find out, since the plant for the separation of isotopes need not be of such a size as to attract attention. Information that could be helpful in this respect would be data about the exploitation of the uranium mines under German control (mainly in Czechoslovakia) and about any recent German purchases of uranium abroad. It is likely that the plant would be controlled by Dr. K. Clusius (Professor of Physical Chemistry in Munich University), the inventor of the best method for separating isotopes, and therefore information as to his whereabouts and status might also give an important clue.

At the same time it is quite possible that nobody in Germany has yet realised that the separation of the uranium isotopes would make the construction of a super-bomb possible. Hence it is of extreme importance to keep this report secret since any rumour about the connection between uranium separation and a super-bomb may set a German scientist thinking along the right lines.

4. If one works on the assumption that Germany is, or will be, in the possession of this weapon, it must be realised that no shelters are available that would be effective and could be used on a large scale. The most effective reply would be a counter-threat with a similar bomb. Therefore it seems to us important to start production as soon and as rapidly as possible, even if it is not intended to use the bomb as a means of attack. Since the separation of the necessary amount of uranium is, in the most favourable circumstances, a matter of several months, it would obviously be too late to start production when such a bomb is known to be in the hands of Germany, and the matter seems, therefore, very urgent.

5. As a measure of precaution, it is important to have detection squads available in order to deal with the radioactive effects of such a bomb. Their task would be to approach the danger zone with measuring instruments, to determine the extent and probable duration of the danger and to prevent people from entering the danger zone. This is vital since the radiations kill

APPENDIX I

instantly only in very strong doses, whereas weaker doses produce delayed effects, and hence near the edges of the danger zone people would have no warning until it were too late.

For their own protection, the detection squads would enter the danger zone in motor-cars or aeroplanes which are armoured with lead plates, which absorb most of the dangerous radiation. The cabin would have to be hermetically sealed and oxygen carried in cylinders because of the danger from contaminated air.

The detection staff would have to know exactly the greatest dose of radiation to which a human being can be exposed safely for a short time. This safety limit is not at present known with sufficient accuracy and further biological research for this purpose is urgently required.

As regards the reliability of the conclusions outlined above, it may be said that they are not based on direct experiments, since nobody has ever yet built a super-bomb, but they are mostly based on facts which, by recent research in nuclear physics, have been very safely established. The only uncertainty concerns the critical size for the bomb. We are fairly confident that the critical size is roughly a pound or so, but for this estimate we have to rely on certain theoretical ideas which have not been positively confirmed. If the critical size were appreciably larger than we believe it to be, the technical difficulties in the way of constructing the bomb would be enhanced. The point can be definitely settled as soon as a small amount of uranium has been separated, and we think that in view of the importance of the matter immediate steps should be taken to reach at least this stage; meanwhile it is also possible to carry out certain experiments which, while they cannot settle the question with absolute finality, could, if their result were positive, give strong support to our conclusions.

O. R. Frisch
R. Peierls

The University,
Birmingham.

APPENDIX II

MEMBERS OF THE BRITISH MISSION TO THE MANHATTAN PROJECT

Director: Sir James Chadwick (Washington)

At Los Alamos *At Berkeley and Oak Ridge*

Otto Frisch	Mark Oliphant
Ernest Titterton	Harrie Massey
Mrs E.W. Titterton	R.H.V. Dawton
Niels Bohr	E.H.S. Burhop
Aage Bohr	J.P. Keene
Rudi Peierls	G. Page
Egon Bretscher	H.W.B. Skinner
Philip Moon	S.C. Curran
Winifred Moon	C.S. Watt
Joseph Rotblat	R.M. Williams
Donald Marshall	S. Rowlands
James Tuck	Jean Sutherland
Sir Geoffrey Taylor	M.J. Moore
William Penney	S.M. Duke
Tony Skyrme	H.J. Morris
Klaus Fuchs	Maurice Wilkins
W.G. Marley	P.P. Starling
Michael Poole	M.E. Haine
Anthony French	D.F. Stanley

MEMBERS OF THE BRITISH MISSION

James Hughes
Derrik Littler
Harold Sheard
George Placzek
Carson Mark

M.P. Edwards
J.D. Craggs
J. Sayers
W.D. Allen
T.E. Allibone
O. Bunemann
H.J. Emeleus
K.J.R. Wilkinson
H.S. Tomlinson
R.R. Nimmo
F. Smith
C.J. Milner
A.G. Jones
H.E. Evans
A.A. Smales
J.P. Baxter (ICI)

GLOSSARY

Alberta: codename for the project to deliver atomic bombs to targets in Japan.

alpha-particle: heavy, positively charged particle (the helium nucleus); component of ionising radiation emitted during radioactive decay.

Alsos: the Scientific Intelligence Mission set up by General Groves in late 1943 to round up German military scientists and acquire their secrets for the USA.

beta-particle: almost weightless, negatively charged particle (the electron) travelling at speed of light; component of ionising radiation emitted during radioactive decay.

boiler: name given by Frédéric Joliot and colleagues to their nuclear reactors.

British Mission to the Manhattan Project: group of sixty British scientists, led by James Chadwick, who left Britain to work with the Americans from 1943 to 1945 at Los Alamos and Berkeley, on both the U-235 and plutonium atomic bombs.

calutron: modified cyclotron which purified U-235 by electromagnetic separation. Hundreds of calutrons were linked to form the 'racetracks' at Oak Ridge.

GLOSSARY

chain reaction: see nuclear chain reaction.

CP-1: codename for Enrico Fermi's nuclear reactor ('Chicago Pile-1'), which first sustained a chain reaction in December 1943.

critical mass: the minimum amount of fissionable material needed to sustain a self-perpetuating nuclear chain reaction.

cyclotron: extremely powerful particle accelerator with a circular path, invented by Ernest Lawrence.

deuterium: isotope of hydrogen with an extra neutron, atomic weight 2; replaces ordinary hydrogen in heavy water.

Deutsche Physik ('German Physics'): a short-lived but influential movement, led by Philipp Lenard and Johannes Stark, to eliminate 'Jewish' and non-Aryan theoretical physics from Germany.

'dragon' experiment: method devised by Otto Frisch to estimate the critical mass of U-235 for Little Boy; after the war, two experimenters died while extending Frisch's experiments.

element: a chemically pure substance consisting entirely of a single type of atom. Ninety-two elements occur naturally.

elements 92, 93 and 94: uranium, neptunium and plutonium.

emanation: Rutherford's original name for radon, element 86.

Farm Hall: the heavily bugged manor house in Godmanchester, Cambridgeshire, where members of the *Uranverein* were interned (and their conversations recorded) during Operation Epsilon, July 1945–January 1946.

fast neutrons: typically produced during fission and travelling close to the speed of light. Poor at inducing fission in natural uranium but essential for detonating an atomic bomb.

Fat Man: nickname for the plutonium bomb which devastated Nagasaki, 9 August 1945.

Frisch–Peierls memorandum: the secret document written in early 1940 by Otto Frisch and Rudi Peierls, envisaging a hugely powerful 'super-bomb' made of a few pounds of pure U-235.

GLOSSARY

gadget: nickname for the implosion device which detonated the plutonium bombs Trinity and Fat Man; also used for the weapons themselves.

gaseous diffusion: method for concentrating U-235 by forcing gaseous hex across a thin metal membrane perforated with millions of tiny pores.

graphite: pure carbon (pencil lead), used to moderate (i.e. slow) neutrons to the optimal speed for a controlled chain reaction in a nuclear reactor such as Fermi's CP-1.

gun detonator: a device, modified from an artillery gun, which fired together two subcritical masses of U-235 to trigger an explosive chain reaction in the Little Boy uranium bomb.

Hanford (Site W): top-secret facility in Washington State where plutonium was made by bombarding uranium with neutrons, for the Trinity and Fat Man bombs.

heavy hydrogen: see deuterium.

heavy water: water containing deuterium (D_2O) instead of ordinary hydrogen (H_2O); used to moderate nuclear reactors. During the 1940s, the only industrial source was the Norsk Hydro plant in Vemork, Norway.

hex: uranium hexafluoride (UF_6), an extremely toxic compound which exists as a liquid and a gas at relatively low temperatures. Hex was subjected to thermal diffusion and gaseous diffusion separation methods to concentrate U-235.

hibakusha: Japanese term for survivors of the atomic bombings.

the Hill: Los Alamos.

ICI (Imperial Chemical Industries): major British chemical company closely involved with both the M.A.U.D. Committee and Tube Alloys.

implosion: the method for detonating the plutonium bomb. The explosion of a large sphere of high explosive is focused to crush a solid plutonium ball into a smaller, supercritical mass, triggering an explosive chain reaction.

GLOSSARY

initiator: a device containing polonium and beryllium which releases neutrons into a critical mass of U-235 or plutonium, igniting an explosive chain reaction.

ionising: radiation powerful enough to knock electrons off atoms; originally detected as the ability of radiation to make air conduct electricity.

isotopes: forms of an element which are chemically identical but, because their nuclei contain variable numbers of neutrons, have different atomic weights and physical properties. Examples are the uranium isotopes U-235 and U-238.

KWI (Kaiser Wilhelm Institute): a series of elite German research centres, including the KWI for Physics and the KWI for Chemistry in Berlin.

Little Boy: the U-235 atomic bomb which devastated Hiroshima on 6 August 1945.

Los Alamos (Site Y): the top-secret nuclear research laboratory in New Mexico which developed both the U-235 and plutonium atomic bombs.

Los Alamos Primer: written by Robert Serber, a top-secret handbook about the U-235 and plutonium bombs available only to those working at Los Alamos.

Manhattan Project: the American-led collaboration with the British to build the atomic bomb. Administered by the US Army Corps of Engineers and named after the site of its headquarters in New York.

M.A.U.D. Committee: top-secret British group of nuclear physicists, chaired by George Thomson, which concluded that the U-235 'super-bomb' envisaged in the Frisch–Peierls memorandum was feasible but could only be built in collaboration with the Americans.

membrane diffusion: see gaseous diffusion.

moderator: material such as graphite or heavy water which slows neutrons to the optimal speed for igniting a controlled nuclear chain reaction in a uranium-fuelled reactor.

GLOSSARY

National Defense Research Committee (NDRC): high-level military research committee established by Roosevelt in 1940 and initially chaired by Vannevar Bush.

neptunium: element 93, the first man-made transuranic element; short-lived, decays into element 94 (plutonium).

neutron: uncharged nuclear particle, discovered by James Chadwick in 1932. Neutrons can smash U-235 and plutonium nuclei into smaller fragments ('fission'), and trigger an explosive nuclear chain reaction.

94: secret codename for plutonium, used in USA and Britain.

Norsk Hydro: Norwegian electricity producer and chemical manufacturer, which owned the heavy-water plant at Vemork.

nuclear chain reaction: self-perpetuating fission reaction, driven by the neutrons generated during fission.

nuclear cross-section: a measure of an element's capacity to capture neutrons and be transformed into another element, or to undergo fission.

nuclear fission: the splitting of a U-235 or plutonium nucleus by a single neutron, which releases immense energy and more neutrons. The term was coined by Otto Frisch in January 1939.

nuclear reactor: machine which converts the heat generated by a controlled nuclear chain reaction into electricity by boiling water to drive a steam turbine. Can also be used to irradiate uranium for the production of plutonium for nuclear weapons.

nucleus: the tiny centre of an atom (1/100,000th of its diameter), containing protons and neutrons and essentially all the atom's mass.

Oak Ridge (Site X): the U-235 production facility in Tennessee which made nuclear explosive for the Little Boy bomb.

Office for Scientific Research and Development (OSRD): top-level military research committee established by Roosevelt in spring 1941 and chaired by Vannevar Bush.

GLOSSARY

Operation Big: Alsos mission to capture members of the *Uranverein* and acquire their nuclear secrets.

Operation Epsilon: internment of *Uranverein* members under covert surveillance at Farm Hall, Godmanchester.

Operation Freshman: the SOE's catastrophic attempt to sabotage the Norsk Hydro heavy-water plant at Vemork, November 1942.

Operation Gunnerside: the successful SOE mission which wrecked the Norsk Hydro heavy-water plant at Vemork, February 1943.

periodic table: spreadsheet of the elements in ascending order of atomic number, organised into vertical groups which possess similar chemical properties. The first and lightest element is hydrogen (H) and the last and heaviest naturally occurring is element 92 (uranium).

pit: the plutonium core of Fat Man, under 4 inches in diameter and weighing only 13 pounds.

pitchblende: principal ore of uranium, discovered in Bohemian silver mines in 1595.

plutonium (Pu): element 94, created artificially by bombarding uranium with neutrons and isolated by Glenn Seaborg in 1943. Even more fissionable than U-235; the nuclear explosive in Fat Man, the bomb dropped on Nagasaki.

polonium (Po): element 84, discovered by Marie and Pierre Curie, and named after Poland. Emits high-energy alpha-particles; used in the urchin initiators to detonate the plutonium and U-235 bombs.

protactinium (Pa): element 91, immediately preceding uranium in the periodic table. Discovered in 1918 by Lise Meitner and Otto Hahn.

Pu: chemical symbol for plutonium.

Quebec Agreement: secret accord, signed in August 1943 by Roosevelt and Churchill, committing Britain and America to collaborate fully on developing the atomic bomb.

racetracks: massive arrays of calutrons at Oak Ridge, used to concentrate U-235 by electromagnetic separation.

GLOSSARY

radioactive decay: spontaneous emission of alpha- and/or beta-particles, transforming an element into a daughter product which differs in atomic number and/or weight and has distinct chemical or physical properties.

radioactivity: contraction of *activité radiante*, coined by Marie Curie, denoting the invisible rays emitted by certain elements.

radioisotopes: isotopes of elements, naturally occurring or artificially produced, which emit ionising radiation.

radium (Ra): element 88, discovered by Marie and Pierre Curie in 1898. A highly radioactive metal, emitting alpha- and beta-particles; decays into radon.

radon (Rn): element 86, discovered in 1899 by Ernest Rutherford (who named it the 'emanation'). An intensely radioactive gas, emitting powerful alpha-particles; decay product of radium, and in turn decays into polonium.

Secret Intelligence Service (SIS): British agency overseeing espionage and counter-espionage, including Military Intelligence Sections 5 and 6 (MI5 and MI6, the latter dealing with foreign activities).

Site W, X, Y: secret codenames for the 'Atomic Cities' of the Manhattan Project – Hanford, Oak Ridge and Los Alamos.

slow neutrons: travelling at less than one-fifth of the speed of a fast neutron; best suited for controlled fission of U-235 in a nuclear reactor, but far too slow to trigger an explosive chain reaction in an atomic bomb.

Smyth Report: official US history of the Manhattan Project, commissioned by General Groves in spring 1944, written by Professor Henry Smyth (Princeton) and published on 12 August 1945, three days after the atomic bombing of Nagasaki. The report focused exclusively on the American input to building the bombs.

S-1 Section and Committee: established in 1941; iterations of the Uranium Advisory Committee set up by Roosevelt in 1939.

Special Operations Executive (SOE): clandestine British organisation set up in 1940 to conduct reconnaissance, espionage and sabotage in

GLOSSARY

German-occupied Europe. Its missions included Operations Freshman and Gunnerside, targeting the Norsk Hydro heavy-water plant at Vemork.

TA: abbreviation for 'Tube Alloys' and by implication the British nuclear programme; also used in America as a codename for uranium.

tamper: heavy shell surrounding the nuclear explosive in an atomic bomb, which reflects escaping neutrons back into the chain reaction and enhances the power of the explosion.

thermal diffusion: method for concentrating U-235 from liquid hex.

Tizard Mission: The British Technical and Scientific Mission led by Sir Henry Tizard to Canada and the USA in 1940, primarily to exchange military scientific secrets and to try to establish free collaboration with the Americans.

transuranic elements: man-made elements, heavier than uranium and therefore lying beyond no. 92 in the periodic table. The first two to be created were elements 93 (neptunium) and 94 (plutonium).

Trinity: the test plutonium bomb, the forerunner of Fat Man, successfully exploded at Alamagordo, New Mexico, on 16 July 1945.

Tube Alloys (TA): the top-secret British organisation, established in November 1941, which took over from the M.A.U.D. Committee to direct the British war effort on uranium fission. This included both the 'fast neutron' programme (the U-235 bomb) and 'slow neutron' work (the boiler reactor developed by French refugees in Cambridge). In the USA, 'tube alloy' (or 'tuballoy') was used as a codename for uranium.

Tube Alloys Directorate: the entire Tube Alloys (TA) organisation. The TA Technical Committee coordinated hands-on research and reported to the TA Consultative Council which directly advised the British War Cabinet.

U: chemical symbol for uranium. U-235, U-238 are the main isotopes of uranium.

GLOSSARY

U-235: rare isotope, comprising only 0.7 per cent of native uranium. Unlike the inert U-238, U-235 is fissionable and can be used in an atomic bomb.

U-238: common isotope (99.2 per cent of native uranium), which does not undergo fission but can be converted into neptunium and then plutonium by bombardment with neutrons.

Uran: German for uranium.

Uranbrenner ('uranium burner'): the various designs of nuclear reactor invented by the *Uranverein*.

Uranium Advisory Committee: set up by Roosevelt in 1939. Chaired by Lyman Briggs, who remained chair of successive iterations of the committee (see S-1 Section) despite having no knowledge of nuclear physics.

Uranverein ('Uranium Club'): network of nuclear researchers scattered across Germany and Austria, established in 1939 to exploit nuclear fission for military purposes, notably reactors to power submarines and atomic bombs.

urchin: nickname for the polonium–beryllium initiators used to detonate the U-235 and plutonium bombs.

Valley: codename for the Special Products factory at Rhydymwyn, North Wales, where prototypes of the British gaseous diffusion U-235 separation equipment were to be tested.

Virus House: nickname for the secret *Uranbrenner* reactor facility built in the grounds of a virology research lab next to the KWI for Physics in Berlin.

ACKNOWLEDGEMENTS

This book took me so deep into alien territory that only an idiot wouldn't have shouted for help. So I did, and I'm extremely grateful to all those who came to my rescue; without them, *The Impossible Bomb* could well have lived up to its title.

Top of my thank-you list is Dr Robin Pitman OBE, alias The Man from Aldermaston. Robin fired me up with his own passion for the 'hidden story' of Tube Alloys, and his collection of memorabilia about the British nuclear weapons programme helped to simplify the task of untangling what actually happened. Robin has always been sparkling company, and he maintained his keen interest in the project despite deteriorating health. We'd both been looking forward to the moment when I could hand him an advance copy of the book, but sadly, that moment never came. Robin died in February 2025.

Next come another two wise men. Dr Richard Moore, who shuttles between the archives of the Atomic Weapons Establishment and the Department of War History at King's College London, generously shared his encyclopaedic knowledge across the field; learning from him has been an absolute pleasure. I also owe much to Alan Samson, whom I first met two books ago and who emerged from pseudo-retirement to provide advice and encouragement for this one. Alan – as always, thank you.

I'm indebted to several critical friends – Dr Jane Dalzell, Dr Bob Spencer, Mr John Rainey FRCS, Jenny Roberts, Alison Bell, Dr Vince

ACKNOWLEDGEMENTS

Smith OBE, Professor Paul Trayhurn FRSE, Professor Jon Arch, Ray and Jeanne Loadman, Dr Michael Bennett and Tracy Spencer – for their invaluable feedback, often by return of email, on interminable drafts of the manuscript. My thanks to them all for their stringent quality control, heroic patience and unflagging good humour. Ray deserves additional thanks and praise for bringing his characteristic clarity and style to the line drawings; this is our fifth book together.

These days, the internet makes light work of finding and checking material, but it will never match the thrill of visiting real archives. Allen Packwood (Director), Andrew Riley (Archives Manager) and their team at the Archives of Churchill College, Cambridge, have again been outstandingly helpful and accommodating. To them, my sincere thanks. I've greatly enjoyed learning about Sir Michael Perrin from Charles and Nicola Perrin, his son and granddaughter, who kindly granted permission to quote extracts from Perrin's diary and papers. I'm similarly grateful to the Niels Bohr Library and Archives at the American Institute of Physics, for allowing me to quote from the Oral Histories recorded by Lew Kowarski, Otto Frisch, Norman Feather, Laura Fermi, Walther Gentner and Hanni Bretscher; and to the Science History Institute, Philadelphia, for permission to quote from Alfred O.C. Nier's Oral History. My thanks also to Dr Christoph Laucht (University of Liverpool) for clarifying the Americanisation of the Manhattan story; Aurélie Lemoine (Musée Curie, Paris), Agence Di (Paris), Marie-Ève Rakuzin and Roger Nougaret (BNP Archives, Paris), Stephen Witkowski (British Council) and Dave Loadman for their expert assistance with elusive illustrations; and the Fitzwilliam Book Group for their feedback and encouragement.

Throughout the five years it took to write this book, my children and friends kept my spirits buoyant with their company, wit and wisdom. Special thanks go to Alison Bell, Jane Dalzell, Bob Spencer, Paul Beck, Tim and Julie Mann, Eryl and Trevor Daniels, Andy and Ainslie Levy, Ernest Woolford, and, for much-appreciated musical therapy, my fellow minstrels in the Rockhampton Wind Quintet, Class Act Quartet and Bristol Concert Orchestra.

Last, but only because they appear late in the life cycle of a book, everyone at Yale University Press London has been absolutely brilliant.

ACKNOWLEDGEMENTS

I'm indebted to Joanna Godfrey, Senior Commissioning Editor, who saw potential in this book while it was still embarrassingly embryonic. Without Jo's encouragement and firm steer, *The Impossible Bomb* would have been a significantly lesser offering. Top marks also to Katie Urquhart (Associate Editor and expert pathfinder through the permissions jungle), Rachael Lonsdale (Managing Editor and Design Manager), Lucy Buchan and Stuart Weir (proofs and indexing), David Watkins (master of the invisible art of copy-editing), Martin Brown (creator of beautiful maps) and Jack Smyth (for his stunning cover design). We all hope that the book does well; the ever-dynamic Chloe Foster and Zara Gillick in Publicity and Marketing have given it the best possible start on its journey. To all the team at Bedford Square, my heartfelt thanks.

This was the first book for which I couldn't share the ups and downs of writing with my wife Caroline (1955–2019). Having sat with me through grim meetings of the Medical Campaign Against Nuclear Weapons during the Cold War years, she'd have been surprised by this one. I believe, though, that she would have found it a thriller as well as a horror story – which is what I hope for you too.

Rockhampton, Gloucestershire
May 2025

NOTES

1 Setting the Scene

1. M. René, 'History of uranium mining in Central Europe' (2018): http://dx.doi.org/10.5772/intechopen.71962
2. René (2018), p. 2.
3. René (2018), pp. 3–4.
4. M. Kaji, 'D.I. Mendeleev's concept of chemical elements and *The Principles of Chemistry*', *Bull Hist Chem*, 27 (2002), pp. 9–10.
5. P. Fournier and J. Fournier, 'Hasard ou mémoire dans la découverte de la radio-activité?', *Rev Hist Sci*, 52 (1999), pp. 51–79; M. Meyer and E. Gonthier. 'Y a-t-il encore polémique autour de la découverte des phénomènes dits radioactifs?', *Sci Trib* (June 1997): http://www.tribunes.com/tribune/art97/meyer.dtm
6. L. Badash, 'Radioactivity before the Curies', *Am J Physics*, 33 (1965), pp. 128–35.
7. È. Curie, *Madame Curie* (London: William Heinemann, 1937), pp. 1–118.
8. Curie (1937), pp. 120–50.
9. Curie (1937), pp. 153–65, 177; P. Curie and Mme P. Curie, 'Sur une substance nouvelle radio-active, contenue dans la pechblende', *C R Acad Sci Paris*, 127 (1898), pp. 175–8; P. Curie, Mme P. Curie and G. Bémont. 'Sur une substance fortement radio-active, contenue dans la pechblende', *C R Acad Sci Paris*, 127 (1898), pp. 1215–17.
10. Curie (1937), pp. 208, 216–23.
11. American Institute of Physics web exhibit, *Marie Curie and the Science of Radioactivity*: https://history.aip.org/exhibits/curie/recdis2.htm
12. A.H. Becquerel, 'On radioactivity, a new property of matter', Nobel Lecture, 11 December 1903: https://www.nobelprize.org/prizes/physics/1903/becquerel/lecture; Fournier and Fournier (1999), pp. 51–79; Curie (1937), pp. 210–11.
13. G.F. Kunz and C. Baskerville, 'The action of radium and Roentgen rays on minerals and gems', *Science*, 18 (1903), pp. 769–83; A.S. Green, 'Notes on two cases of rodent ulcer treated by radium', *Lancet*, 163 (1904), p. 794; F. Soddy, *The Interpretation of Radium and the Structure of the Atom* (London: John Murray, 1909), p. 16.

14. Becquerel (1903), pp. 56–63.
15. A.S. Eve and J. Chadwick, 'Lord Rutherford, 1871–1937', *Biogr Mem Fell Roy Soc*, 2 (1938), pp. 395–8.
16. E. Rutherford, 'Uranium radiation and the electrical conduction produced by it', *Philos Mag*, 47 (1899), pp. 109–63.
17. E. Rutherford and T. Royds, 'The nature of the alpha particle from radioactive substances', *Philos Mag*, 17 (1909), pp. 281–6.
18. Soddy (1909), p. 105.
19. E. Rutherford and F. Soddy, 'The cause and nature of radioactivity, part 1', *Philos Mag*, 4 (1902), pp. 370–96; Soddy (1909), pp. 142–51.
20. M. Gowing. *Britain and Atomic Energy, 1935–1945* (London: Macmillan Publishing, 1964), pp. 5, 10; Eve and Chadwick (1938), pp. 405–6.
21. E. Rutherford. 'The scattering of alpha and beta particles by matter and the structure of the atom', *Philos Mag*, 21 (1911), pp. 669–88.
22. Soddy (1909), pp. 78–9.
23. Soddy (1909), pp. 227–33, 244–8.
24. Soddy (1909), pp. 241–5; F. Soddy. 'The atomic weight of "thorium" lead', *Nature*, 98 (1917), p. 469.
25. Soddy (1909), pp. 78–9.
26. Soddy (1909), pp. 168–77, 181–5.
27. H.G. Wells, *The World Set Free: A Story of Mankind* (London: Macmillan & Co., 1914), pp. 106–7.

2 War Games

1. M. von Laue, 'Fritz Haber, 1868–1933', *Naturwissenschaften*, 22 (1934), p. 97.
2. Chronik des Kaiser-Wilhelm-Max-Planck-Institut–für Chemie: https://www.mpic.de/de/geschichte; R. Spence, 'Otto Hahn, 1879–1968', *Biogr Mem Fell Roy Soc*, 16 (1970), pp. 282–3.
3. Spence (1970), pp. 284–6; O.R. Frisch, 'Lise Meitner, 1878–1968', *Biogr Mem Fell Roy Soc*, 16 (1970), pp. 405–6; Chronik des Kaiser-Wilhelm-Max-Planck-Institut–für Chemie: https://www.mpic.de/de/geschichte.
4. Spence (1970), p. 287.
5. J. Campbell, 'Rutherford's war', *Physics World* (February 2016), pp. 1–12.
6. Curie (1937), pp. 295, 304.
7. Curie (1937), pp. 297, 303–5.
8. B. Friedrich, 'Fritz Haber (1868–1934)', *angewandte Chemie* (Intl Edn), 44 (2005), pp. 3967–8.
9. Friedrich (2005), pp. 3969–70.
10. Curie (1937), p. 311.
11. Frisch (1970), p. 407; L. Meitner, 'Ueber das protactinium', *Naturwissenschaften*, 6 (1918), pp. 324–7.
12. J. Navarro, *Ether and Modernity* (Oxford: Oxford University Press, 2018), p. 115.
13. J.F. Mulligan, 'Heinrich Hertz and Philipp Lenard: two distinguished physicists, two disparate men', *Phys in Perspect*, 1 (1999), pp. 350–2, 361; P. Ball, *Serving the Reich: The Struggle for the Soul of Physics under Hitler* (Chicago: University of Chicago Press, 2014), pp. 21–42; B.R. Wheaton, 'Philipp Lenard and the photoelectric effect, 1889–1911', *Hist Studies Physical Sci*, 9 (1978), pp. 299–322.

14. A. Einstein, 'Zur Elektrodynamik bewegter Körper', *Annalen der Physik*, 17 (1905), pp. 891–916; T. Jayaraman, 'Albert Einstein: radical pacifist and democrat', *Current Sci*, 89 (2005), pp. 2141–6; Mulligan (1999), p. 358.
15. Jayaraman (2005), p. 2142.
16. Ball (2014), pp. 40–2.
17. Friedrich (2005), p. 3968; M. Dunikowska and L. Turko, 'Fritz Haber, the damned scientist', *angewandte Chemie*, 50 (2011), p. 10057.
18. Citation for Haber's Nobel Prize in Chemistry 1918: https://www.nobelprize.org/prizes/chemistry/1918/ceremony-speech/
19. Wells (1914), pp. 106–7.
20. P. Curie, 'Radioactive substances, especially radium', Nobel Lecture, 6 June 1905, p. 78: https://www.nobelprize.org/prizes/physics/1903/pierre-curie/lecture/

3 Pure Physics

1. A. Gall, 'Otto Baumbach – Rutherford's glassblower', *Hist Physics Newsletter*, 223 (2008), pp. 47–8: https://www.physicshistory.org.uk/NewsLTR/n23/n23p44.htm
2. A. Brown, *The Neutron and the Bomb: A Biography of Sir James Chadwick* (Oxford: Oxford University Press, 1997), pp. 8–11.
3. Brown (1997), pp. 14–18, 21–2, 27–34, 39.
4. Eve and Chadwick (1938), p. 411; N. Todd, 'The centenary of transmutation', *Inst Physics Hist Group Newsletter* (October 2021), pp. 23–36.
5. J.D. Cockcroft, 'Niels Henrik David Bohr, 1885–1962', *Biogr Mem Fell Roy Soc*, 9 (1963), pp. 41–2; G. Hartcup and T.E. Allibone, *Cockcroft and the Atom* (Bristol: Adam Hilger (Institute of Physics), 1984), pp. 121–2.
6. Cockcroft (1963), p. 42.
7. Brown (1997), p. 50; E. Rutherford (1919), 'Collision of alpha-particle with light atoms. IV. An anomalous effect in nitrogen', *Philos Mag*, 37 (1919), pp. 581–8; E. Rutherford and J. Chadwick, 'The disintegration of elements by alpha-particles', *Philos Mag*, 44 (1922), pp. 417–32.
8. Eve and Chadwick (1938), p. 412.
9. Curie (1937), pp. 310–11.
10. Curie (1937), pp. 330–3.
11. Curie (1937), pp. 311, 354–6.
12. Curie (1937), p. 363.
13. Curie (1937), pp. 340, 346.
14. Curie (1937), pp. 354–6.
15. G. Gilmore and G. Tausch-Pebody, 'The 1919 eclipse results that verified general relativity, and their later detractors: a story re-told', *Notes Rec Roy Soc*, 76 (2021), pp. 155–80: http://doi.org/10.1098/rsnr.2020.0040
16. D.E. Rowe, 'Einstein and relativity: what price fame?', *Science in Context*, 25 (2012), p. 21: doi.10.1017/SO26988971200004X
17. Citation for Einstein's Nobel Prize in Physics 1921: https://nobelprize.org/prizes/physics/1921/summary
18. Rowe (2012), pp. 225–7; J. van Dongen, 'Reactionaries and Einstein's fame: "German Scientists for the Preservation of Pure Science", relativity, and the Bad Neuheim meeting', *Physics in Perspective*, 4 (2007), pp. 212–16: https://doi.org/10.1007/s00016-006-0318-y

19. Van Dongen (2007), pp. 217–22.
20. Van Dongen (2007), pp. 223–4; Rowe (2012), pp. 224, 227.
21. M. Walker, *Nazi Science: Myth, Truth, and the German Atomic Bomb* (New York: Plenum Press, 1995), pp. 7–9, 11–13.
22. Walker (1995), pp. 12–13.
23. Walker (1995), pp. 14–16.
24. M. Janssen et al. (eds), *The Collected Papers of Albert Einstein*, Vol. 7: The Berlin Years (Princeton, NJ: Princeton University Press, 2002), p. 113.
25. N. Mott and R. Peierls, 'Werner Heisenberg, 1901–1976', *Biogr Mem Fell Roy Soc*, 23 (1977), pp. 219–25.
26. R. Peierls, *Bird of Passage* (Princeton, NJ: Princeton University Press, 1979), pp. 32–48.
27. Peierls (1979), pp. 55–9; O.R. Frisch, *What Little I Remember* (Cambridge: Cambridge University Press, 1979), pp. 90–2.

4 Years of Wonder

1. S. Cockburn and D. Ellyard, *Oliphant: The Life and Times of Sir Mark Oliphant* (Adelaide: Axiom Books, 1981), pp. 21–30.
2. Brown (1997), pp. 103–4.
3. J. Chadwick, 'Possible existence of a neutron', *Nature*, 129 (1932), p. 312; J. Chadwick, 'The existence of a neutron', *Proc Roy Soc A*, 136 (1932), pp. 692–708.
4. Hartcup and Allibone (1984), pp. 9–16, 21–40.
5. Hartcup and Allibone (1984), pp. 41–6; M.L.E. Oliphant and W. Penney, 'John Douglas Cockcroft, 1897–1967', *Biogr Mem Fell Roy Soc*, 14 (1968), pp. 148–9; E.C. Finch, 'E.T.S. Walton: atom splitter and man of peace', *Dublin Hist Rec*, 60 (2007), pp. 111–23.
6. Hartcup and Allibone (1984), pp. 9–16, 21–40.
7. Hartcup and Allibone (1984), pp. 52–3; Oliphant and Penney (1968), pp. 149–50.
8. J.D. Cockcroft and E.T.S. Walton, 'Disintegration of lithium by swift protons', *Nature*, 129 (1932), p. 649; B. Cathcart, 'Rutherford's resonance: responses to discoveries in 1911 and 1932', *Inst Physics Hist Phys Group Newsletter* (December 2012), pp. 39–48.
9. Cockcroft and Walton (1932), pp. 39–48.
10. T. Levenson (2017), 'The scientist and the fascist', *Atlantic* (9 June 2017), pp. 2–4: https://www.theatlantic.com/science/archive/2017/06/einstein-germany-and-the-bomb/528534/
11. Levenson (2017), pp. 6–8; Letter from Einstein to his son Eduard, from SS *Belgenland*, 28 March 1933, Shapell Manuscript Foundation, SMC 1792: https://www.shapell.org/manuscript/einstein-renounces-german-citizenship-becomes-outlaw-in-nazi-germany/
12. Levenson (2017), pp. 6–8; Letter from Einstein to his son Eduard (1933).
13. M. Planck, 'Mein Besuch mit Adolf Hitler', *Physikalische Blätter*, 3 (1947), p. 143: https://doi.org/10.1002/phbl.19470030502; Levenson (2017), p. 6.
14. D. Ferguson, 'Einstein on the run: how the world's greatest scientist hid from Nazis in a Norfolk hut', *Observer* (10 February 2024); G. Hitchings and D. Styan, *Locker-Lampson: Einstein's Protector* (Cromer, Norfolk: Norfolk Museums, 2010).

15. Walker (1995), pp. 20–1; E. Rutherford, 'Makers of Science', *Nature*, 132 (1933), p. 367.
16. R.J. Evans, *The Coming of the Third Reich* (London: Penguin, 2005), pp. 532–3.
17. M.G. Manning, *When Books Went to War* (Boston: Houghton Mifflin Harcourt, 2014), pp. 1–8; M. Richardson, 'Philanthropy and the internationality of learning', *Minerva*, 28 (1990), p. 38.
18. A list of authors and publishers banned by the Nazis can be found at: https://verbrannte-und-verbannte.de
19. Manning (2014), pp. 1–8.
20. Walker (1995), pp. 18–19; P. Ball, 'How 2 pro-Nazi Nobelists attacked Einstein's "Jewish Science"', *Sci American* (13 February 2015): https://www.scientificamerican.com/article/how-2-pro-nazi-nobelists-attacked-einstein-s-jewish-science-excerpt1/
21. Walker (1995), p. 16.
22. H. Heine, from his play *Almansor* (1821): 'Dort, wo man Bücher verbrennt, verbrennt man am Ende auch Menschen.'
23. Ball, 'How 2 pro-Nazi Nobelists attacked Einstein's "Jewish Science"' (2015), p. 10.
24. J.L. Heilbron, *The Dilemmas of an Upright Man: Max Planck as Spokesman for German Science* (Cambridge, MA: Harvard University Press, 1996), p. 190.
25. L. Kowarski, interview by Charles Weiner, New York, Session I, 20 March 1969, p. 6, American Institute of Physics: https://www.aip.org/history-programs/niels-bohr-library/oral-histories/4717-1
26. 'A second generation of Curies: Jean-Frédéric Joliot and Irène Curie', American Institute of Physics: https://history.aip.org/exhibits/curie/2ndgen1.htm
27. Brown (1997), pp. 103–4.
28. 'Carl D. Anderson: Facts', The Nobel Prize: https://www.nobelprize.org/prizes/physics/1936/anderson/facts/
29. The time taken for the strength of emitted radiation to fall by 50 per cent; the unique 'signature' of a radioisotope.
30. F. Joliot and I. Curie, 'Artificial production of a new type of radioactive element', *Nature*, 133 (1934), pp. 201–3.
31. I. Joliot-Curie, 'Artificial production of radioactive elements', Nobel Lecture, 12 December 1935: https://www.nobelprize.org/prizes/chemistry/1935/joliot-curie/lecture/
32. Joliot-Curie, 'Artificial production of radioactive elements', Nobel Lecture (1935).
33. H. Massey and N. Feather, 'James Chadwick, 1891–1974', *Biogr Mem Fell Roy Soc*, 22 (1976), p. 25.
34. Joliot-Curie, 'Artificial production of radioactive elements', Nobel Lecture (1935); F. Joliot, 'Chemical evidence of the transmutation of elements', Nobel Lecture, 12 December 1935: https://www.nobelprize.org/prizes/chemistry/1935/joliot-fred/lecture/
35. Curie (1937), pp. 385–6.
36. Anon., 'Philipp-Lenard-Institut at Heidelberg', *Nature*, 137 (1936), pp. 93–4.
37. A. Grant, 'The scientific exodus from Nazi Germany', *Physics Today* (2018): https://doi.org/10.1063/PT.6.4.20180926a
38. Anon., 'Heavy water in chemistry', *Nature*, 134 (1934), p. 843: https://doi.org/10.1038/134843a0

39. Heilbron (1996), p. 163; M. Oliphant, P. Harteck and E. Rutherford, 'Transmutation effects observed with heavy hydrogen', *Proc Roy Soc A*, 144 (1934), pp. 692–703.
40. P.F. Dahl, *Heavy Water and the Wartime Race for Nuclear Energy* (Philadelphia: CRC Press, 1999), pp. 41–8.

5 Beyond Nature?

1. O.R. Frisch, 'Induced radioactivity of sodium and phosphorus', *Nature*, 133 (1934), pp. 721–2.
2. Frisch was generally called 'Robert' but became 'Otto' at Los Alamos, to distinguish himself from the many other Roberts there.
3. Frisch (1979), pp. 1–11; O.R. Frisch, interview by Charles Weiner, American Institute of Physics, 3 May 1967, p. 1: https://www.aip.org/history-programs/niels-bohr-library/oral-histories/4616
4. Frisch (1979), pp. 31–41; Frisch (1967), pp. 1–2.
5. Frisch (1979), pp. 41–52; Frisch (1967), pp. 1–2.
6. Frisch (1979), pp. 53–6; Frisch (1967), p. 2.
7. Frisch (1979), pp. 75–6; Frisch (1967), p. 2.
8. Frisch (1979), pp. 76–80; Frisch (1967), p. 2.
9. Frisch (1979), pp. 74, 77–80.
10. Brown (1997), p. 125; S.K. Allison, 'Enrico Fermi, 1901–1954', *Biograph Memoirs Nat Acad Sci*, 30 (1957), p. 134; S.M. DiScala, 'Science and Fascism: the case of Enrico Fermi', *Totalit Movts Polit Relig*, 6 (2011), pp. 200–2.
11. E. Fermi, E. Amaldi et al., 'Artificial radioactivity produced by neutron bombardment', *Proc Roy Soc A*, 146 (1934), pp. 483–500; E. Fermi, E. Amaldi et al., 'Possible production of elements of atomic number higher than 92', *Nature*, 133 (1934), pp. 898–9; DiScala (2011), p. 203.
12. DiScala (2011), p. 203.
13. I. Noddack, 'Über das Element 93', *Zeitschr für angewandte Chemie*, 47 (1934), pp. 653–5; G. Hermann, 'Five decades ago: from the transuranics to nuclear fission', *angewandte Chemie* (Intl Edn), 29 (1990), pp. 483–4; M. Gowing, *Britain and Atomic Energy, 1935–1945* (London: Macmillan Publishing, 1964), p. 26n.
14. Noddack (1934), pp. 653–5; Hermann (1990), pp. 483–4; Gowing (1964), p. 26n.
15. Hermann (1990), pp. 483–5.
16. Hermann (1990), pp. 486–7.
17. R.L. Sime, *Lise Meitner: A Life in Physics* (Berkeley, CA: University of California Press, 1996), p. 285.
18. Hermann (1990), pp. 488–9.
19. Hermann (1990), p. 488.
20. Allison (1957), p. 129.
21. E. Bretscher and J.D. Cockcroft, 'Enrico Fermi, 1901–1954', *Biogr Mem Fell Roy Soc*, 1 (1955), pp. 71–2.
22. L. Bonolis, 'Enrico Fermi's scientific work', in C. Bernardini and L. Bonolis (eds), *Enrico Fermi: His Work and Legacy* (Bologna: Società Italiana di Fisica/Springer, 2001), p. 321; A.J. Dempster, 'Isotopic constitution of uranium', *Nature*, 136 (1935), p. 180.
23. Massey and Feather (1976), pp. 22–5; Dahl (1999), p. 79.

24. Brown (1997), p. 123.
25. Brown (1997) pp. 23, 158–9.
26. J.J. Thomson quoted in *Time Magazine*, 1 November 1937; Eve and Chadwick (1938), p. 422; Ernest Rutherford quotes: https://www.goodreads.com/author/quotes/437411.Ernest_Rutherford
27. J.G. Jenkin, 'Atomic energy is "moonshine": what did Rutherford *really* mean?', *Physics in Perspect*, 13 (2011), pp. 128–45: https://doi.org/10.1007/s00016-010-0038-1
28. Walker (1995), pp. 36–9, 43–7.
29. Walker (1995), pp. 42–3, 47.
30. Ball (2015).

6 To the Brink

1. A. Grant, 'The scientific exodus from Nazi Germany', *Physics Today* (26 September 2018): https://pubs.aip.org/physicstoday/online/5299/The-scientific-exodus-from-Nazi-Germany
2. Sime (1996), pp. 184–94; Frisch (1970), pp. 410–1.
3. Sime (1996), pp. 195–204.
4. Frisch (1970), pp. 411–12.
5. DiScala (2011), pp. 204–5.
6. 'To Fermi – with love – Part 2', interview with Laura Fermi, 12 September 1971, Voices of the Manhattan Project, Atomic Heritage Foundation: https://ahf.nuclearmuseum.org/voices/oral-histories/fermi-love-part-2/
7. Frisch (1970), p. 413; Hermann (1990), p. 494.
8. Hermann (1990), p. 490.
9. O. Hahn and F. Strassmann, 'Über den Nachweis und das Verhalten der bei der Bestrahlung des Urans mittels Neutronen entstehenden Erdalkalimetalle' [Concerning the existence of alkali metals resulting from neutron irradiation of uranium], *Naturwissenschaften*, 27 (1939), pp. 11–15.
10. Frisch (1979), pp. 83–4, 90–2.
11. Frisch (1979) pp. 98, 114.
12. Frisch (1979), pp. 114–16; Frisch (1967), p. 3; Frisch (1970), p. 413.
13. Frisch (1967), p. 3; Frisch (1979), pp. 115–16.
14. Frisch (1979), pp. 116–17; R.H. Stuewer, 'Bringing the news of fission to America', *Physics Today*, 38 (1985), p. 51: https://doi.org/10.1063/1.881016
15. Frisch (1979), p. 117; Frisch (1967), p. 3.
16. Stuewer (1985), pp. 49, 51–2.
17. Stuewer (1985), p. 52.
18. Stuewer (1985), pp. 54–5.
19. L. Meitner and O.R. Frisch, 'Disintegration of uranium by neutrons: a new form of nuclear reaction', *Nature*, 143 (1939), pp. 239–40; O.R. Frisch, 'Physical evidence for the division of heavy nuclei under neutron bombardment', *Nature*, 143 (1939), p. 276.
20. Frisch (1979), p. 118.
21. N. Halasz and R. Halasz, 'Leo Szilard: the reluctant father of the atomic bomb', *New Hungarian Quarterly*, 15 (1974), pp. 165–6; R. Rhodes, *The Making of the Atomic Bomb* (New York: Simon & Schuster, 1986), p. 215.

22. T. Frank, 'Ever ready to go: the multiple exiles of Leo Szilard', *Physics in Perspective*, 7 (2005), pp. 204–52.
23. Rhodes (1986), pp. 203–4; L. Badash, E. Hodes and A. Tiddens, 'Nuclear fission: reaction to the discovery in 1939', *Proc Am Philosoph Soc*, 130 (1986), p. 209.
24. W. Lanouette, 'The odd couple and the bomb', *Sci Am*, 283 (2000), pp. 105–6.
25. Rhodes (1986), pp. 291–2.
26. L. Szilard and W.H. Zinn, 'Instantaneous emission of fast neutrons in the interaction of slow neutrons with uranium', *Physical Rev*, 55 (1939), pp. 799–800.
27. O. Hahn and F. Strassmann, 'Nachweis der Entstehung activer Bariumisotoper aus Uran und Thorium durch Neutronenbestrahlung' [Proof of the formation of active isotopes of barium from uranium and thorium irradiated with neutrons], *Naturwissenschaften*, 27 (1939), pp. 89–95.
28. F. Joliot, 'Preuve expérimentale de la rupture explosive des noyaux d'uranium et de thorium sous l'action des neutrons' [Experimental evidence of the explosive rupture of uranium and thorium nuclei under the action of neutrons], *C R Acad Sci*, 208 (1939), pp. 341–2.
29. F. Joliot, H. Halban, L. Kowarski and F. Perrin, 'Mise en évidence d'une réaction nucléaire en chaîne au sein d'une masse uranifère' [Evidence of a nuclear chain reaction within a mass of uranium], *J Phys*, 10 (1939), pp. 428–31.
30. F. Joliot, 'Chemical evidence of the transmutation of the elements', Nobel Lecture, 10 December 1935.
31. P.M.S. Blackett, 'Fréderic Joliot, 1900–1958', *Biogr Mem Fell Roy Soc*, 6 (1960), p. 94; D. Mongin, 'Aux origines du programme atomique militaire français', *Matériaux pour l'Histoire de Notre Temps*, 31 (1993), p. 13.
32. Rhodes (1986), p. 259. No citation is provided and the source has proved elusive. See https://hsm.stackexchange.com/questions/13903/whats-the-origin-of-the-claim-that-a-single-uranium-atom-fissioning-would-relea
33. S. Flügge, 'Kann der Energieinhalt der Atomkerne technisch nutzbar gemacht werden?' [Can the energy content of the nucleus be practically harnessed?], *Naturwissenschaften*, 27 (1939), pp. 402–10.
34. Badash et al. (1986), p. 214.
35. Badash et al. (1986), pp. 214–15.
36. S.R. Weart, 'Scientists with a secret', *Physics Today*, 29 (1978), p. 26: https://doi.org/10.1063/1.3023312
37. Halasz and Halasz (1974), p. 163.
38. Letter from Albert Einstein to President F.D. Roosevelt, 2 August 1939. Facsimile in Franklin Delano Roosevelt Presidential Library and Museum: https://www.fdrlibrary.org/documents/356632/390886/document007.pdf/3483329d-7b68-442d-953d-eb91e0c5c9b1
39. Lanouette (2000), p. 108.
40. Spence (1970), p. 296.
41. J.L. Logan et al., 'Heisenberg, Goudsmidt and the German "A" bomb', *Physics Today*, 44 (1991), pp. 13–14.
42. Brown (1997), p. 174.
43. N. Bohr and J.A. Wheeler, 'The mechanism of nuclear fission', *Phys Rev*, 56 (1939), pp. 426–50: https://doi.org/10.1103/PhysRev.56.426
44. A. Nier, 'The isotopic constitution of uranium and the half-lives of the uranium isotopes', *Phys Rev*, 55 (1939), pp. 150–3.

45. Rhodes (1986), p. 286; B.C. Reed, 'From fission to censorship: 18 months on the road to the bomb', *Ann Phys (Berlin)*, 530 (2018), p. 1700455 (2/5).
46. Interview with John Wheeler, 12 September 1965, Voices of the Manhattan Project, Atomic Heritage Foundation: https://ahf.nuclearmuseum.org/voices/oral-histories/john-wheeler
47. Frisch (1979), pp. 121–2.
48. Reed (2018), p. 1700455 (3/5).
49. Anon., 'The possibility of producing an atomic bomb. A review of the position', 3 May 1939, The National Archives, Kew, PRO AB1/9.
50. Reed (2018), p. 1700455 (3–4/5).

7 Warm-Up

1. Frisch (1979), pp. 108, 120–2.
2. Frisch (1979) p. 122; Frisch (1967), p. 4.
3. Frisch (1979), pp. 123–4.
4. Frisch (1979), pp. 118–19, 122, 124–5.
5. Peierls (1979), pp. 1, 16–23, 32, 40–6, 55.
6. Peierls (1979), pp. 61–70.
7. Peierls (1979), pp. 84–6, 90–6, 100–1, 114–20, 127–35; S. Lee, 'Sir Rudolf Ernst Peierls, 1907–1995', *Biogr Mem Fell Roy Soc*, 53 (2007), p. 270.
8. Peierls (1979), pp. 140–3.
9. Peierls (1979), p. 153; R. Clark, *Tizard* (London: Methuen, 1965), p. 43.
10. R. Clark, *The Birth of the Bomb: The Untold Story of Britain's Part in the Weapon That Changed the World* (London: Phoenix House, 1961), p. 33; A. Whitaker, 'Henry Tizard, 1885–1959', *Inst of Physics Newsletter* (October 2019), pp. 46–7, 51–4.
11. Whitaker, 'Henry Tizard, 1885–1959' (2019), p. 54; Anon., obituary of Dr H.E. Wimperis, *Nature*, 188 (1960), pp. 622–3.
12. B. Katz, 'Archibald Vivian Hill, 1886–1977', *Biogr Mem Fell Roy Soc*, 24 (1978), pp. 87–9, 106–7, 109–10; Clark (1961), pp. 33–4.
13. B. Lovell, 'Patrick Maynard Stuart Blackett, 1897–1974', *Biogr Mem Fell Roy Soc*, 21 (1975), pp. 3–10, 22–9, 50.
14. Clark (1961), p. 116.
15. Clark (1961), pp. 33–4, 63–4.
16. Lovell (1975), pp. 51–2; Whitaker, 'Henry Tizard, 1885–1959' (2019), p. 54.
17. Gowing (1964), pp. 34–5; Clark (1965), pp. 35–6.
18. Gowing (1964), pp. 35–6.
19. Gowing (1964), pp. 35–7.
20. Gowing (1964), p. 36.
21. Clark (1961), pp. 33–4.
22. Clark (1961), pp. 35–6; H.A.H. Boot and J.T. Randall, 'Historical notes on the cavity magnetron', *IEEE Transactions on Electron Devices*, 23 (1976), pp. 724–9.
23. Brown (1997), pp. 177–83.
24. Brown (1997), p. 179; G. Farmelo, *Churchill's Bomb* (New York: Basic Books, 2013), pp. 110–11.
25. Gowing (1964), p. 39.
26. G.P. Thomson, 'Frederick Alexander Lindemann, Viscount Cherwell, 1886–1957', *Biogr Mem Fell Roy Soc*, 4 (1958), p. 58.

27. Lanouette (2000), pp. 105–6; Halasz and Halasz (1974), p. 167.
28. Lanouette (2000), p. 107.
29. 'Fermi on Chicago Pile-1', Manhattan Project History, Atomic Heritage Foundation: https://ahf.nuclearmuseum.org/ahf/key-documents/fermi-chicago-pile-1/#:~:text=Italian%20physicist%20Enrico%20Fermi%20directed,1%2C%20in%20the%20excerpt%20below
30. W. Lanouette and B. Silard, *Genius in the Shadows: A Biography of Leo Szilard* (New York: Skyhorse Publishing, 1992), pp. 194–5; E. Fermi, 'The development of the first chain-reacting pile', *Proc Am Philosoph Soc*, 90 (1946), pp. 20–4.
31. L. Szilard and W.H. Zinn, 'Instantaneous emission of fast neutrons in the interaction of slow neutrons with uranium', *Phys Rev*, 55 (1939), pp. 799–800.
32. Rhodes (1986), pp. 309–13.
33. Rhodes (1986), p. 314; R. Jungk, *Brighter Than a Thousand Suns: A Personal History of the Atomic Scientists* (New York: Harcourt, Brace, 1956), pp. 109–11.
34. Anon., obituary of Lyman J. Briggs, *Physics Today*, 16 (1963), p. 104: https://doi.org/10.1063/1.3050957

8 Liaisons Dangereuses

1. Rhodes (1986), p. 315; R.G. Hewlett and O.E. Anderson, *The New World, 1939–1946* (University Park, PA: Pennsylvania State University Press, 1962), pp. 19–20.
2. Rhodes (1986), pp. 315–17; Hewlett and Anderson (1962), pp. 20–1.
3. Rhodes (1986), p. 317.
4. Hewlett and Anderson (1962), p. 23.
5. Weart (1978), pp. 25, 29.
6. L.A. Turner, 'Nuclear fission', *Rev Mod Physics*, 12 (1940), pp. 1–29.
7. J.A. Wheeler, 'Fission in 1939', *Ann Rev Nucl Part*, 39 (1989), p. xxvii; A.O. Nier, E.T. Booth, J.R. Dunning and A.V. Grosse, 'Nuclear fission of separated uranium isotopes', *Phys Rev*, 57 (1940), p. 546; Rhodes (1986), pp. 322–3.
8. Rhodes (1986), p. 297.
9. Weart (1978), p. 28.
10. Reed (2018), p. 1700455 (4/5); Weart (1978), p. 29.
11. E. McMillan and P.H. Abelson, 'Radioactive element 93', *Phys Rev*, 57 (1940), pp. 1185–6; Reed (2018), p. 1700455 (4/5).
12. Brown (1997), p. 206.
13. Reed (2018), p. 1700455 (3–4/5); H.H. McAllister, obituary of Gregory Breit, *Physics Today*, 36 (1983), pp. 103–4.
14. S. Goldberg, 'Inventing a climate of opinion: Vannevar Bush and the decision to build the bomb', *Isis*, 83 (1992), p. 430; Rhodes (1986), pp. 331–3, 338; A.M. Weinberg, 'Eugene Wigner, nuclear engineer', *Physics Today*, 55 (2002), pp. 42–6.
15. Hewlett and Anderson (1962), pp. 24–5; H.D. Smyth, *Atomic Energy for Military Purposes* (Princeton, NJ: Princeton University Press, 1945), pp. 48–9. Full text at: https://www.osti.gov/opennet/manhattan-project-history/publications/smyth_report.pdf
16. Hewlett and Anderson (1962), pp. 24–5.
17. J. Chadwick, obituary of Mme Irène Joliot-Curie, *Nature*, 177 (1956), pp. 964–5.
18. P. Craig, 'Frédéric Joliot and France's nuclear heritage', *New Sci* (7 February 1985), pp. 16–19.

19. Blackett (1960), pp. 94–6; C. Waltham, 'An early history of heavy water', *Physics in Canada*, 49 (1993), pp. 81–6; B. Goldschmidt, 'France's contribution to the discovery of the chain reaction', *Internat Atomic Energy Agency Bulletin*, 4-0 (1963), pp. 22–3: https://www.iaea.org/publications/magazines/bulletin/4-0/frances-contribution-discovery-chain-reaction
20. F. Joliot, H. Halban and L. Kowarski (1949), 'Sur la possibilité de produire dans un milieu uranifère des réactions nucléaires en chaîne illimitée', *C R Acad Sci Paris*, 299, 19 (1949). Deposited in a sealed envelope (No. 11.620) at the academy on 30 October 1939; opened 18 August 1948; presented at the session of 7 November 1949.
21. Clark (1961), p. 23; Dahl (1999), pp. 101–4.
22. Dahl (1999), pp. 104–5.
23. Dahl (1999), pp. 104–7.
24. Dahl (1999), pp. 105, 107–8.
25. Dahl (1999), pp. 108–9.
26. Dahl (1999), p. 109.
27. Dahl (1999), p. 109.
28. A. Encrevé, 'Jacques Allier', in P. Cabanel and A. Encrevé (eds), *Dictionnaire Biographique des Protestants Français de 1787 à nos jours* (Paris: Editions Max Chaleil, 2015), vol. 1, p. 45; H. Bhys, 'The perilous mission of a Paris banker during the "Phoney War"', *Historia* (2022), BNP Paribas: https://histoire.staging.bnpparibas/en/the-perilous-mission-of-a-paris-banker-during-the-phoney-war/
29. Gowing (1964), p. 39; P.B. Moon, 'George Paget Thomson, 1892–1975', *Biogr Mem Fell Roy Soc*, 23 (1977), pp. 542–3.
30. A. Marsh, 'From World War II radar to microwave popcorn', *IEEE Spectrum* (31 October 2018).
31. Brown (1997), pp. 178, 186.
32. Brown (1997), p. 181. For a concise explanation of neutrons and uranium-bomb physics, see J. Bernstein, *Hitler's Uranium Club: The Secret Recordings at Farm Hall* (New York: Springer Verlag, 2001), p. 23.
33. Brown (1997), p. 181.
34. Brown (1997), pp. 183, 186.
35. Brown (1997), p. 186.
36. Peierls (1979), p. 146.
37. Frisch (1979), p. 127.
38. Frisch (1979), pp. 125–6.
39. O.R. Frisch, 'Nuclear fission', *Annual Reports on Progress in Chemistry*, 1939, 36 (1940), p. 25.

9 Memorandum of Understanding

1. Peierls (1979), pp. 153–4.
2. O.R. Frisch and R. Peierls, 'Memorandum on the properties of a radioactive "super-bomb"' (1940), UK Public Record Office, AB 1/210. The full text is available at: https://www.atomicarchive.com/resources/documents/beginnings/frisch-peierls-2.html; Peierls (1979), pp. 146, 154. See Appendix I, p. 351.
3. Frisch and Peierls (1940), pp. 1, 4–6.
4. Frisch and Peierls (1940), p. 3.

5. Frisch and Peierls (1940), pp. 1–2; Clark (1961), p. 51.
6. Clark (1961), pp. 58–9; Gowing (1964), pp. 43–4.
7. Rhodes (1986), pp. 328–9; Dahl (1999), pp. 110–13; Waltham (1993), p. 7.
8. G. Hevesy, *Adventures in Radioisotope Research* (New York: Pergamon, 1962), vol. 1, p. 27.
9. Dahl (1999), p. 119.
10. Dahl (1999), pp. 113–14.
11. Gowing (1964), pp. 43–5.
12. Gowing (1964), pp. 43–5; Hartcup and Allibone (1984), p. 121; Brown (1997), p. 190.
13. Brown (1997), p. 189.
14. Brown (1997), pp. 189–90; Clark (1961), p. 62.
15. Clark (1961), p. 61.
16. Clark (1961), p. 65; Gowing (1964), p. 45.
17. Hartcup and Allibone (1984), pp. 121–2, 122n; Peierls (1979), p. 156; Gowing (1964), p. 45.
18. Peierls (1979), p. 155.
19. Frisch (1979), pp. 128–9.
20. Peierls (1979), pp. 155–6; Gowing (1964), pp. 46–7.
21. K. Freeman, *The Civilian Bomb-Disposing Earl* (Barnsley, South Yorks.: Pen & Sword Military, 2015), pp. 54–6, 64.
22. Freeman (2015), pp. 56, 70–87.
23. Freeman (2015), pp. 70–1.
24. Blackett (1960), p. 94; Clark (1961), p. 23; Dahl (1979), p. 105; Gowing (1964), p. 50; L. Kowarski, interview by Charles Wiener, New York, Session III, 20 October 1969, p. 1, American Institute of Physics: https://www.aip.org/history-programs/niels-bohr-library/oral-histories/4717-3
25. Kowarski (1969), Session III, p. 2; Waltham (1993), p. 7; Freeman (2015), pp. 79–82.
26. Freeman (2015), p. 82.
27. Kowarski (1969), Session III, p. 3; Freeman (2015), pp. 82–3.
28. Kowarski (1969), Session III, pp. 3–4; Freeman (2015), p. 83.
29. Freeman (2015), pp. 85–8, 96.
30. Freeman (2015), pp. 89, 91–2, 96–8, 113, 126–8.
31. Kowarski (1969), Session III, pp. 5–6; Dahl (1999), p. 129; Freeman (2015), pp. 100–7.
32. Kowarski (1969), Session III, p. 6; Freeman (2015), pp. 108–17.
33. Kowarski (1969), Session III, p. 7; Freeman (2015), p. 123.
34. J.-L. Pasquier, 'L'actinium d'Irène dans la cité sanitaire de Clairvivre', 6bisruedemessine, 14 April 2013: https://6bisruedemessine.wordpress.com/2013/04/14/lactinium-direne-dans-la-cite-sanitaire-de-clairvivre-juste-une-histoire-vraie-bien-quincroyable/
35. Brown (1997), pp. 191–2; Lovell (1975), p. 72.

10 Separation Anxieties

1. Peierls (1979), pp. 145–7, 156–7.
2. J. Medawar and D. Pyke, *Hitler's Gift: Scientists Who Fled Nazi Germany* (London: Blake Publishing, 2000), pp. 195–201.
3. R.C. Williams, *Klaus Fuchs, Atom Spy* (Cambridge, MA: Harvard University Press, 1987), pp. 32–3.

4. Peierls (1979), pp. 150–1.
5. Peierls (1979), pp. 148–9.
6. Frisch (1979), p. 131.
7. Frisch (1979), pp. 135–6.
8. Frisch (1979), pp. 133–4, 140.
9. Frisch (1979), pp. 138–9; Brown (1997), pp. 178, 186–7; R.A. Hinde and J.L. Finney, 'Sir Joseph (Józef) Rotblat, 1908–2005', *Biogr Mem Fell Roy Soc*, 53 (2007), p. 313.
10. Frisch (1979), p. 139.
11. Frisch (1979), pp. 137–8; F.N. Flakus, 'Detecting and measuring ionizing radiation: a short history', *IAEA Bulletin*, 23 (1981), p. 34.
12. Clark (1961), p. 85; Gowing (1964), p. 56.
13. Gowing (1964), pp. 48, 56–7, 61–3; Clark (1961), p. 91.
14. N. Kurti, 'Franz Eugen Simon, 1893–1956', *Biogr Mem Fell Roy Soc*, 4 (1958), pp. 225–31; Gowing (1964), pp. 46–7; Clark (1961), pp. 88, 91.
15. Kurti (1958), pp. 242, 246, 248, 250.
16. Gowing (1964), pp. 46–7; Kurti (1958), pp. 231, 242; Clark (1961), pp. 90–1.
17. Clark (1961), p. 92; Gowing (1964), pp. 54–7; Peierls (1979), pp. 158–9.
18. Clark (1961), pp. 93–4; Gowing (1964), p. 47.
19. Kowarski (1969), Session III, pp. 8–9; Clark (1961), p. 101.
20. Kowarski (1969), Session III, p. 8; Gowing (1964), p. 59; Clark (1961), p. 102.
21. Kowarski (1969), Session III, pp. 8–10; Gowing (1964), p. 51; Clark (1961), pp. 102–3.
22. Kowarski (1969), Session III, p. 15; Gowing (1964), p. 51 and note.
23. W. Cochran and S. Devons, 'Norman Feather, 1904–1978', *Biogr Mem Fell Roy Soc*, 27 (1981), pp. 258–60; N. Feather, interview by Charles Weiner, Edinburgh, Session I, 25 February 1971, American Institute of Physics, pp. 24–5, 31: https://www.aip.org/history-programs/niels-bohr-library/oral-histories/4599-1; N. Feather and E. Bretscher, 'Atomic numbers of the so-called transuranic elements', *Nature*, 143 (1939), p. 516.
24. Peierls (1979), pp. 159–60; Frisch (1979), pp. 142–3; N. Feather, interview by Charles Weiner, Edinburgh, Session II, 5 November 1971, p. 35: https://www.aip.org/history-programs/niels-bohr-library/oral-histories/4599-2; Clark (1961), p. 103; Kowarski (1969), Session III, p. 21.
25. Clark (1961), p. 120.
26. Gowing (1964), pp. 48–9, 51–4.
27. Feather (1971), Session II, p. 33; Gowing (1964), pp. 59–60.
28. Brown (1997), p. 206.
29. Freeman (2015), pp. 139–40.
30. A.B. Hartley, *Unexploded Bomb* (London: Cassell, 1958); Freeman (2015), pp. 158–9.
31. Freeman (2015), pp. 141–4, 147–53, 180.

11 Men with Missions

1. M. Goldsmith, *The Curie Family* (Geneva: Edito-Service, 1971), pp. 145–6.
2. Blackett (1960), p. 96; W. Gentner, interview by Charles Weiner, Heidelberg, 15 November 1971, p. 88, American Institute of Physics: https://www.aip.org/history-programs/niels-bohr-library/oral-histories/5080

3. Gentner (1971), pp. 8–11, 16, 23, 34, 52; V.L. Telegdi and V.F. Weisskopf, obituary of Wolfgang Gentner, *Physics Today*, 34 (1981), pp. 91–2.
4. Dahl (1999), p. 143; Goldsmith (1971), pp. 146–7.
5. Goldsmith (1971), p. 147; Gentner (1971), p. 80; Dahl (1999), pp. 143–4.
6. Gentner (1971), p. 88; Goldsmith (1971), p. 148; Dahl (1999), pp. 144–5; Blackett (1960), p. 96.
7. Clark (1965), p. 258.
8. Clark (1961), pp. 160–2; S. Phelps, *The Tizard Mission: The Top-Secret Operation That Changed the Course of World War II* (Yardley, PA: Westholme, 2010), pp. 44, 58, 134, 144.
9. Phelps (2010), p. 65; Clark (1961), p. 250; Clark (1965), p. 161.
10. Clark (1965), pp. 250, 256, 266; Clark (1961), p. 161; Phelps (2010), p. 65.
11. Gowing (1964), pp. 43–4; Clark (1961), p. 161.
12. Phelps (2010), pp. 123, 130; Clark (1965), pp. 255–8.
13. Clark (1965), p. 248.
14. Phelps (2010), pp. 39–41, 97; Clark (1965), p. 258.
15. Clark (1965), p. 260.
16. Clark (1961), p. 162; Clark (1965), pp. 254, 258.
17. Phelps (2010), p. 70; Clark (1965), p. 251.
18. Phelps (2010), p. 32; Clark (1965), p. 269.
19. Clark (1965), p. 269.
20. Clark (1965), p. 260.
21. Clark (1965), pp. 261, 263.
22. Clark (1965), pp. 264–5; Phelps (2010), p. 112.
23. Clark (1965), pp. 259–60, 264; Phelps (2010), pp. 163–6.
24. J.P. Baxter, *Scientists Against Time* (Boston, MA: Little, Brown & Co., 1946), p. 142.
25. Hartcup and Allibone (1984), p. 99.
26. Phelps (2010), p. 168.
27. Phelps (2010), pp. 169, 171–6, 181–4; Clark (1965), pp. 267–8; Hartcup and Allibone (1984), pp. 99–100.
28. Phelps (2010), p. 181; Hartcup and Allibone (1984), p. 102; Clark (1965), pp. 269–70.
29. Clark (1965), p. 269.
30. Phelps (2010), p. 181; Clark (1965), p. 269; Hartcup and Allibone (1984), p. 102.
31. Hartcup and Allibone (1984), p. 102; Gowing (1964), p. 65.
32. Hartcup and Allibone (1984), p. 102.
33. Clark (1961), pp. 118–19; Gowing (1964), pp. 57–8.
34. Kowarski (1969), Session III, p. 23; Blackett (1960), p. 94.
35. Feather (1971), Session II, p. 32; Gowing (1964), pp. 59–60.
36. G. Seaborg, *The Transuranic Elements* (New Haven, CT: Yale University Press, 1958), p. 7.
37. Dahl (1999), p. 145.
38. Dahl (1999), p. 146; Goldsmith (1971), p. 180.
39. Dahl (1999), pp. 145–6.

12 In the Dark

1. Gowing (1964), pp. 65-6.
2. Gowing (1964), p. 66.
3. G.T. Seaborg and E. Segrè, 'The trans-uranium elements', *Nature*, 159 (1947), pp. 863-5; G.T. Seaborg, 'Transuranium elements: a half century', *Proceedings of the American Chemical Society Symposium to Commemorate the 50th Anniversary of Transuranium Elements*, Washington DC, 27 August 1990, pp. 3-6, 8: https://orau.org/health-physics-museum/files/library/transuranicsseaborg.pdf
4. Seaborg (1990), pp. 6-7.
5. Smyth (1945), p. 48; Peierls (1979), pp. 168-9; A. Kramish, *The Griffin* (London: Macmillan, 1986), p. 121; D. Irving, *The Virus House* (London: William Kimber, 1967), p. 35.
6. Peierls (1979), p. 168; Dahl (1999), p. 89; Gowing (1964), p. 43.
7. Kramish (1986), p. 52; M. Smith, *Foley: The Spy Who Saved 10,000 Jews* (London: Hodder & Stoughton, 1999), pp. 58-9.
8. Kramish (1986), pp. 15-17; Smith (1999), p. 184.
9. Kramish (1986), p. 50.
10. Kramish (1986), pp. 22-6.
11. Kramish (1986), pp. 52-4; Dahl (1999), p. 89.
12. Kramish (1986), pp. 48-9.
13. Dahl (1999), pp. 130-2.
14. Kramish (1986), p. 54.
15. Dahl (1999), pp. 89, 130-2; Irving (1967), pp. 29-31; Kramish (1986), p. 118.
16. Dahl (1999), pp. 135, 137.
17. Dahl (1999), pp. 136-7, 173; Irving (1967), pp. 44-5.
18. Irving (1967), pp. 131-2, 135-7; Irving (1967), p. 32.
19. Dahl (1999), pp. 132-3, 136.
20. Dahl (1999), pp. 136-7; Kramish (1986), p. 173; Irving (1967), pp. 37-42.
21. Dahl (1999), pp. 136-7, 173; Irving (1967), pp. 44-5.
22. Dahl (1999), pp. 137-8; Bernstein (2001), p. 137.
23. Irving (1967), p. 53.
24. Dahl (1999), pp. 188-9.
25. Smith (1999), p. 147.
26. Hartcup and Allibone (1984), p. 120.
27. Clark (1961), p. 51.

13 Minority Reports

1. Oliphant and Penney (1968), pp. 142-4; Hartcup and Allibone (1984), p. 104; Clark (1961), p. 86.
2. Hartcup and Allibone (1984), p. 104; Clark (1961), p. 86.
3. Clark (1961), pp. 111-12; Frisch (1979), p. 142; Gowing (1964), pp. 63-4, 82.
4. Gowing (1964), p. 181.
5. Gowing (1964), pp. 68-9; Clark (1961), p. 127.
6. Gowing (1964), pp. 61-2, 68-9, 100.
7. Gowing (1964), pp. 68-9; Clark (1961), p. 127.
8. H. Bretscher, interview by John Bennett and Anna Shepherd, 10 July 1984, p. 17, American Institute of Physics: https://www.aip.org/history-programs/niels-bohr-library/oral-histories/4536

9. Gowing (1964), p. 82.
10. Gowing (1964), pp. 66n, 70-1, 74-5; Seaborg (1990), pp. 4-8.
11. Gowing (1964), pp. 55-6; Clark (1961), p. 163.
12. Gowing (1964), pp. 66, 84.
13. Gowing (1964), pp. 67-8.
14. Gowing (1964), p. 68.
15. Peierls (1979), pp. 163-4; Williams (1987), p. 40.
16. Williams (1987), pp. 13-17, 20-2; Clark (1961), pp. 82-3.
17. Williams (1987), p. 40; Clark (1961), pp. 82-3; Peierls (1979), pp. 163-4.
18. Gowing (1964), pp. 73, 76, 80, 85; Brown (1997), pp. 210-13.
19. Gowing (1964), pp. 80, 91.
20. M.A.U.D. Committee, 'Report by M.A.U.D. Committee on the use of uranium for a bomb', Ministry of Aircraft Production, London, July 1941. Reproduced in full as Appendix 2 in Gowing (1964), pp. 394-438. Page nos in text correspond to the facsimile at: https://fissilematerials.org/library/1941/07/report_by_maud_committee_on_th.html
21. M.A.U.D. Committee (1941), pp. 1, 3.
22. M.A.U.D. Committee (1941), p. 6.
23. M.A.U.D. Committee (1941), pp. 6, 8-9.
24. M.A.U.D. Committee (1941), pp. 2, 9-10.
25. M.A.U.D. Committee (1941), pp. 15, 20-1.
26. M.A.U.D. Committee (1941), pp. 1, 14.
27. M.A.U.D. Committee (1941), pp. 10, 11-13, 16.
28. M.A.U.D. Committee (1941), p. 15.
29. M.A.U.D. Committee (1941), p. 3.
30. Lovell (1975), p. 70.
31. Clark (1961), pp. 105-6.
32. Gowing (1964), pp. 72-5.
33. Gowing (1964), p. 72.
34. Gowing (1964), pp. 75-6.
35. L. Kowarski, interview by Charles Wiener, New York, 19 October 1969, Session II, p. 61, American Institute of Physics: https://www.aip.org/history-programs/niels-bohr-library/oral-histories/4717-2
36. Kowarski (1969), Session I, pp. 6-7; Kowarski (1969), Session II, pp. 7-8, 11, 47.
37. Kowarski (1969), Session III, pp. 4-6.
38. Peierls (1979), pp. 163-4.
39. Gowing (1964), p. 87.
40. Kramish (1986), p. 119; Walker (1995), p. 148; C. Brown, *Operation Big: The Race to Stop Hitler's A-Bomb* (Stroud: Amberley, 2016), pp. 133-4.
41. Brown (2016), p. 133; Kramish (1986), p. 44.
42. Walker (1995), p. 149.
43. Aage Bohr, letter to Sir Michael Perrin, 8 October 1984, Perrin Papers, PERR 3/16; Kramish (1986), p. 120; Walker (1995), p. 149; Brown (2016), pp. 133-7.
44. Brown (2016), p. 174; Kramish (1986), pp. 120-1.
45. Kramish (1986), p. 103; Smith (1999), p. 173.
46. Smith (1999), pp. 18-21, 115, 193-9; Kramish (1986), pp. 84, 88.
47. M.S. Goodman, 'MI6's atomic spy: the rise and fall of Eric Welsh', *War in History*, 23 (2016), pp. 101-2.

48. Goodman (2016), pp. 102–4.
49. Dahl (1999), pp. 149–51.
50. Dahl (1999), pp. 115–16, 156.
51. H.A.H. Boot and J.T. Randall, 'Historical notes on the cavity magnetron', *IEEE Transactions on Electron Devices*, 23 (1976), pp. 724–9.
52. Gowing (1964), pp. 83–4.
53. Hartcup and Allibone (1984), p. 102; Gowing (1964), pp. 83–4.
54. Goldberg (1992), p. 441.
55. Clark (1961), pp. 131–2, 163; Goldberg (1992), pp. 435–6; F.M. Szasz, *British Scientists and the Manhattan Project: The Los Alamos Years* (New York: St Martin's Press, 1992), p. 3.
56. Gowing (1964), p. 85; Clark (1961), p. 135.
57. Clark (1961), p. 139.
58. Hewlett and Anderson (1962), p. 24; Goldberg (1992), pp. 430–1; Gowing (1964), pp. 55–6.
59. Goldberg (1992), p. 438.
60. Alfred O.C. Nier, interview by Michael A. Grayson and Thomas Krick, 7–10 April 1989, University of Minnesota, Minneapolis (Philadelphia: Chemical Heritage Foundation, Oral History Transcript #0112), pp. 93–4; Goldberg (1992), pp. 430, 438.
61. Gowing (1964), p. 94.
62. Goldberg (1992), pp. 436–7.
63. Goldberg (1992), pp. 432, 438, 440.
64. Farmelo (2013), p. 197; Szasz (1992), p. 13; Brown (2016), p. 224.
65. Farmelo (2013), p. 197; Brown (2016), p. 224.
66. Goldberg (1992), pp. 444–5; Farmelo (2013), p. 198; M. Oliphant, Notes on conversations with E.O. Lawrence, 23–4 September 1941, The National Archives, Kew, PRO AB1/495.
67. Gowing (1964), pp. 91, 110.

14 Tube Alloys

1. Gowing (1964), pp. 91–2; Clark (1961), p. 135.
2. Gowing (1964), p. 93.
3. Gowing (1964), pp. 93–4; Farmelo (2013), p. 186.
4. Gowing (1964), pp. 97–9, 102–5; Clark (1961), pp. 153–6; Brown (1997), p. 221.
5. Clark (1961), p. 195; E.B. Bridges, 'John Anderson, Viscount Waverley, 1882–1958', *Biogr Mem Fell Roy Soc*, 4 (1958), pp. 322–3; Gowing (1964), pp. 107–8; Brown (1997), p. 218.
6. Gowing (1964), pp. 95–6; Clark (1961), pp. 154–5.
7. A. Fleck, 'Wallace Alan Akers, 1888–1954', *Biogr Mem Fell Roy Soc*, 1 (1955), pp. 1–4; Gowing (1964), pp. 108–9; Clark (1961), pp. 155–7; Farmelo (2013), p. 211.
8. Anon., 'Michael Perrin: the man to whom Fuchs confessed', *New Scientist*, 24 January 1957.
9. M. Perrin diaries, 1941, multiple entries, Perrin Papers, 3/1/3.
10. Gowing (1964), p. 109; Clark (1961), pp. 156–8; Farmelo (2013), p. 200.

NOTES TO PP. 158–166

11. Gowing (1964), p. 109.
12. Gowing (1964), pp. 109–10, 126; Clark (1961), pp. 157–8; Brown (1997), pp. 225–6.
13. Clark (1961), pp. 157–8; Gowing (1964), pp. 110–11; Brown (1997), pp. 225–6; Farmelo (2013), p. 200; Hartcup and Allibone (1984), p. 125; Kowarski (1969), Session III, pp. 25–6.
14. Clark (1961), pp. 156–7; Farmelo (2013), pp. 201, 211.
15. Tube Alloys Consultative Committee minutes, 15 and 16 January 1942, 4 June 1942; M. Perrin memo to J. Rickett, War Cabinet Office, 20 January 1942: both Chadwick Papers, GBR/0014/CHAD I 12/2.
16. Clark (1961), p. 173.
17. M. Wilkins, *The Third Man of the Double Helix: The Autobiography of Maurice Wilkins* (Oxford: Oxford University Press, 2003), pp. 76–7; M.H.F. Wilkins, 'Recollections of the 1930s', in H. Rose and S. Rose (eds), *Science at the Crossroads: A Socialist View of Science, Technology and Medicine*, 51 (1982), pp. 10–11.
18. Tube Alloys Technical Committee minutes, 5 and 6 June 1942, Chadwick Papers GBR/0014/CHAD I 12/2.
19. Clark (1961), p. 174; T. Jones, *The X-Site: Britain's Most Mysterious Government Facility* (Rhyl: Gwasg Helgain, 2000), pp. 6–10.
20. Clark (1961), pp. 128–9, 172–4.
21. Clark (1961), p. 175.
22. S. Oldfield, *The Black Book: Britons on the Nazi Hit List* (London: Profile Books, 2022), p. 88.
23. Clark (1961), p. 172; Tube Alloys Technical Committee minutes, 20 January 1942, Chadwick Papers GBR/0014/CHAD I 12/2.
24. Gowing (1964), pp. 93–4; Farmelo (2013), p. 186.
25. Goldberg (1992), pp. 436–7.
26. Goldberg (1992), p. 433.
27. Goldberg (1992), p. 438.
28. Goldberg (1992), pp. 438–9; Gowing (1964), p. 115; Clark (1961), pp. 164–5.
29. A.H. Compton, Report of National Academy of Sciences Committee on Atomic Fission, 17 May 1941, US National Archives, Records of the Office of Scientific Research and Development, Record Group 227, Bush-Conant Papers microfilm collection, Roll 1, Target 2, Folder 1, S-1 Historical File, Section A (1940–1941).
30. Goldberg (1992), pp. 440–2.
31. Goldberg (1992), pp. 442–3; Clark (1961), pp. 164–5.
32. Clark (1961), pp. 166–7; Gowing (1964), p. 122.
33. Clark (1961), p. 160.
34. Clark (1961), pp. 166–7; Farmelo (2013), p. 198.
35. Gowing (1964), p. 123; Farmelo (2013), p. 188.
36. Farmelo (2013), pp. 172–3.
37. Gowing (1964), pp. 96–7, 106; Farmelo (2013), p. 193.
38. Farmelo (2013), p. 189–91; Gowing (1964), pp. 106–7.
39. Clark (1961), pp. 168–9.
40. Clark (1961), pp. 85, 119; Clark (1961), pp. 168–9; Gowing (1964), pp. 85, 119.
41. Gowing (1964), p. 120.

42. M. Perrin diary entry for 13 November 1941, Perrin Papers, 3/1/1.
43. Clark (1961), pp. 168–70.
44. Goldberg (1992), p. 445; Gowing (1964), p. 117; Clark (1961), pp. 167, 170.
45. Goldberg (1992), p. 449; Gowing (1964), pp. 122–3; Clark (1961), p. 170; Brown (1997), pp. 225–6.
46. Farmelo (2013), p. 194.
47. Farmelo (2013), pp. 203–4; Gowing (1964), p. 123.
48. Farmelo (2013), p. 203.
49. Gowing (1964), pp. 126–7; Clark (1961), p. 182.
50. Farmelo (2013), p. 201.
51. Brown (1997), pp. 226–7.

15 Über Alles

1. M. Perrin memo to D.H.F. Rickett, War Cabinet Office, 20 January 1942, Chadwick Papers, GBR/0014/CHAD I 12/2; Gowing (1964), p. 87.
2. Dahl (1999), pp. 151–5; Irving (1967), pp. 94–6, 125.
3. Walker (1995), pp. 151–5.
4. Clark (1961), p. 142.
5. Perrin memo to Rickett (1942), Chadwick Papers, GBR/0014/CHAD I 12/2; Gowing (1964), p. 87.
6. Irving (1967), pp. 46, 53.
7. Irving (1967), pp. 45, 48, 56–7, 82, 95–6.
8. Irving (1967), pp. 57, 82–4.
9. Irving (1967), pp. 100–1.
10. Irving (1967), pp. 65–9; Dahl (1999), p. 150.
11. W. Bothe and P. Jensen, 'Die Absorption thermischer Neutronen in Kohlenstoff', *Zeit Phys*, 122 (1944), p. 749; Irving (1967), pp. 92–3; Dahl (1999), pp. 139–41; Kramish (1986), p. 174.
12. Dahl (1999), p. 158.
13. Dahl (1999), pp. 188–9; Irving (1967), pp. 58, 67, 70, 87–9.
14. Dahl (1999), pp. 189–90.
15. J. Schintlmeister and F. Hernegger, 'Über ein bisher unbekanntes alphastrahlendes chemisches Element' (1940), ADDM FA 002/Vorl No. 0055. First of four secret papers, declassified 1971. See K. Hentschel and A.M. Hentschel, *Physics and National Socialism: An Anthology of Primary Sources* (Basel: Birkhäuser, 1996), Appendix E, 'Kernphysikalische Forschungsberichte'; Irving (1967), pp. 79–80.
16. F. Strassmann and O. Hahn, 'Über die Isolierung und einige Eigenschaften des Elements 93', *Naturwissenschaften*, 30 (1942), pp. 256–60; O. Hahn and F. Strassmann, 'Zur Folge des Enstehung des 2.3 Tage Isotops des Elements 93', *Naturwissenschaften*, 30 (1942), pp. 260–4; Irving (1967), p. 45.
17. Bernstein (2001), p. 34.
18. C.F. von Weizsäcker, 'Eine Möglichkeit der Energiewinnung aus Uran 238', KPF G-59 (1940), in D. Irving (ed.), *Third Reich Documents*, Group 11, German Atomic Research, microfilm DJ-29, pp. 451–5.
19. Bernstein (2001), p. 32.
20. Irving (1967), pp. 116–17.

21. Irving (1967), p. 43; Bernstein (2001), p. 46.
22. Irving (1967), pp. 38, 77–8.
23. Bernstein (2001), p. 33.
24. Dahl (1999), pp. 186–7; Irving (1967), pp. 118–19; Bernstein (2001), pp. 36, 40.
25. W. Heisenberg, 'A lecture on bomb physics: February 1942' [English translation], *Physics Today*, 48 (1995), pp. 27–34: https://doi.org/10.1063/1.881468
26. Dahl (1999), pp. 187–8.
27. Clark (1961), pp. 98–9; Goldsmith (1971), p. 156; Gentner (1971), p. 91.
28. Gentner (1971), p. 89.
29. Gentner (1971), pp. 90–1; Spence (1970), p. 298.
30. Goldsmith (1971), p. 151; Dahl (1999), p. 145; Anon., 'Irène Joliot-Curie', *Current Biography: Who's News and Why* (New York: H.W. Wilson, 1940), pp. 78–83.
31. Gentner (1971), p. 39.
32. Dahl (1999), p. 146.
33. Goldsmith (1971), p. 156.
34. Kramish (1986), p. 192; Dahl (1999), pp. 158–9.
35. Dahl (1999), pp. 156–7, 165–6.
36. Dahl (1999), pp. 152–8.
37. Dahl (1999), pp. 164–6.
38. Dahl (1999), pp. 157, 167–8.
39. Kramish (1986), p. 98.
40. Kramish (1986), pp. 111–14.

16 Double Dealing

1. Farmelo (2013), pp. 205–6.
2. Farmelo (2013), pp. 205–6.
3. Hewlett and Anderson (1962), pp. 49–51; Gowing (1964), p. 127; Farmelo (2013), pp. 206, 217.
4. Hewlett and Anderson (1962), pp. 51–6.
5. Clark (1961), p. 182; Gowing (1964), p. 126.
6. Peierls (1979), pp. 169–70.
7. Gowing (1964), pp. 126–8.
8. Gowing (1964), pp. 128–9; Peierls (1979), p. 171.
9. Gowing (1964), p. 128; Peierls (1979), p. 172.
10. Hewlett and Anderson (1962), pp. 56–9; Gowing (1964), p. 128.
11. Gowing (1964), p. 131.
12. M Perrin, note for Tube Alloys Directorate, 4 June 1942, Chadwick Papers, GBR/0014/CHAD I 12/2; Gowing (1964), pp. 130, 134–5; Kowarski (1969), Session III, pp. 27–8.
13. M Perrin, note for Tube Alloys Directorate, 4 June 1942, Chadwick Papers, GBR/0014/CHAD I 12/2; Hewlett and Anderson (1962), pp. 89–90.
14. Peierls (1979), pp. 171–2.
15. Peierls (1979), p. 173.
16. Clark (1961), p. 182; Gowing (1964), p. 131.
17. Gowing (1964), p. 137.
18. Farmelo (2013), pp. 207–8.
19. Gowing (1964), pp. 132–3.

20. Farmelo (2013), p. 214.
21. Dahl (1999), pp. 189–90.
22. Kramish (1986), p. 126; J.L. Lohan, H. Rechenberg et al., 'Heisenberg, Goudsmit and the German "A"-bomb', *Physics Today*, 44 (1991), pp. 13–15, 90–6: https://doi.org/10.1063/1.2810103
23. Kramish (1986), p. 126.
24. Kramish (1986), p. 127.
25. Kramish (1986), p. 128.
26. Kramish (1986), p. 129.
27. Dahl (1999), p. 192; Kramish (1986), p. 128.
28. Kramish (1986), pp. 129–30.
29. Dahl (1999), pp. 163–4.
30. Dahl (1999), p. 164; Kramish (1986), pp. 131–2.
31. Kramish (1986), pp. 131, 162–4; Dahl (1999), pp. 193–4.
32. Clark (1961), pp. 182–3; Gowing (1964), p. 145.
33. Personal communication, Nicola Perrin to GW, 3 May 2021; High Commissioner Canada MOST SECRET telegram to Dominions Office, London, 16 June 1942, Chadwick Papers, GBR/0014/CHAD I 12/6; Farmelo (2013), p. 209.
34. M. Perrin diary entries from 3 June to 10 July 1942, Perrin Papers, PERR 3/4 [TA].
35. Personal communication, Nicola Perrin to GW, 3 May 2021.
36. M. Perrin diary entries from 3 June to 10 July 1942, Perrin Papers, PERR 3/4 [TA]; Gowing (1964), pp. 138–9; Farmelo (2013), p. 214.
37. Gowing (1964), pp. 137, 144–5; Farmelo (2013), pp. 214–15.
38. Gowing (1964), pp. 144; Farmelo (2013), p. 210.
39. Hewlett and Anderson (1962), pp. 89–90; Seaborg (1990), pp. 8–9.
40. Hewlett and Anderson (1962), p. 103.
41. Hewlett and Anderson (1962), pp. 71–83; Gowing (1964), p. 141.
42. Hewlett and Anderson (1962), pp. 105–6.
43. Hewlett and Anderson (1962), p. 193; Kramish (1986), p. 162.
44. Dahl (1999), pp. 156–8, 192–4.
45. Dahl (1999), p. 192.
46. Dahl (1999), pp. 194–5.
47. Dahl (1999), pp. 167–9.
48. Dahl (1999), pp. 194–5.
49. T. Gallagher, *Assault in Norway: Sabotaging the Nazi Nuclear Program* (Guilford, CT: Lyons Press, 1975), pp. 33–4; Dahl (1999), pp. 195–6.
50. Gallagher (1975), pp. 37–8; Dahl (1999), pp. 196–7.

17 Critical Masses

1. C. Allardice and E.R. Trapnell, 'The first pile' (1946), reissued by US Department of Energy, December 1982, p. 27: https://www.iaea.org/sites/default/files/publications/magazines/bulletin/bull4-0/04005004147su.pdf; T.M. McAndrew, 'Subterfuge in the city' (2019): https://will.illinois.edu/news/story/subterfuge-in-the-city-how-illinois-helped-create-the-nuclear-age
2. Goldberg (1992), p. 435.
3. Goodman (2016), p. 105; Kramish (1986), p. 132.
4. McAndrew (2019), p. 6.

5. L. Groves, *Now It Can Be Told: The Story of the Manhattan Project* (New York: Harper & Row, 1962), p. 13; Rhodes (1986), pp. 425–6; Goldberg (1992), p. 122.
6. Gowing (1964), p. 150; Brown (1997), p. 233; Hewlett and Anderson (1962), pp. 227–9.
7. V. Bush memo to W.A. Akers, Tube Alloys Directorate, 10 July 1942, Chadwick Papers GBR/0014/CHAD I 12/6; Clark (1961), p. 184.
8. Gowing (1964), pp. 148–50.
9. Gowing (1964), p. 145; Brown (1997), pp. 228–9.
10. Gowing (1964), p. 150.
11. M. Perrin diary entry for 10 October 1942, Perrin Papers, PERR 3/1/4; Hewlett and Anderson (1962), p. 269.
12. Gowing (1964), pp. 150–3; Hewlett and Anderson (1962), pp. 264–5; Brown (1997), p. 229.
13. Gowing (1964), pp. 150–3, 157.
14. Fermi (1946), pp. 21–30; Rhodes (1986), pp. 435–7; Allardice and Trapnell (1946), p. 44.
15. Allardice and Trapnell (1946), p. 44.
16. Gowing (1964), p. 154.
17. Gowing (1964), p. 135; Clark (1961), p. 186; Memo from W.A. Akers, Tube Alloys Directorate, 6 June 1942, Chadwick Papers, GBR/0014/CHAD I 12/6; Peierls (1979), pp. 174–5; Kowarski (1969), Session III, p. 26.
18. High Commission Canada telegram to Tube Alloys Directorate, 12 July 1942, Chadwick Papers GBR/0014/CHAD I 12/6; Gowing (1964), pp. 192–4.
19. Kowarski (1969), Session II, p. 62; V. Bush telegram to W.A. Akers, Tube Alloys Directorate, 12 October 1942, Chadwick Papers, GBR/0014/CHAD I 12/6.
20. Hewlett and Anderson (1962), pp. 268–70; Brown (1997), p. 234.
21. Hewlett and Anderson (1962), p. 267; Brown (1997), p. 239; Gowing (1964), p. 175.
22. Hewlett and Anderson (1962), p. 268.
23. Hewlett and Anderson (1962), p. 268; Gowing (1964), pp. 155–7.
24. Gowing (1964), p. 157; Farmelo (2013), pp. 218–22; Brown (1997), p. 235.
25. Hewlett and Anderson (1962), pp. 269–70; Brown (1997), p. 235; Gowing (1964), pp. 158, 160.
26. Allardice and Trapnell (1946), p. 45.
27. Allardice and Trapnell (1946), pp. 44–5; Fermi (1946), pp. 21, 23; Rhodes (1986), pp. 436–8.
28. Allardice and Trapnell (1946), pp. 45–6; Rhodes (1986), pp. 439–40.
29. Allardice and Trapnell (1946), pp. vii, 46–7.
30. Rhodes (1986), p. 442.
31. Allardice and Trapnell (1946), p. 47; Rhodes (1986), p. 442.
32. Allardice and Trapnell (1946), pp. 28–31.
33. Fermi (1946), p. 24; Rhodes (1986), p. 436.
34. Hewlett and Anderson (1962), pp. 188–9, 213–15; D. Harvey, 'History of the Hanford Site, 1943–1990', Pacific Northwest National Laboratory/US Department of Energy (2000), pp. 2–6: https://doi.org/10.2172/887452
35. Harvey (2000), pp. 3–7.
36. Harvey (2000), pp. 3–5.
37. Harvey (2000), p. 3.
38. Goldberg (1992), pp. 122–3.

39. Goldberg (1992), p. 123; Harvey (2000), pp. 2–6; Hewlett and Anderson (1962), pp. 213–15.
40. Groves (1962), p. 25; Hewlett and Anderson (1962), pp. 116–19; Harvey (2000), pp. 2–4.
41. Bernstein (2001), p. 28.
42. Brown (1997), p. 228; C. Edmonson, 'A reporter's journey into how the US funded the bomb', *New York Times*, 18 January 2024: https://www.nytimes.com/U.S./politics/
43. Hewlett and Anderson (1962), pp. 229–30.
44. Groves–Conant letter to Oppenheimer, 25 February 1943, Nuclear Museum: https://ahf.nuclearmuseum.org/ahf/key-documents/groves-conant-letter

18 Breakdown and Repair

1. Gowing (1964), pp. 161–2.
2. Goldberg (1992), p. 124; Gowing (1964), pp. 155, 157, 168, 174.
3. Hewlett and Anderson (1962), pp. 261–3; Gowing (1964), pp. 175–6; Bretscher (1984), p. 27.
4. Gowing (1964), pp. 165, 167; Brown (1997), p. 244.
5. Hewlett and Anderson (1962), pp. 267–8; Gowing (1964), p. 175.
6. Gowing (1964), pp. 193–4.
7. Gowing (1964), pp. 164, 184–5, 192–4.
8. Gowing (1964), pp. 161–2.
9. Gowing (1964), p. 247.
10. Gowing (1964), p. 246; Brown (1997), p. 242; Kramish (1986), pp. 192–3.
11. Brown (1997), pp. 242–3.
12. Dahl (1999), pp. 23, 198.
13. Dahl (1999), pp. 198–9.
14. Dahl (1999), p. 191; R.D. McFadden, 'Joachim Rønneberg, leader of a raid that thwarted a Nazi atomic bomb, dies at 99', *New York Times*, 22 October 2018.
15. Dahl (1999), pp. 200–2.
16. Dahl (1999), pp. 201–2.
17. Dahl (1999), pp. 202–3.
18. Dahl (1999), pp. 203–5.
19. Dahl (1999), pp. 208–9.
20. Dahl (1999), p. 208.
21. Gowing (1964), pp. 157–8, 160, 162–3.
22. Gowing (1964), pp. 159–60, 164–5.
23. Gowing (1964), pp. 152–3; Brown (1997), p. 253.
24. Gowing (1964), pp. 165–6.
25. Gowing (1964), pp. 154, 159, 167; Hewlett and Anderson (1962), pp. 262–3, 272–4.
26. Gowing (1964), p. 164.
27. Hewlett and Anderson (1962), pp. 272–4.
28. Gowing (1964), pp. 168–70.
29. Gowing (1964), pp. 171–3; Hewlett and Anderson (1962), pp. 278–80.
30. Gowing (1964), pp. 171–2.
31. Clark (1961), p. 177; Gowing (1964), p. 247; Kramish (1986), pp. 193–4.
32. Clark (1961), pp. 177–8; Kramish (1986), p. 194.

33. Goodman (2016), p. 105; Kramish (1986), p. 194.
34. Clark (1961), p. 178; Kramish (1986), p. 195.
35. Clark (1961), p. 177; Gowing (1964), pp. 247–8.
36. Clark (1961), p. 179; Gowing (1964), pp. 248–9; Kramish (1986), p. 195.
37. Kramish (1986), p. 198.
38. Clark (1961), p. 179; Goodman (2016), p. 105; Charles Perrin, personal communication to GW, 20 May 2021.
39. Clark (1961), p. 179; Gowing (1964), pp. 249–50.
40. Kramish (1986), pp. 195–6.
41. Kramish (1986), p. 194; Brown (1999), p. 252.
42. Letter from Viktor Weisskopf to Robert Oppenheimer, 28 October 1942, Oppenheimer Papers, Box 77, Weisskopf folder, Library of Congress archives.
43. Kramish (1986), pp. 180, 205; Goodman (2016), p. 105.
44. Kramish (1986), p. 147.
45. Kramish (1986), p. 197.
46. Hewlett and Anderson (1962), p. 263.
47. D. Stafford, *Secret Agent: The True Story of the Special Operations Executive* (London: BBC Books, 2000), p. 122; Dahl (1999), p. 209.
48. Dahl (1999), p. 165; Clark (1961), p. 143; Kramish (1986), p. 159.
49. Dahl (1999), pp. 211–12.
50. Dahl (1999), pp. 159, 225–6.
51. Kramish (1986), p. 197.
52. Kramish (1986), pp. 172, 182; Dahl (1999), p. 193.
53. Bernstein (2001), p. 28; Kramish (1986), pp. 172–4; Dahl (1999), pp. 213–18.
54. Bernstein (2001), p. 28; Dahl (1999), pp. 224–6.

19 Over There

1. Szasz (1992), pp. xix–xx, 148; Brown (1997), p. 276.
2. Peierls (1979), p. 178.
3. Brown (1997), pp. 245–7, 249.
4. Brown (1997), pp. 248–50; Hewlett and Anderson (1962), p. 282.
5. Brown (1997), pp. 247–8; Gowing (1964), pp. 154, 175–6; Clark (1961), p. 187.
6. Telegrams from Air Ministry to Webster (assistant to Chadwick), 20 January 1944, and from Webster to M.W. Perrin, 20 January 1944, Chadwick Papers, GBR/0014/CHAD I 12/6; Brown (1997), pp. 251–3.
7. Frisch (1979), pp. 145–6; Clark (1961), p. 192.
8. Frisch (1979), pp. 146–7; Peierls (1979), pp. 182–3.
9. Peierls (1979), pp. 184–5.
10. Frisch (1979), p. 148.
11. D. Hawkins, *Project Y: The Los Alamos Story* (Los Angeles, CA: Tomash Publishers, 1961), pp. 18, 34–6; Brown (1997), p. 200; Frisch (1979), p. 149.
12. 'Stories of Displacement', *Manhattan Project*, US National Park Service: https://www.nps.gov/mapr/learn/historyculture/displacement.htm#:~:text=The%20Manhattan%20Project%20required%20great,three%20main%20centers%20of%20operation
13. Frisch (1979), pp. 151–3; Peierls (1979), p. 193; Brown (1997), p. 260.
14. Telegrams from Air Ministry to Webster and from Webster to Perrin (1944), Chadwick Papers, GBR/0014/CHAD I 12/6; Brown (1997), pp. 251–3.

15. Brown (1997), pp. 254, 260–1; Szasz (1992), p. 149.
16. Peierls (1979), pp. 187–9; Szasz (1992), p. 149.
17. Gowing (1964), p. 249; Brown (1997), pp. 261–3; Hewlett and Anderson (1962), pp. 310–11; Farmelo (2013), p. 257.
18. Brown (1997), pp. 261–3; Farmelo (2013), pp. 248, 258–60.
19. Medawar and Pyke (2000), pp. 226–7; Brown (1997), pp. 252, 254, 261–2; Szasz (1992), p. 149.
20. Kramish (1986), p. 197; Groves (1962), pp. 187–90.
21. Groves (1962), pp. 191–3.
22. B. Pash, *The Alsos Mission* (New York: Charter Books, 1969), pp. 22, 29–32.
23. 'Morris "Moe" Berg', Atomic Heritage Foundation: https://ahf.nuclearmuseum.org/ahf/profile/morris-moe-berg-1/
24. Kramish (1986), pp. 199–200.
25. Goodman (2016), p. 105; Kramish (1986), p. 132; Gowing (1964), p. 87; Clark (1961), pp. 47, 80; Tube Alloys Consultative Council memo to Lord President, War Cabinet Office, 8 September 1943, Chadwick Papers, GBR/0014/CHAD I 12/6.
26. Kramish (1986), p. 221.
27. Clark (1961), pp. 143–4; Dahl (1999), pp. 225–32.
28. Clark (1961), p. 144; Dahl (1999), pp. 235–80.

20 Missionaries

1. Szasz (1992), pp. xix–xx, 148–9; Gowing (1964), pp. 260–2.
2. Hawkins (1961), pp. 34, 36–8, 65; Szasz (1992), p. 14; Peierls (1979), p. 195.
3. Brown (1997), p. 255.
4. Hawkins (1961), p. 65; Peierls (1979), pp. 194, 198.
5. R. Serber, *The Los Alamos Primer: The First Lectures on How to Build an Atomic Bomb* (Berkeley, CA: University of California Press, 1992); Szasz (1992), p. 17; B.C. Reed, 'Revisiting the *Los Alamos Primer*', *Physics Today*, 70 (2017), pp. 42–9: https://doi.org/10.1063/PT.3.3692
6. Hawkins (1961), pp. 18, 60, 62, 68, 75, 78; Hewlett and Anderson (1962), p. 310; Peierls (1979), p. 193; S. Goldberg, 'Racing to the finish: the decision to bomb Hiroshima and Nagasaki', *J Am-East Asian Relns*, 4 (1995), p. 120.
7. Peierls (1979), p. 201; Gowing (1964), pp. 262–3.
8. E. Teller in Groves (1962), p. v; Frisch (1979), p. 154; Hawkins (1961), pp. 68–70; Peierls (1979), pp. 196–7; R. Feynman, 'Los Alamos from below' (1975), lecture transcript at: https://www.mathpax.com/lecture-by-richard-feynman
9. Peierls (1979), pp. 193–4; Brown (1997), pp. 260–1; Frisch (1979), pp. 156–70.
10. Frisch (1979), p. 151; Gowing (1964), p. 262.
11. Hawkins (1961), p. 17; Szasz (1992), p. xvi; Brown (1997), p. 261; Frisch (1979), p. 150; I. Gallagher, 'James Chadwick', *Daily Mail*, 20 July 2023: https://www.dailymail.co.uk/news/article-12327581/Oppenheimer-remember-atom-bomb-test-Odd-fellow-not-clever-father-say-twin-daughters-96-humble-British-scientist-vital-role-Manhattan-Project-airbrushed-movie.html
12. Smyth (1945), p. 207; Peierls (1979), p. 199; Gowing (1964), p. 261.
13. Brown (1997), p. 250.

14. Brown (1997), pp. 265–7.
15. M.W. Perrin, note for Tube Alloys Technical Committee, 11 April 1944, Chadwick Papers, GBR/0014/CHAD I 12/6.
16. Gowing (1964), pp. 311–12.
17. Brown (1997), pp. 263–6; Minutes of Tube Alloys Technical Committee, 11 April 1944, Chadwick Papers, GBR/0014/CHAD I 12/6.
18. Gowing (1964), pp. 321–4, 334–5.
19. Brown (2016), p. 152.
20. Williams (1987), p. 69.
21. Farmelo (2013), pp. 256, 258; Brown (1997), pp. 261–3.
22. Farmelo (2013), pp. 248, 260; Brown (1997), p. 262.
23. Farmelo (2013), p. 248.
24. Farmelo (2013), p. 261; Brown (2016), p. 140.
25. Brown (1997), p. 269; Farmelo (2013), pp. 247, 266.
26. Brown (1997), p. 270; Farmelo (2013), pp. 259–60.
27. Farmelo (2013), pp. 260–1.
28. Farmelo (2013), p. 266.

21 Liberation

1. Groves (1962), pp. 207–8; M. Goldhaber, obituary of Samuel A. Goudsmit, *Physics Today*, 32 (1979), pp. 71–2. https://doi.org/10.1063/1.2995511
2. Groves (1962), p. 208; Brown (2016), pp. 156–8.
3. Brown (2016), pp. 154, 157–8.
4. Report of Alsos Mission operations in Paris, August–September 1944, Perrin Papers, PERR 3/24; Brown (2016), pp. 159–64.
5. Alsos Mission, transcript of interviews with Frédéric Joliot, Paris, 28 August 1944, Perrin Papers, PERR 3/24; Brown (2016), pp. 165–6; Brown (1997), p. 268.
6. Goldsmith (1971), pp. 151–3, 155; Clark (1961), p. 99.
7. Alsos Mission, transcript of interviews with Frédéric Joliot, London, 5 and 7 September 1944, with notes by Michael Perrin and Wallace Akers, Perrin Papers, PERR 3/23; Brown (1997), pp. 268–9, 271–6.
8. Minutes of 1st meeting of TA Intelligence Committee, London, 30 January 1945, Perrin Papers, PERR 3/23; Brown (2016), pp. 167–8.
9. Hewlett and Anderson (1962), p. 233; Peierls (1979), p. 195; Goldberg (1995), p. 124; Brown (1997), p. 249.
10. Reed (2017), pp. 44–6.
11. Frisch (1979), pp. 159–60.
12. Reed (2017), p. 46.
13. Reed (2017), pp. 45–7; Brown (1997), pp. 233, 248.
14. Hewlett and Anderson (1962), pp. 310–12; Hawkins (1961), pp. 114–16.
15. Reed (2017), p. 46; Hawkins (1961), p. 103.
16. Hawkins (1961), p. 47.
17. Hewlett and Anderson (1962), pp. 116–19.
18. Hewlett and Anderson (1962), pp. 282, 298–9; Brown (1997), pp. 253, 317; Gowing (1964), pp. 251–4, 256–9; Clark (1961), pp. 250–4.
19. Gowing (1964), pp. 118, 256.
20. Hewlett and Anderson (1962), p. 153; Brown (1997), p. 254.

21. 'Calutron Girls', *Tennessee Encyclopedia* (2019), Tennessee Historical Society: http://tennesseeencyclopedia.net/entries/calutron-girls
22. Hewlett and Anderson (1962), p. 298; 'Y-12: Construction, 1943', *The Manhattan Project: An Interactive History*: https://www.osti.gov/opennet/manhattan-project-history/Events/1942-1944_ur/y-12_construction.htm
23. Hewlett and Anderson (1962), pp. 168–73, 296–7.
24. Hewlett and Anderson (1962), pp. 289–90.
25. Hawkins (1961), pp. 114–17.
26. Reed (2017), pp. 47–9.
27. Hawkins (1961), pp. 66, 95, 103, 108.
28. Hawkins (1961), p. 74; Hewlett and Anderson (1962), p. 311; Goldberg (1995), p. 125.
29. Hawkins (1961), p. 82; Hewlett and Anderson (1962), p. 247.
30. Hewlett and Anderson (1962), p. 246; Hawkins (1961), pp. 124–5.
31. Hawkins (1961), pp. 80, 125; Peierls (1979), pp. 192, 194.
32. Hewlett and Anderson (1962), pp. 247, 311–12; F. Dainton, 'George Bogdan Kistiakowski, 1900–1982', *Biogr Mem Fell Roy Soc*, 31 (1985), pp. 376–408.
33. Peierls (1979), p. 200.
34. Peierls (1979), p. 200; Hewlett and Anderson (1962), pp. 249, 311; Hawkins (1961), pp. 69–70, 81, 98–9.
35. Hewlett and Anderson (1962), pp. 312; R. Moore and E.N. Brown, 'Woolwich, Bruceton, Los Alamos: Munroe jets and the Trinity gadget', *Nuclear Technology*, 207 (2021), supplement 1, pp. S222–30: https://doi.org/10.1080/00295450.2021.1905463
36. Hawkins (1961), pp. 132–3; 'Little Boy and Fat Man', Atomic Heritage Foundation: https://ahf.nuclearmuseum.org/ahf/history/little-boy-and-fat-man
37. R. Serber and R.P. Crease, *Peace and War: Reminiscences of a Life on the Frontiers of Science* (New York: Columbia University Press, 1998), p. 78.
38. Hawkins (1961), pp. 130–2; Hewlett and Anderson (1962), pp. 250–1.
39. Hawkins (1961), pp. 132–3.
40. Hawkins (1961), pp. 118–20.
41. Groves (1962), p. 254; Hawkins (1961), pp. 130–3; Szasz (1992), p. 27.
42. Hewlett and Anderson (1962), pp. 321–3; Hawkins (1961), p. 21.
43. Hawkins (1961), pp. 128, 169.
44. K. Bird and M.J. Sherwin, *American Prometheus: The Triumph and Tragedy of J. Robert Oppenheimer* (New York: Alfred A. Knopf, 2005), pp. 250–2.

22 The Giving of All Help

1. Gowing (1964), pp. 236–40; Brown (1997), pp. 263–4, 279.
2. Minutes of a special meeting of the Tube Alloys Technical Committee, Washington, 13–17 May 1944, Chadwick Papers, GBR/0014/CHAD I 12/6.
3. Gowing (1964), pp. 251–4, 256–9; Clark (1961), pp. 250–4.
4. Gowing (1964), pp. 258–60; Hewlett and Anderson (1962), pp. 281–2.
5. Frisch (1979), pp. 159–60.
6. Hewlett and Anderson (1962), p. 311; Szasz (1992), p. 29.
7. Szasz (1992), pp. 18, 149; Hawkins (1961), p. 60.
8. Szasz (1992), pp. 23–4; Moore and Brown (2021), pp. S223–5; Hawkins (1961), pp. 49, 70, 99.

9. Moore and Brown (2021), pp. S223–4, S226–8; Hawkins (1961), pp. 70, 82, 126.
10. Hawkins (1961), p. 82; Peierls (1979), p. 201; Szasz (1992), pp. 22, 149; Moore and Brown (2021), pp. S223–5.
11. Peierls (1979), p. 201; Szasz (1992), pp. 22, 149.
12. Williams (1987), pp. 75–8.
13. J. Chadwick, memo on position of the TA projects in the USA, 30 August 1944, Chadwick Papers, GBR/0014/CHAD I 12/6; Hewlett and Anderson (1962), pp. 288, 312.
14. J. Chadwick, memo on position of the TA projects in the USA, 30 August 1944, Chadwick Papers, GBR/0014/CHAD I 12/6; Hawkins (1961), pp. 70–1.
15. Hewlett and Anderson (1962), pp. 195, 208–9, 211–12.
16. Harvey (2000), pp. 2–8.
17. Hewlett and Anderson (1962), pp. 306–7.
18. Frisch (1979), pp. 152, 158; Szasz (1992), p. xxx.
19. Brown (1997), p. 283.
20. Frisch (1979), p. 169; Brown (1997), p. 261.
21. Williams (1987), p. 77.
22. Brown (1997), pp. 270–2.
23. Brown (1997), pp. 271–5.
24. Hewlett and Anderson (1962), pp. 296–7.
25. Hewlett and Anderson (1962), p. 312.
26. Hawkins (1961), p. 98.
27. A. Rimmer, *Between Heaven and Hell* (n.p.: Lulu.com, 2012), pp. 11–20.
28. Brown (1997), p. 269.
29. Brown (1997), p. 269.
30. Brown (1997), pp. 174, 176–7; Kramish (1986), p. 236.
31. Alsos preliminary report from Strasbourg Mission, 8 December 1944, Perrin Papers, PERR 3/24; Kramish (1986), pp. 175–6.
32. M. Perrin and R. Furman, Alsos report, 'TA Project Intelligence Targets: Germany', Perrin Papers, PERR 3/23; Kramish (1986), pp. 176–7.
33. Hewlett and Anderson (1962), pp. 312–13.
34. Hewlett and Anderson (1962), p. 248.
35. Hewlett and Anderson (1962), pp. 245, 312.
36. Hewlett and Anderson (1962), p. 313.
37. W. Tobey, 'Nuclear scientists as assassination targets', *Bull Atomic Scientists*, 68 (2012), pp. 61–2.
38. Brown (2016), pp. 178–9.
39. Brown (2016), p. 180.
40. Brown (1997), p. 262; J. Rotblat, 'Leaving the bomb project', *Bull Atomic Scientists*, 41 (1985), pp. 16–18.
41. Rotblat (1985), p. 19.

23 Countdown

1. Hewlett and Anderson (1962), p. 316; Goldberg (1995), p. 124.
2. Hewlett and Anderson (1962), p. 319; Szasz (1992), p. xviii.
3. Brown (1997), p. 279.
4. Hawkins (1961), pp. 147, 193–5; Hewlett and Anderson (1962), p. 317.
5. Hewlett and Anderson (1962), pp. 108, 168, 298–301.

6. Hewlett and Anderson (1962), pp. 300, 302.
7. Frisch (1979), pp. 159–62; Hawkins (1961), pp. 198–9; Rhodes (1986), p. 611; Hewlett and Anderson (1962), p. 320; Szasz (1992), p. 23.
8. Hewlett and Anderson (1962), p. 317; Hawkins (1961), pp. 182, 199.
9. Harry Daghlian and Louis Slotin died from acute neutron irradiation following two separate accidents with the dragon assembly after the war, in August 1945 and May 1946. See Atomic Accidents at: https://ahf.nuclearmuseum.org/ahf/history/atomic-accidents
10. T.A. Chadwick and M.B. Chadwick, 'Who invented the Trinity nuclear test's Christy gadget?', *Crit Rev, Tech Papers & Tech Notes*, 207 (2021), supplement 1, pp. S356–73: https://doi.org/10.1080/00295450.2021.1903300
11. Chadwick and Chadwick (2021), pp. S356–73.
12. Hewlett and Anderson (1962), pp. 309–10, 319; Hawkins (1961), pp. 24–5, 45, 72–3, 75.
13. Hewlett and Anderson (1962), p. 317; Hawkins (1961), p. 55; Reed (2017), p. 45.
14. Hawkins (1961), pp. 126, 128–30, 168–9.
15. Hewlett and Anderson (1962), p. 317; Hawkins (1961), p. 112, 158; 'Ernest William Titterton, 1916–1990', Australian Academy of Science: https://www.science.org.au/fellowship/fellows/biographical-memoirs-1/ernest-william-titteron-1916-1990
16. Hawkins (1961), pp. 123, 128–9, 168–9; Szasz (1992), p. 24.
17. Details of the 'urchin' (and possibly Bohr's input) were redacted when the Los Alamos files were declassified, and remain unknown.
18. Hewlett and Anderson (1962), pp. 317–18; Reed (2017), pp. 47–8; Hawkins (1961), pp. 68, 168, 205; Chadwick and Chadwick (2021).
19. Hewlett and Anderson (1962), p. 318; Hawkins (1961), p. 158.
20. K.T. Bainbridge, 'Trinity' (official report) (Los Alamos, NM: Los Alamos Scientific Laboratory, University of California, 1976), pp. 3–5; Hewlett and Anderson (1962), p. 377; Hawkins (1961), pp. 170, 241.
21. Bainbridge (1976), pp. 1, 3–5, 13; Hawkins (1961), pp. 168, 170; Hewlett and Anderson (1962), p. 377; Rhodes (1986), p. 654.
22. Bainbridge (1976), pp. 1–2; Reed (2017), pp. 46–7; Szasz (1992), pp. 22–4, 149; Hawkins (1961), pp. 46, 48, 88, 238; Hewlett and Anderson (1962), p. 319.
23. Hewlett and Anderson (1962), p. 319; Hawkins (1961), pp. 75, 80, 200, 238; Szasz (1992), pp. 22, 24; Bainbridge (1976), pp. 1–2.
24. Hawkins (1961), pp. 158, 168–70; Hewlett and Anderson (1962), p. 318.
25. J. Chadwick, memo for Tube Alloys Consultative Council on current state and activities of British TA staff in USA, 30 April 1945, Chadwick Papers, GBR/0014/CHAD I 12/6; Brown (1997), pp. 279–81; Szasz (1992), p. 40.
26. Gowing (1964), pp. 279–80.
27. Groves (1962), pp. 348–9.
28. Hewlett and Anderson (1962), p. 368.
29. Szasz (1992), pp. 29, 37.
30. Rhodes (1986), pp. 598–600; B. Lenon and J. Lozuka, 'History's deadliest air raid happened in Tokyo during World War II and you've probably never

heard of it', CNN, 7 March 2020: https://edition.cnn.com/2020/03/07/asia/japan-tokyo-fire-raids-operation-meetinghouse-intl-hnk/index.html
31. Hewlett and Anderson (1962), pp. 351, 363–4, 376.
32. Hewlett and Anderson (1962), pp. 365–6.
33. Rimmer, *Between Heaven and Hell* (2012), p. 20.

24 On the Run

1. Hewlett and Anderson (1962), pp. 334–5.
2. Hewlett and Anderson (1962), pp. 340, 349.
3. Hewlett and Anderson (1962), pp. 343, 345.
4. Hewlett and Anderson (1962), pp. 344–5, 354, 360.
5. Hewlett and Anderson (1962), p. 365.
6. Dahl (1999), pp. 253, 260–1; Brown (2016), p. 186.
7. E. Teller in Groves (1962), p. vii.
8. M. Perrin and R. Furman, Alsos report, 'TA Project Intelligence Targets: Germany', Perrin Papers, PERR 3/23; Minutes of 4th meeting of TA Intelligence Committee, London, 28 February 1945, Perrin Papers, PERR 3/23; Dahl (1999), pp. 252–4.
9. Dahl (1999), pp. 252, 254–5.
10. Brown (2016), p. 183; Dahl (1999), p. 259.
11. Minutes of 6th meeting of TA Intelligence Committee, Paris, 9 April 1945, Perrin Papers, PERR 3/23; Alsos report of mission operations in Strasbourg (14 February 1945), Munich (12 May 1945) and South-West Germany (14 May), Perrin Papers, PERR 3/23; Dahl (1999), pp. 260–1; Brown (2016), pp. 186–7.
12. Dahl (1999), pp. 260, 263.
13. Alsos report of mission operations in South-West Germany, 14 May 1945, Perrin Papers, PERR 3/23; Alsos photos and plans of Haigerloch laboratory, April 1945, Perrin Papers, PERR 3/25-27; Brown (2016), pp. 181, 189–91, 193; Dahl (1999), pp. 260–1.
14. Dahl (1999), p. 262; Brown (2016), pp. 198–9.
15. Alsos report of mission operations in South-West Germany, 14 May 1945, Perrin Papers, PERR 3/23; Alsos photos and plans of Haigerloch laboratory, April 1945. Perrin Papers, PERR 3/25-27; Dahl (1999), p. 262; Brown (2016), pp. 182, 199.
16. Alsos report of mission operations in South-West Germany, 14 May 1945, Perrin Papers, PERR 3/23; Alsos photos and plans of Haigerloch laboratory, April 1945, Perrin Papers, PERR 3/25-27; Interview with Edwin Seaver, 9 March 1947, Northwestern University, Evanston IL, Samuel A. Goudsmith Papers, box 1, folder 04; Dahl (1999), p. 262.
17. Alsos report of Alpine mission, 18 May 1945, Perrin Papers, PERR 3/24; Dahl (1999), pp. 253, 263; Brown (2016), pp. 180, 200, 205–6, 208.
18. Brown (2016), p. 200.
19. Eric Welsh, memo to Chancellor Sir John Anderson, 'On the status and treatment of ten scientists detained late April 1945 by the Alsos Mission on behalf of the TA Organisation', London, 8 June 1945, with covering letter by Michael Perrin, Perrin Papers, PERR 3/23; Dahl (1999), pp. 262, 271–2; Brown (1997), p. 291.

25 Proof of Concept

1. Hawkins (1961), pp. 238–9.
2. Bainbridge (1976), pp. 7–9, 12; Hawkins (1961), p. 170.
3. Hawkins (1961), p. 168; C. Laucht, *Elemental Germans: Klaus Fuchs, Rudolf Peierls and the Making of British Nuclear Culture, 1939–1959* (Basingstoke: Palgrave Macmillan, 2012), p. 40; 'Trinity and beyond', Los Alamos National Laboratory: https://www.lanl.gov/media/publications/national-security-science/0824-trinity-and-beyond
4. S. Hassan, 'No other love: heart-wrenching letters from Richard Feynman to his late wife, Arline', *Science & Pop Cult*, 27 April 2021.
5. Hewlett and Anderson (1962), pp. 353–5.
6. Hewlett and Anderson (1962), p. 365; A. Wellerstein, 'The Kyoto misconception: what Truman did, and didn't know about Hiroshima', in M.D. Gordin and G.J. Ikenberry (eds), *The Age of Hiroshima* (Princeton, NJ: Princeton University Press, 2020), pp. 38–40.
7. Hewlett and Anderson (1962), pp. 342, 355–8, 365–6; Farmelo (2013), p. 288.
8. Hewlett and Anderson (1962), p. 366.
9. Hewlett and Anderson (1962), pp. 366–7; Szasz (1992), pp. 29–30.
10. Bainbridge (1976), p. 15; Hewlett and Anderson (1962), pp. 351, 354, 361–4, 368–73.
11. Landeshauptstadt Potsdam: https://en.potsdam.de/content/babelsberg-palace-and-park
12. Hewlett and Anderson (1962), p. 373.
13. Farmelo (2013), pp. 293–4, 297.
14. Farmelo (2013), pp. 285–6, 293–5.
15. Farmelo (2013), pp. 287, 298.
16. Hewlett and Anderson (1962), pp. 238, 249, 269; Hawkins (1961), pp. 235, 238–40; Bainbridge (1976), pp. 15, 28.
17. Bainbridge (1976), pp. 28, 29, 40.
18. Bainbridge (1976), pp. 39–41.
19. Bainbridge (1976), pp. 39–40; Hewlett and Anderson (1962), pp. 377–8.
20. Hawkins (1961), p. 168; Bainbridge (1976), p. 33; Hewlett and Anderson (1962), pp. 377–8.
21. Hawkins (1961), p. 168.
22. Hawkins (1961), pp. 240–1; Hewlett and Anderson (1962), p. 378.
23. Bainbridge (1976), pp. 15, 20, 25, 36; Hawkins (1961), pp. 235, 238–40; Hewlett and Anderson (1962), p. 378; Brown (1997), p. 292; Peierls (1979), pp. 201–2; Frisch (1979), pp. 163–4.
24. Rhodes (1986), p. 661.
25. Bainbridge (1976), p. 31.
26. Bainbridge (1976), pp. 20, 29, 30–1, 42; Hawkins (1961), p. 241; Hewlett and Anderson (1962), p. 379; Brown (1997), p. 293; Rhodes (1986), p. 666.
27. Bainbridge (1976), p. 31; Hawkins (1961), pp. 240–1.
28. Hawkins (1961), p. 241; Bainbridge (1976), p. 15; Szasz (1992), p. 71; Brown (1997), p. 292.
29. Peierls (1979), p. 202.
30. Frisch (1979), p. 157; Gowing (1964), pp. 441–2.
31. Szasz (1992), p. 25; Farmelo (2013), p. 289.

32. Hawkins (1961), p. 271; Hewlett and Anderson (1962), p. 379; Szasz (1992), p. 24.
33. Smyth (1945), p. 253; Hewlett and Anderson (1962), p. 379; Szasz (1992), p. 25; Peierls (1979), p. 202; Farmelo (2013), p. 289.
34. Bainbridge (1976), p. 32; Hawkins (1961), pp. 272–3; Hewlett and Anderson (1962), p. 378; Peierls (1979), p. 203.
35. Bretscher (1984), p. 22; Hawkins (1961), p. 272.
36. Hewlett and Anderson (1962), p. 379.
37. Bird and Sherwin (2005), p. 309.

26 Instrument of War

1. Hewlett and Anderson (1962), pp. 382–3.
2. Hewlett and Anderson (1962), pp. 380, 383–6; Farmelo (2013), pp. 298–300.
3. Hewlett and Anderson (1962), pp. 386.
4. Hewlett and Anderson (1962), pp. 385–8.
5. Hewlett and Anderson (1962), pp. 380, 389.
6. Hewlett and Anderson (1962), pp. 389–91; Farmelo (2013), pp. 301–2.
7. Hewlett and Anderson (1962), pp. 391–2.
8. Hewlett and Anderson (1962), pp. 392–4.
9. Hewlett and Anderson (1962), p. 394.
10. Farmelo (2013), pp. 303–4.
11. Farmelo (2013), p. 304; Gowing (1964), pp. 448–9.
12. Hawkins (1961), p. 243.
13. Hewlett and Anderson (1962), pp. 379–80, 389; Hawkins (1961), pp. 249; Szasz (1992), p. 24.
14. A.B. Carr, 'Thirty minutes before the dawn', *Nuclear Technology*, 207 (2021), supplement 1, p. S11.
15. Lord Sherfield, 'William George Penney, Baron Penney of East Hendred', *Biogr Mem Fell Roy Soc*, 39 (1993), pp. 285–6; Szasz (1992), p. 30.
16. P. Morrison, interview by Studs Terkel for *The Good War: An Oral History of World War II* (New York: The New Press, 1985).
17. Goldberg (1995), pp. 399–400.
18. 'Safety and the Trinity Test', *The Manhattan Project: An Interactive History*: https://www.osti.gov/opennet/manhattan-project-history/Events/1945/trinity_safety.htm
19. Hewlett and Anderson (1962), p. 319; Hawkins (1961), pp. 248–9; 'Tinian Island', Atomic Heritage Foundation: https://ahf.nuclearmuseum.org/ahf/location/tinian-island/
20. P.J. Kutnick, 'Defending the indefensible: a meditation on the life of Hiroshima pilot Paul Tibbets', *Asia-Pacific Journal/Japan Focus*, 6 (2008), pp. 1–3; 'Paul Tibbets', Atomic Heritage Foundation: https://ahf.nuclearmuseum.org/ahf/profile/paul-tibbets/
21. Hawkins (1961), pp. 240, 251–3; 'Tinian Island', Atomic Heritage Foundation.
22. Hawkins (1961), pp. 248–50; Hewlett and Anderson (1962), p. 401.
23. Hewlett and Anderson (1962), pp. 248–9.
24. Hewlett and Anderson (1962), pp. 380, 388; Hawkins (1961), pp. 248–53; 'Tinian Island', Atomic Heritage Foundation.
25. Hawkins (1961), p. 253; 'Paul Tibbets', Atomic Heritage Foundation.

26. Kutnick (2008), pp. 4–5.
27. Kutnick (2008), pp. 4–5; Hawkins (1961), p. 253.
28. Kutnick (2008), p. 5.
29. Kutnick (2008), pp. 5–6; Hawkins (1961), p. 253.
30. Kutnick (2008), p. 6; 'Paul Tibbets', Atomic Heritage Foundation.
31. Kutnick (2008), p. 6; Hawkins (1961), p. 253.
32. Kutnick (2008), pp. 6–8.
33. Hawkins (1961), p. 253.
34. Kutnick (2008), p. 10.
35. P.M.S. Blackett, *Military and Political Consequences of Atomic Energy* (London: Turnstile Press, 1948), pp. 36–9; Hawkins (1961), p. 253; 'Bombings of Hiroshima and Nagasaki – 1945', Atomic Heritage Foundation: https://ahf.nuclearmuseum.org/ahf/history/bombings-hiroshima-and-nagasaki-1945/. For discussion about differing estimates of numbers killed, see A. Wellerstein, 'Counting the dead at Hiroshima and Nagasaki', *Bull At Sci*, 4 August 2020: https://thebulletin.org/2020/08/counting-the-dead-at-hiroshima-and-nagasaki/
36. 'Bombings of Hiroshima and Nagasaki – 1945', Atomic Heritage Foundation: https://ahf.nuclearmuseum.org/ahf/history/bombings-hiroshima-and-nagasaki-1945/
37. 'Bombings of Hiroshima and Nagasaki – 1945', Atomic Heritage Foundation: https://ahf.nuclearmuseum.org/ahf/history/bombings-hiroshima-and-nagasaki-1945/; 'An eyewitness account of Hiroshima', World Council of Churches: https://www.oikoumene.org/resources/documents/an-eye-witness-account-of-hiroshima
38. J.A. Slemes, 'Eyewitness account of Hiroshima', Atomic Archive: https://www.atomicarchive.com/resources/documents/hiroshima-nagasaki/hiroshima-siemes.html
39. Kutnick (2008), p. 10.

27 Aftermaths

1. Hawkins (1961), p. 503; Kutnick (2008), p. 90.
2. Hawkins (1961), pp. 501–3.
3. Szasz (1992), pp. 30–1; Frisch (1979), p. 176; Peierls (1979), p. 203.
4. Hewlett and Anderson (1962), pp. 402–3.
5. Hewlett and Anderson (1962), p. 402.
6. Hewlett and Anderson (1962), p. 402. Full text of Truman's address is at: https://millercenter.org/the-presidency/presidential-speeches/august-6-1945-statement-president-announcing-use-bomb
7. Harvey (2000), p. 23.
8. *New York Times*, 7 August 1945, front page; Harvey (2000), p. 18.
9. E. Bradbury and S. Blakeslee, 'The harrowing story of the Nagasaki bombing mission', *Bull Atomic Sci*, 4 August 2015, p. 3: https://thebulletin.org/2015/08/the-harrowing-story-of-the-nagasaki-bombing-mission/
10. Bradbury and Blakeslee (2015), pp. 3, 7; Hawkins (1961), p. 250; Hewlett and Anderson (1962), p. 380.
11. Hewlett and Anderson (1962), pp. 402–3; 'Propaganda post-Hiroshima', Atomic Archive: https://www.atomicarchive.com/resources/documents/med/med_chp2.html

12. Hewlett and Anderson (1962), pp. 403–4.
13. Bradbury and Blakeslee (2015), pp. 3, 7. An image of Norman Ramsey (1989 Nobel Prize in Physics) signing Fat Man is at: https://ahf.nuclearmuseum.org/ahf/location/tinian-island/
14. Bradbury and Blakeslee (2015), pp. 8–10.
15. Bradbury and Blakeslee (2015), p. 7.
16. Bradbury and Blakeslee (2015), p. 10.
17. Bradbury and Blakeslee (2015), pp. 10–11.
18. Bradbury and Blakeslee (2015), pp. 12–14.
19. Bradbury and Blakeslee (2015), pp. 14–16.
20. Hawkins (1961), p. 250.
21. Hewlett and Anderson (1962), pp. 403–5.
22. Hewlett and Anderson (1962), pp. 402–3; Bradbury and Blakeslee (2015), p. 17.

28 Winners and Losers

1. Szasz (1992), pp. xvi, 30–1; Frisch (1979), p. 176; Peierls (1979), p. 203.
2. Szasz (1992), p. 42.
3. Szasz (1992), pp. 44–6.
4. Gowing (1964), pp. 322–34; Bernstein (2001), p. 135; Brown (1997), p. 294.
5. Szasz (1992), p. 43; Peierls (1979), p. 206; Brown (1997), p. 244.
6. Szasz (1992), pp. 43–4.
7. Szasz (1992), pp. 44, 48–9, 51–2; Brown (1997), p. 301.
8. Peierls (1979), p. 205.
9. Bird and Sherwin (2005), pp. 328–9.
10. Peierls (1979), pp. 206–8; Szasz (1992), pp. 89, 149.
11. M.J. Neufeld, 'Wernher von Braun and the Nazis', American Experience, PBS (2019): https://www.pbs.org/americanexperience/features/wernher-von-braun
12. Bernstein (2001), pp. xxi, 53.
13. J. Bernstein and D. Cassidy, 'Bomb apologetics: Farm Hall, August 1945', *Physics Today*, 48 (1995), p. 32.
14. Brown (2016), pp. 278, 283.
15. Bernstein (2001), pp. 66–71, 74; Brown (2016), p. 234.
16. Bernstein (2001), pp. xvii, 73–4; Brown (2016), p. 234.
17. Brown (2016), pp. 236–7; Bernstein (2001), p. 53.
18. Brown (2016), pp. 209, 228; Bernstein (2001), pp. xviii, 54; Bernstein and Cassidy (1995), p. 32.
19. Clark (1961), p. 149; Bernstein (2001), pp. 77, 87, 89, 91, 97, 99, 102, 107; Brown (2016), p. 238.
20. Brown (2016), pp. 114–15; Medawar and Pyke (2000), p. 173.
21. Bernstein (2001), pp. 117–19; Bernstein and Cassidy (1995), p. 32.
22. Bernstein (2001), pp. 120, 357–9; Bernstein and Cassidy (1995), p. 33.
23. Bernstein (2001), pp. 35, 117, 120, 122, 127–9; Bernstein and Cassidy (1995), pp. 34–5.
24. Bernstein (2001), pp. 126, 132–3, 137, 146, 352.
25. Bernstein (2001), p. 151.
26. Bernstein (2001), pp. 137, 157, 162, 169–85, 254.
27. Bernstein (2001), pp. 163–70.
28. Bernstein (2001), pp. 210, 225, 265, 270.

29. Bernstein (2001), p. 281.
30. Bernstein (2001), pp. 282–3.
31. Clark (1961), p. 144; Bernstein (2001), pp. 121, 127, 137–8, 352.
32. Bernstein (2001), pp. 140–1.
33. Clark (1961), pp. 145–8; A.A. Lucas, 'Revisiting Farm Hall', *Europhysics News*, 38 (2007), p. 28.
34. Bernstein (2001), p. 36.
35. Bernstein (2001), p. 36.
36. Gowing (1964), p. 39.

29 Whodunnit

1. Full text of Truman's address is at: https://millercenter.org/the-presidency/presidential-speeches/august-6-1945-statement-president-announcing-use-bomb
2. Smyth (1945).
3. Smyth (1945), pp. vi, 52–4, 70–1, 74, 80, 104, 173, 181, 202, 210, 214–16, 226, 248.
4. Szasz (1992), p. xiii; M.W. Perrin, 'Statements relating to the Atomic Bomb', *Rev Modern Physics*, 17 (1945), pp. 472–89.
5. Perrin (1945), pp. 472–4.
6. Perrin (1945), pp. 475–89; Clark (1961), p. 197.
7. C. Laucht, 'An extraordinary achievement of the "American way": Hollywood and the Americanization of the making of the atom bomb in *Fat Man & Little Boy*', *Eur J Am Culture*, 28 (2009), p. 43.
8. Smyth (1945), pp. 226, 248; Groves (1962), pp. 349–52.
9. Baxter (1946), pp. 142, 425–8, 438–9, 444.
10. Anon., *Manhattan Project: Official History and Documents*, Book VIII, Los Alamos Project, part 16, Diplomatic History of the Manhattan Project, US National Archives microfilm 1947 [1977], reel 10, pp. 42–5.
11. Hawkins (1961), pp. 26–8, 198, 294–303; Groves (1962), p. xiv.
12. Hewlett and Anderson (1962), pp. 37–43, 249, 260, 281–3, 310, 313.
13. Groves (1962), pp. iii, xvii, 2, 7, 23.
14. Groves (1962), pp. 7, 117–19, 125–37, 196–7, 408.
15. Bird and Sherwin (2005), p. 284.
16. Szasz (1992), p. xiv; Laucht, 'An extraordinary achievement of the "American way"' (2009), p. 43.
17. Laucht, 'An extraordinary achievement of the "American way"' (2009), pp. 42–3, 49–52.
18. Review by Paul Schrader, *Variety*, 18 July 2023: https://variety.com/film/news/oppenheimer-best-film
19. Groves (1962), pp. 125, 129–31; Gowing (1964), p. 237.
20. Groves (1962), pp. 348–51.
21. Hartcup and Allibone (1984), pp. 156, 160, 162, 272–4, 286.
22. Groves (1962), pp. viii, 143; Szasz (1992), pp. xviii, 9–10; Farmelo (2013), p. 287.
23. Groves (1962), p. 408; Gowing (1964), p. 258, 266–7; Szasz (1992), pp. 148–9.
24. Gowing (1964), pp. 158–62.
25. Gowing (1964), pp. 234, 236, 239, 240, 243, 256, 263.
26. Cockburn and Ellyard (1981), pp. 193–5; Gowing (1964), pp. 256–9.

27. Gowing (1964), pp. 260–1; Szasz (1992), p. 198.
28. Frisch (1979), pp. 159–62.
29. Gowing (1964), pp. 250–4; Szasz (1992), p. 87; Hewlett and Anderson (1962), p. 26.
30. M. Perrin diary entries for 8 September, 21 October and 30 December 1943, Perrin Papers, PERR 3/1/5; Gowing (1964), p. 256.
31. Peierls (1979), pp. 187, 199; Szasz (1992), p. 89; Gowing (1964), pp. 264–5; C. Laucht, '"Los Alamos in a way was a city of foreigners": German-speaking émigré scientists and the making of the atomic bomb at Los Alamos, New Mexico, 1943-6', New Mexico Historical Review, 86 (2011), p. 233.
32. Moore and Brown (2021), pp. S224–6; Szasz (1992), p. 24.
33. Brown (1997), p. 292; J.D. Newton, 'Ernest William Titterton, 1916–1990', Australian Academy of Science: https://www.science.org.au/fellowship/fellows/biographical-memoirs/ernest-william-titterton-1916-1990
34. Gowing (1964), pp. 266–7, 287.
35. Szasz (1992), p. xxvi.
36. Rhodes (1986), pp. 359–62, 368.
37. Rhodes (1986), p. 372.
38. Letter from W.A. Akers to Munro, 14 January 1943, Perrin Papers, PERR 3/7.
39. Rhodes (1986), p. 372.
40. Szasz (1992), pp. xviii–xx; Clark (1961), p. 197; Frisch (1979), pp. 176–7; Peierls (1979), pp. 203–5.
41. Clark didn't cover the plutonium bomb, then still classified top secret.
42. Clark (1961), pp. xi, xiv, 138, 159–60, 163–6, 168–70, 192–3.
43. Gowing (1964), pp. xiii, 85; R. MacLeod, 'Margaret Gowing, 1921–1998', Biogr Mem Fell Roy Soc, 58 (2012), pp. 74–8; Szasz (1992), p. xv.
44. MacLeod (2012), pp. 78–9; Gowing (1964), pp. 364–5; Szasz (1992), p. xv.
45. Szasz (1992), p. xv.
46. Szasz (1992), p. xv.

30 Follow-Up

1. Gowing (1964), p. 328; Hartcup and Allibone (1984), pp. 130, 136–9, 147.
2. Groves (1962), pp. 349–51.
3. Groves (1962), pp. 253, 261, 286.
4. Groves (1962), p. 140; Szasz (1992), p. 137; Sherfield (1993), pp. 288, 291, 298.
5. Lord Waverley and A. Fleck, 'Wallace Allan Akers, 1888–1954', Biogr Mem Fell Roy Soc, 1 (1955), pp. 1–4; A. Fleck, 'Akers, Sir Wallace A.', Oxford Dictionary of National Biography: https://doi.org/10.1093/ref:odnb/30359
6. Anon., 'Michael Perrin: the man to whom Fuchs confessed', New Scientist, 24 January 1957.
7. Diaries and papers of Michael Perrin, Perrin Papers, PERR 3/1/3, 3/1/5, 3/25.
8. Szasz (1992), pp. 47, 89.
9. Szasz (1992), pp. 47–9; J.O. Newton, 'Ernest William Titterton, 1916–1990', Hist Record Australian Science, 9 (1992), pp. 167–87.
10. Moore and Brown (2021), p. S228; Szasz (1992), pp. 105, 140.
11. Moon (1977), pp. 544, 547, 551–3.

12. Kurti (1958), pp. 225–6, 231–3, 245–8; P.W. Bridgman, 'Sir Francis Simon', *Science*, 131 (1960), pp. 1649–50.
13. Szasz (1992), pp. 82–4, 88; Williams (1987), pp. 184–6, 188.
14. Szasz (1992), pp. 84–6; Williams (1987), pp. 135, 186, 188.
15. Williams (1987), pp. 184, 192, 199.
16. Brown (1997), pp. 291–4, 323, 358–9; Gowing (1964), p. 234.
17. Brown (1997), pp. 343–8, 355–7, 361–2.
18. Brown (1997), p. 363; I. Gallagher, 'James Chadwick, humble British scientist who played vital role in Manhattan Project', *Daily Mail*, 20 July 2023: https://www.dailymail.co.uk/news/article-12327581
19. B. Bleaney, 'Sir Mark Oliphant, 1901–2000', *Biogr Mem Fell Roy Soc*, 47 (2001), pp. 390–1.
20. Bleaney (2001), p. 392.
21. Szasz (1992), p. 8.
22. Lee (2007), pp. 275–81; Peierls (1979), pp. 223–5, 278, 324; M. Shifman, *Love and Physics: The Peierlses* (Singapore: World Scientific, 2019), p. 243.
23. Lee (2007), pp. 275, 279–80; Peierls (1979), pp. 282, 284.
24. Lee (2007), pp. 270, 280–1; Peierls (1979), pp. 326, 339.
25. Frisch (1979), pp. 192–3, 197, 200–2; R. Peierls, 'Otto Robert Frisch, 1904–1979', *Biogr Mem Fell Roy Soc*, 27 (1981), p. 294.
26. Frisch (1979), pp. 195, 199, 201, 203–8, 214–16; Peierls, 'Otto Robert Frisch' (1981), pp. 299, 301; L. Meitner, interview by O.R. Frisch and T.S. Kuhn, Cambridge, MA, 12 May 1963, American Institute of Physics: https://repository.aip.org/object/nbla:270959
27. Frisch (1979), pp. 218–19.
28. L.M.M. Blume, 'The elusive horror of Hiroshima', *National Geographic*, 5 August 2020; K. Hignett, 'The devastation of Nagasaki', *Newsweek*, 24 November 2022.
29. Blume (2020).
30. Hiroshima Peace Memorial Museum: https://hpmmuseum.jp/?lang=eng
31. Trinity Site, White Sands National Park: https://www.nps.gov/whsa/learn/historyculture/trinity-site.htm; R.E. Hermes and W.D. Strickfadden, 'A new look at trinitite', *Nucl Weapons J*, 2 (2005), pp. 2–5; D. Burge, 'Have a blast: Trinity site allows public to visit where atom bomb was first tested', *El Paso Times*, 4 April 2018.
32. Einstein–Szilard letter, 1939. Atomic Heritage Foundation. https://ahf.nuclearmuseum.org/key-documents/einstein-szilard; A. Einstein, quoted in Jungk (1956), p. 87.
33. Rhydymwyn Valley Works. *Newcomen*, International Society for the History of Engineering and Technology: https://www.newcomen.com/activity/visit-to-rhydymwyn
34. Bernstein (2001), pp. xix, 32, 55; Bernstein and Cassidy (1995), p. 36; Gowing (1964), p. vi, xiii, xv; A.A. Lucas, 'Revisiting Farm Hall', *Europhysics News*, 38 (2007), pp. 25–7.
35. W. Atkins, 'The bomb that killed my grandfather', *FT Magazine*, 5 April 2019; Freeman (2015), pp. 197–208.
36. *La Bataille de l'Eau Lourde* (1948), directed by J. Dréville and T. Vibe-Müller (English version, *Operation Swallow*): see https://doi.org/10.4000/chrhc.16298

37. The Saboteurs' Trail, Rjukan, Norway: https://www.visitnorway.com/listings/the-saboteurs-trail/1469/
38. Mayer et al., 'Uranium from German nuclear power projects of the 1940s – a nuclear forensic investigation', *angewandte Chemie*, 54 (2015), pp. 13452–6: https://doi.org/10.1002/anie.201504874

SELECT BIBLIOGRAPHY

Allardice, C., and E.R. Trapnell (1946). 'The first pile', reissued by US Department of Energy, December 1982: https://www.iaea.org/sites/default/files/publications/magazines/bulletin/bull4-0/04005004147su.pdf

Allison, S.K. 'Enrico Fermi, 1901–1954', *Biograph Memoirs Nat Acad Sci*, 30 (1957).

Badash, L., E. Hodes and A. Tiddens. 'Nuclear fission: reaction to the discovery in 1939', *Proc Am Philosoph Soc*, 130 (1986).

Bainbridge, K.T. 'Trinity' (official report) (Los Alamos, NM: Los Alamos Scientific Laboratory, University of California, 1976).

Baxter, J.P. *Scientists Against Time* (Boston, MA: Little, Brown & Co., 1946).

Becquerel, A.H. 'On radioactivity, a new property of matter', Nobel Lecture, 11 December 1903: https://www.nobelprize.org/prizes/physics/1903/becquerel/lecture

Bernstein, J. *Hitler's Uranium Club: The Secret Recordings at Farm Hall* (New York: Springer Verlag, 2001).

Bernstein, J., and D. Cassidy. 'Bomb apologetics: Farm Hall, August 1945', *Physics Today*, 48 (1995).

Bird, K., and M.J. Sherwin. *American Prometheus: The Triumph and Tragedy of J. Robert Oppenheimer* (New York: Alfred A. Knopf, 2005).

Blackett, P.M.S. 'Fréderic Joliot, 1900–1958', *Biogr Mem Fell Roy Soc*, 6 (1960).

Bradbury, E. and S. Blakeslee. 'The harrowing story of the Nagasaki bombing mission', *Bull Atomic Sci*, 4 August 2015: https://thebulletin.org/2015/08/the-harrowing-story-of-the-nagasaki-bombing-mission/

Bretscher, H. Interview by John Bennett and Anna Shepherd, 10 July 1984, American Institute of Physics: https://www.aip.org/history-programs/niels-bohr-library/oral-histories/4536

Brown, A. *The Neutron and the Bomb: A Biography of Sir James Chadwick* (Oxford: Oxford University Press, 1997).

Brown, C. *Operation Big: The Race to Stop Hitler's A-Bomb* (Stroud: Amberley, 2016).

Chadwick Papers: papers and correspondence of Sir James Chadwick, Archives of Churchill College, Cambridge, GBR/0014/CHAD.

Clark, R. *The Birth of the Bomb: The Untold Story of Britain's Part in the Weapon That Changed the World* (London: Phoenix House, 1961).

SELECT BIBLIOGRAPHY

Clark, R. *Tizard* (London: Methuen, 1965).

Cochran, W., and S. Devons. 'Norman Feather, 1904–1978', *Biogr Mem Fell Roy Soc*, 27 (1981).

Cockburn, S., and D. Ellyard. *Oliphant: The Life and Times of Sir Mark Oliphant* (Adelaide: Axiom Books, 1981).

Curie, È. *Madame Curie* (London: William Heinemann, 1937).

Dahl, P.F. *Heavy Water and the Wartime Race for Nuclear Energy* (Philadelphia: CRC Press, 1999).

DiScala, S.M. 'Science and Fascism: the case of Enrico Fermi', *Totalit Movts Polit Relig*, 6 (2011).

Eve, A.S. and J. Chadwick. 'Lord Rutherford, 1871–1937', *Biogr Mem Fell Roy Soc*, 2 (1938).

Farmelo, G. *Churchill's Bomb* (New York: Basic Books, 2013).

Feather, N. Interviews by Charles Weiner, Edinburgh, 1971, American Institute of Physics: Session I, 25 February 1971: https://www.aip.org/history-programs/niels-bohr-library/oral-histories/4599-1; Session II, 5 November 1971: https://www.aip.org/history-programs/niels-bohr-library/oral-histories/4599-2

Fermi, E. 'The development of the first chain-reacting pile', *Proc Am Phil Soc*, 90 (1946).

Fermi, L. 'To Fermi – with love – Part 2', interview with Laura Fermi, 12 September 1971, Voices of the Manhattan Project, Atomic Heritage Foundation: https://ahf.nuclearmuseum.org/voices/oral-histories/fermi-love-part-2/

Freeman, K. *The Civilian Bomb-Disposing Earl* (Barnsley, South Yorks.: Pen & Sword Military, 2015).

Friedrich, B. 'Fritz Haber (1868–1934)', *angewandte Chemie* (Intl Edn), 44 (2005).

Frisch, O.R. Interview by Charles Weiner, American Institute of Physics, 3 May 1967: https://www.aip.history-programs/niels-bohr-library/oral-histories/4616

Frisch, O.R. 'Lise Meitner, 1878–1968, *Biogr Mem Fell Roy Soc*, 16 (1970).

Frisch, O.R. *What Little I Remember* (Cambridge: Cambridge University Press, 1979).

Frisch, O.R., and R. Peierls. 'Memorandum on the properties of a radioactive "super-bomb"' (1940), UK Public Record Office, AB 1/210. Full text available at: https://www.atomicarchive.com/resources/documents/beginnings/frisch-peierls-2.html

Gentner, W. Interview by Charles Weiner, Heidelberg, 15 November 1971, American Institute of Physics: https://www.aip.org/history-programs/niels-bohr-library/oral-histories/5080

Goldberg, S. 'Inventing a climate of opinion: Vannevar Bush and the decision to build the bomb', *Isis*, 83 (1992).

Goldberg, S. 'Racing to the finish: the decision to bomb Hiroshima and Nagasaki', *J Am-East Asian Relns*, 4 (1995).

Goldsmith, M. *The Curie Family* (Geneva: Edito-Service, 1971).

Goodman, M.S. 'MI6's atomic spy: the rise and fall of Eric Welsh', *War in History*, 23 (2016).

Gowing, M. *Britain and Atomic Energy, 1935–1945* (London: Macmillan Publishing, 1964).

Groves, L. *Now It Can Be Told: The Story of the Manhattan Project* (New York: Harper & Row, 1962).

Halasz, N, and R. Halasz. 'Leo Szilard: the reluctant father of the atomic bomb', *New Hungarian Quarterly*, 15 (1974).

Hartcup, G., and T.E. Allibone. *Cockcroft and the Atom* (Bristol: Adam Hilger (Institute of Physics), 1984).

Harvey, D. 'History of the Hanford Site, 1943–1990', Pacific Northwest National Laboratory/US Department of Energy (2000): https://doi.org/10.2172/887452

Hawkins, D. *Project Y: The Los Alamos Story* (Los Angeles, CA: Tomash Publishers, 1961).

Heilbron, J.L. *The Dilemmas of an Upright Man: Max Planck as Spokesman for German Science* (Cambridge, MA: Harvard University Press, 1996).

Heisenberg, W. 'A lecture on bomb physics: February 1942', *Physics Today*, 48 (1995).

Hermann, G. 'Five decades ago: from the transuranics to nuclear fission', *angewandte Chemie* (Intl Edn), 29 (1990).

Hewlett, R.G. and O.E. Anderson. *The New World, 1939–1946* (University Park, PA: Pennsylvania State University Press, 1962).

Hinde, R.A. and J.L. Finney. 'Sir Joseph (Józef) Rotblat, 1908–2005', *Biogr Mem Fell Roy Soc*, 53 (2007).

Irving, D. *The Virus House* (London: William Kimber, 1967).

Jungk, R. *Brighter Than a Thousand Suns: A Personal History of the Atomic Scientists* (New York: Harcourt, Brace, 1956).

Kowarski, L. Interviews by Charles Weiner, New York, 1969, American Institute of Physics: Session I, 20 March 1969: https://www.aip.org/history-programs/niels-bohr-library/oral-histories/4717-1; Session II, 19 October 1969: https://www.aip.org/history-programs/niels-bohr-library/oral-histories/4717-2; Session III, 20 October 1969: https://www.aip.org/history-programs/niels-bohr-library/oral-histories/4717-3

Kramish, A. *The Griffin* (London: Macmillan, 1986).

Kurti, N. 'Franz Eugen Simon, 1893–1956', *Biogr Mem Fell Roy Soc*, 4 (1958).

Kutnick, P.J. 'Defending the indefensible: a meditation on the life of Hiroshima pilot Paul Tibbets', *Asia-Pacific Journal/Japan Focus*, 6 (2008).

Lanouette, W. 'The odd couple and the bomb', *Sci Am*, 283 (2000).

Lee, S. 'Sir Rudolf Ernst Peierls, 1907–1995', *Biogr Mem Fell Roy Soc*, 53 (2007).

Lovell, B. 'Patrick Maynard Stuart Blackett, 1897–1974', *Biogr Mem Fell Roy Soc*, 21 (1975).

Massey, H. and N. Feather. 'James Chadwick, 1891–1974', *Biogr Mem Fell Roy Soc*, 22 (1976).

M.A.U.D. Committee. 'Report by M.A.U.D. Committee on the use of uranium for a bomb', Ministry of Aircraft Production, London, July 1941. Reproduced in full as Appendix 2 in Gowing (1964), pp. 394–438.

Medawar, J. and D. Pyke. *Hitler's Gift: Scientists Who Fled Nazi Germany* (London: Blake Publishing, 2000).

Moon, P.B. 'George Paget Thomson, 1892–1975', *Biogr Mem Fell Roy Soc*, 23 (1977).

Moore, R. and E.N Brown. 'Woolwich, Bruceton, Los Alamos: Munroe jets and the Trinity gadget', *Nuclear Technology*, 207 (2021), supplement 1, pp. S222–S230: https://doi.org/10.1080/00295450.2021.1905463

Mott, N, and R. Peierls. 'Werner Heisenberg, 1901–1976', *Biogr Mem Fell Roy Soc*, 23 (1977).

Mulligan, J.F. 'Heinrich Hertz and Philipp Lenard: two distinguished physicists, two disparate men', *Phys in Perspect*, 1 (1999).

Oliphant, M.L.E. and W. Penney. 'John Douglas Cockcroft, 1897–1967', *Biogr Mem Fell Roy Soc*, 14 (1968).

SELECT BIBLIOGRAPHY

Pash, B. *The Alsos Mission* (New York: Charter Books, 1969).

Peierls, R. *Bird of Passage* (Princeton, NJ: Princeton University Press, 1979).

Perrin Papers: papers and correspondence of Sir Michael Perrin, Archives of Churchill College, Cambridge, partially catalogued.

Perrin, M.W. *Statements Relating to the Atomic Bomb* (London: British Information Services, HMSO, 1945); reprinted in *Rev Modern Physics*, 17 (1945).

Phelps, S. *The Tizard Mission: The Top-Secret Operation That Changed the Course of World War II* (Yardley, PA: Westholme, 2010).

Reed, B.C. 'From fission to censorship: 18 months on the road to the bomb', *Ann Phys (Berlin)*, 530 (2018).

Reed, B.C. 'Revisiting the *Los Alamos Primer*', *Physics Today*, 70 (2017): https://doi.org/10.1063/PT.3.3692

René, M. 'History of uranium mining in Central Europe' (2018): http://dx.doi.org/10.5772/intechopen.71962

Rhodes, R. *The Making of the Atomic Bomb* (New York: Simon & Schuster, 1986).

Seaborg, G. *The Transuranic Elements* (New Haven, CT: Yale University Press, 1958).

Seaborg, G.T. 'Transuranium elements: a half century', *Proceedings of the American Chemical Society Symposium to Commemorate the 50th Anniversary of Transuranium Elements*, Washington DC, 27 August 1990: ttps://orau.org/health-physics-museum/files/library/transuranicsseaborg.pdf

Serber, R. *The Los Alamos Primer: The First Lectures on How to Build an Atomic Bomb* (Berkeley, CA: University of California Press, 1992).

Sherfield, Lord. 'William George Penney, Baron Penney of East Hendred', *Biogr Mem Fell Roy Soc*, 39 (1993).

Sime, R.L. *Lise Meitner: A Life in Physics* (Berkeley, CA: University of California Press, 1996).

Smith, M. *Foley: The Spy Who Saved 10,000 Jews* (London: Hodder & Stoughton, 1999).

Smyth, H.D. *Atomic Energy for Military Purposes* (Princeton, NJ: Princeton University Press, 1945). Full text at: https://www.osti.gov/opennet/manhattan-project-history/publications/smyth_report.pdf

Soddy, F. *The Interpretation of Radium and the Structure of the Atom* (London: John Murray, 1909).

Spence, R. 'Otto Hahn, 1879–1968', *Biogr Mem Fell Roy Soc*, 16 (1970).

Stuewer, R.H. 'Bringing the news of fission to America', *Physics Today*, 38 (1985): https://doi.org/10.1063/1.881016

Szasz, F.M. *British Scientists and the Manhattan Project: The Los Alamos Years* (New York: St Martin's Press, 1992).

Walker, M. *Nazi Science: Myth, Truth, and the German Atomic Bomb* (New York: Plenum Press, 1995).

Waltham, C. 'An early history of heavy water', *Physics in Canada*, 49 (1993).

Weart, S.R. 'Scientists with a secret', *Physics Today*, 29 (1978): https://doi.org/10.1063/1.3023312

Wells, H.G. *The World Set Free: A Story of Mankind* (London: Macmillan & Co., 1914).

Williams, R.C. *Klaus Fuchs, Atom Spy* (Cambridge, MA: Harvard University Press (1987).

INDEX

Page numbers in *italic* refer to figures; page numbers in **bold** refer to maps. Plate numbers are indicated by *p*.

'94' (codename for plutonium) 131
509th Composite Group 291

Abelson, Philip 82, 83, 116, 128, 201, 236–7
Abwehr (German military intelligence) 86, 87, 88–9, 129, 138
accidents
    Chadwick laboratory 91, 154
    Heisenberg laboratory 183
    Manhattan Project 257
Advisory Committee on Uranium *see* Uranium Committee (S-1 Committee)
AERE *see* Atomic Energy Research Establishment
Air Defence, Committee for the Scientific Survey of 72, 73
Air Warfare, Committee for the Scientific Survey of 74, 75, 89, 98
Akers, Wallace xx, 157–8, 159, 186, 187, *p*17
    British Mission to the Manhattan Project 209
    and Chadwick 226–7
    exclusion from Manhattan Project 216
    and Groves 194–5
    and ICI 157–8, 204, 332
    knighthood awarded to 332
    and Manhattan Project 194–7, 209, 216
    post-war activities 332
    on shortcomings of Bush and Conant 325
    Tube Alloys tour of US 180, 182
Alamagordo Bombing/Missile Range, New Mexico 242, 342–3
Alberta (Project A) 262, 290–5
    see also Fat Man (plutonium bomb); Little Boy (uranium bomb)
Aldermaston 331
Allier, Jacques 86, 87, 88, 89, 97, 102, 129
    in *La Bataille de l'Eau Lourde* 345, *p*11
Allison, Samuel 154
alpha-particles 6, 7, 8, 21, 30, 31, 36
Alsos Mission 220–1, 230–2, 251, 252, *p*22
    pursuit of *Uranverein* scientists and uranium 269–73, 309–10, *p*23
altimeters 241, 255
aluminium foil 260, 261
Alvarez, Luis 259

INDEX

American Prometheus: The Triumph and Tragedy of J. Robert Oppenheimer (Bird & Sherwin) 318
Anchor Ranch, Los Alamos 234
Anderson, John 156–7, 167, 182–3, 187
   Bush's letter to 193–4
   Conant memorandum 197
   and German atomic bomb threat 315
   and ICI 204
   sharing of atomic bomb secrets with Russia 228
   Tube Alloys–Manhattan Project accord 209
Andes, RMS 216
Anglo-American military and scientific collaboration 121–6, 130–1, 142–3, 155–6, 161, 166, 167–8, 187
   American terms of engagement 196–7
   British resentment at 'predatory' American behaviour 204
   British security concerns 167
   Churchill's attitude towards 164, 165, 183, 185–6, 208
   Conant memorandum 196–7
   Hyde Park Aide-Mémoire 250
   minimisation of British contribution to Manhattan Project 316–19, 326–7
   rapprochement between Britain and US 208–9
Anglo-Russian Treaty 196
Annual Report of Progress in Chemistry 92
anti-Semitism 16, 25, 32, 42, 53
Ashworth, Frederick 302, 303
atomic bomb
   altitude of detonation 241, 242, 245, 294
   American scepticism/indifference to concept of 122, 126, 151–4, 161–2, 323–5
   British contribution to development of 320–3, 326–9
   design of 145
   destructive impact of 290, 295–7
   detonation of 233–4, 237–40, 241, 244, 246, 249, 255, 259
   etymology 10
   exclusion of US from British super-bomb development 165
   explosive and destructive potential of 94, 112, 145, 252
   feasibility of 92, 94, 95, 128, 166, 210
   fireball description 295
   flash and blast damage to people 290, 296
   German atomic bomb, feasibility/threat of 62–4, 66, 75, 76, 94, 117, 131–2, 135–40
   height of detonation 241, 242, 245, 294
   Japan, decision to use atomic bomb against 275
   manufacturing cost and timescale 146
   mushroom cloud 295
   non-military/'harmless demonstration' use of 276
   nose-impact bomb fuses 241
   nuclear chain reaction 234
   in science fiction 17
   self-destruction contingency 241
   target selection 266, 275, 287
   U-235 requirement for uranium bomb 233
   see also Fat Man (plutonium bomb); Little Boy (uranium bomb); nuclear energy/power
'Atomic Cities', Manhattan Project **xxxiv**, 200–2
Atomic Energy Research Establishment (AERE) 330
atomic number 21, 30
atomic structure
   Bohr's model 20
   Rutherford's model 8, 9
   see also nuclear physics
Atomic Weapons Research Establishment, Aldermaston 331
atomic weight 21, 30
Atomkeller (Atomic Cellar) Museum, Haigerloch 346
Attlee, Clement 277, 286, 288

INDEX

Aubert, Axel 87, 96, 138
Augusta, USS 299
'ausonium' ('element 93') 44
Australian National University, Canberra 337

B-I reactor 172
B-VIII *Uranbrenner* reactor 271, *272*, 346, *p*22, *p*23
B-29 long-range bombers (Superfortresses) 241–2, 291, 292
Babelsberg 277
Babes in the Wood 307–8, *p*29
Bad Nauheim, 'Einstein debate' (1920) 24
Bagge, Erich 269, 310, 311, 314
Bainbridge, Kenneth 152, 153, 162, 284
barium 46, 54, 56
Bataille de l'Eau Lourde, La (*The Battle of the Heavy Water*) (film) 345, *p*8, *p*11
Baumbach, Otto 19
Baxter, J.P. 320–1
Baxter, John 317
Beams, Jesse 171, 181
Becquerel, Henri 2–3, 4, 5–6
Beginning or the End, The (film) 318
Belgian Congo, uranium mines 73, 226
Belgium, German invasion/occupation of 98
Bell Telephone Company 125
Berg, Morris (Moe) 221, 252–3
Berkeley, University of California
  '94' (codename for plutonium) 131
  Lawrence's lab 48, 83, 181, 182, 216, 235, 243, *p*16
  Oliphant's visits to 154, 163, 166, 236, 320
  Seaborg's research at 142
  Wilkins's research at 320
Berlin
  'bibliocaust' 34
  Kaiser Wilhelm Institutes 11, 14, 134, 137, 170, 313
  Virus House (*Uranbrenner* reactor facility) 137, 138, 139, 170, 221
beryllium 55, 99, 261
beta-particles 6
Bethe, Hans 239–40

Bhagavad Gita 285
'bibliocaust' (book-burning), Nazi Germany 33–4
Big Stink (photographic aircraft, Nagasaki bombing) 302, 303
Bird, Kai 318
Bird of Passage, A (Peierls) 339
Birmingham, Blitz 108
Birmingham University 67–8, 70, *p*20
Birth of the Bomb, The (Clark) 327, 328
'Black Box' (Tizard Mission) 123, 124–5, 126, 161
Blackett, Patrick 41, 42, 71–2, 106, 146, 313
Blau, Ulla 340
Bock, Frederick 302
Bockscar, Nagasaki bombing 302–4
Bohr, Niels xx, 20, 26, 27, 42, *p*3
  ambivalence about the atomic bomb 219
  atomic bomb feasibility 210
  British Mission to the Manhattan Project 219
  Churchill meeting 228, 229, 251
  Copenhagen conferences 50, *p*7
  extraction from Denmark to US, via Sweden and Britain 209–11
  friendship with Welsh 211
  Frisch's description of 55
  and German invasion/occupation of Denmark 95
  Heisenberg meetings in Copenhagen (September 1941) 148–9, 211
  and initiator for plutonium bomb 260
  invitation to relocate to Britain 205
  at Los Alamos 248
  and Meitner 51, 52
  nuclear chain reaction in uranium 91
  and nuclear fission discovery 56, 57–9, 65
  sharing of atomic bomb secrets with Russia 228
  Tube Alloys and Manhattan Project briefing 210
  viewed as security risk by Churchill 251

416

INDEX

'boilers' (nuclear reactors), Joliot–Halban–Kowarski research 85–6, 97, 113–16, 127, 147
bomb disposal 117–18, 344–5
book-burning ('bibliocaust'), Nazi Germany 33–4
Boot, Harry 74
Bordeaux 103, 105
Born, Max 148
Bothe, Walther 119, 120, 128–9, 134, 137, 138, 139, 176, 270
　Alsos Mission hunt for 231, 232
　graphite 'blind spot' 172
Bowen, Edward ('Taffy') 122–3, 124, 125
Bragg, Lawrence 73
Braun, Wernher von 309, 315
Breit, Gregory 83, 84
Bretscher, Egon 114–15, 127–8, 142
Briggs, Lyman xx–xxi, 79, 80, 85, 126, 153, 324
　and approval for American atomic bomb 180
　confrontation with Oliphant 154
　and M.A.U.D. Committee final report 152, 154, 163
Brighter Than a Thousand Suns (Jungk) 344
Britain
　Anglo-Russian Treaty 196
　Chain Home radar network 74
　Committee for the Scientific Survey of Air Defence 72, 73
　Committee for the Scientific Survey of Air Warfare 74, 75, 89, 98
　contribution to atomic bomb development 320–3, 326–9
　military intelligence sharing/not sharing with US 185, 192–3, 212
　minimisation of British contribution to Manhattan Project by Americans 316–19, 326–7
　nuclear programme 263, 307, 323, 327–8, 330, 331
　radar development 72, 73, 74–5
　Technical and Scientific (Tizard) Mission to the United States 122–3
　see also Anglo-American military and scientific collaboration
Britain and Atomic Energy 1939–1945 (Gowing) 328
British Central Scientific Office, Washington DC 130
British Mission to the Manhattan Project 209, 215–20, 223, 225, 243–6, 248, 264, 306–9
　contributions to atomic bomb development 320–3, 326–9
　farewell party 307–8
　members of 358–9
　post-war activities of members 333
British Scientists and the Manhattan Project: The Los Alamos Years (Szasz) 328–9
Broompark, SS 104, 105, 150
Brun, Jomar 169, 170, 185, 213
　escape from Norway 189–90
　sabotage of Vemork plant 177
Brussels
　Solvay Physics Conference (1927) 26
　Union Minière Company 73, 74, 232
Bush, Vannevar xxi, 126, 152, 153, 162, 163, 165, 183, *p*16
　activation of American atomic bomb programme 166–7
　Akers's assessment of 325
　Anglo-American cooperation 208, 319
　dismissal of atomic bomb programme 324
　exclusion of British from Manhattan Project 203
　and Groves 200
　Manhattan Project 188, 193–4, 203
　and Perrin 186
　Senate Appropriations Committee, appearance before 237
　and Tizard 124
　volte-face on atomic bomb programme 325

cadmium control rods 77, 197, 198
Calder Hall reactor 331

417

INDEX

calutron (modified cyclotron) 235, *236*, 244, 300, 320
   'calutron girls' 236, *p*19
   etymology 181
   racetracks 236
Cambridge 115
   Cavendish Laboratory 20–1, 29–32, 114, 115, 127–8, 340, *p*5
   Science Park 340
cameras
   Nagasaki bombing photographs 302
   Trinity test monitoring 262
Canada
   Eldorado Mining Company 160, 186, 195, 204
   heavy-water reactor, Anglo-Franco-Canadian 264
   nuclear energy programme 195
   Tizard's visit to 124
cathode rays 15
Cavendish Laboratory, Cambridge 20–1, 29–32, 114, 115, 127–8, 340, *p*5
cavity magnetron 74, 89, 125, 151
Chadwick, Aileen 218, 337
Chadwick, James xxi, 19–20, 21, 28, 29, 37, 47–8, 65, *p*2, *p*30
   Anglo-American military and scientific collaboration 167
   and Bohr 210
   British Mission to the Manhattan Project 243
   contribution to atomic bomb development 320
   cyclotron use 48, 89, 91, 106
   death of 337
   and Frisch 108
   German atomic bomb feasibility 75
   and Groves 215–16, 243, 320
   invitation to Bohr to relocate to Britain 205
   and Joliot 232
   knighthood awarded to 254–5
   and Manhattan Project 218, 226–7, 263–4, 318
   M.A.U.D. Committee 98, 105–6, 141, 144
   Pegram and Urey meeting 166
   Peierls's assessment of 70
   post-war activities 336–7
   and Rotblat 89–90, 253
   Smyth Report 317, 319
   Trinity test 283–4, 285
   Tube Alloys Technical Committee 158
Chadwick Building, Liverpool 343
Chain Home network 74
chain reaction *see* nuclear chain reaction
Charlton, Wiltshire 344, 345
chemical warfare 14, 17
Cherwell, Lord *see* Lindemann, Frederick
chlorine gas
   in chemical warfare 14
   thermal diffusion technique for isotope separation 68
Christie Cancer Hospital, Manchester 99
Christy, Robert 258
Churchill, Winston 72–3, 98, 141, 227
   1945 General Election 278, 288
   and Anglo-American military and scientific collaboration 122, 123, 183, 187, 208
   Anglo-Russian Treaty 196
   Bohr meeting 228, 229, 251
   British super-bomb development 164–5
   German atomic bomb feasibility 76
   Hyde Park Aide-Mémoire 250–1
   Potsdam Conference 277–8, 286, 287, 288
   and Roosevelt 121, 179, 185–6, 208–9, 250
   Roosevelt's letter to (October 1941) 163–4, 165, 167
   sharing of atomic bomb secrets with Russia 228, 268
   Smyth Report 317
   and Trinity test success 287
   trips to United States 179, 185–6
   Yalta Conference 267
Clairvivre 105
Clarendon Laboratory, Oxford 111, 142
Clark, Ronald 327, 328
Clermont-Ferrand 101–2, 105

INDEX

Clusius, Klaus 68, 94, 170
Clusius tubes 109, 170, 171
Cockcroft, John xxi–xxii, 30–1, 113, 114, *p*5, *p*31
   Anglo-American military and scientific collaboration 122, 123, 125, 126
   'Black Box' 123
   Britain's nuclear programme 263, 330–1
   death of 331
   German atomic bomb feasibility 139–40
   M.A.U.D. Committee 141
   Meitner telegram 99
   military intelligence work 133
   Nobel Prize 331
   Rosbaud meeting 133, 134, 135
   and 'super-bomb' prospect 98
   and Tube Alloys 159
Cockcroft–Walton accelerator 30–1
Collège de France, Paris 85, 88, 101, *p*8
   Alsos Mission arrival at 231
   under German control 119, 120, 121, 128–9, 175–7
Columbia River 199, 200
Columbia University, New York 76, 84, 181
Committee for the Scientific Survey of Air Defence 72, 73
Committee for the Scientific Survey of Air Warfare 74, 75, 89, 98
Compton, Arthur 162, 163, 166, 198, *p*16
   and approval for American atomic bomb 180
   military use of atomic bomb 276
Conant, James xxii, 151–3, 162, 163, 198, 325, *p*16
   and Akers 196–7
   exclusion of British from Manhattan Project 203
   and Russian atomic bomb programme 268
Congo, Belgian, uranium mines 73, 226
control rods, cadmium 77, 197, 198
Copenhagen
   Bohr's annual conferences 50, *p*7
   Bohr's meetings with Heisenberg (September, 1941) 148–9, 211
   German occupation of 95
   Institute of Theoretical Physics 27
Coster, Dirk 52
'Cowpuncher Committee' 263, 265, 276
CP-1 (Chicago Pile-1) (Fermi's nuclear reactor) 195, 197–9
critical mass
   'dragon' experiment, Frisch 233
   plutonium 258
   uranium 62, 70, 74, 91–2, 94, 100, 174
'CS' (Capital Ship) bomb 245
Curie, Marie 3–5, 13–14, 15, 22–3, 35, 37–8, 69–70
Curie, Pierre 3, 4–5, 10, 13, 18, 37
cyanide capsules 190, 206, 293
cyclotron
   Bohr's cyclotron 95
   Chadwick's use of 48, 89, 91, 106
   German access to 119, 120, 121, 129
   Joliot's sabotage of 176
   Lawrence's work on 154, 181
   Rutherford's rift with Chadwick over 48
   see also calutron (modified cyclotron)
Czechoslovakia, German threat to/invasion of 53, 59

Daladier, Édouard 87
Dale, Henry 313
Darwin, Charles 153, 154, 155, 161, 183, 193
Dautry, Raoul 86, 101
de Gaulle, Charles 248–9
Dempster, Arthur 47
Denmark, German invasion/occupation of 95–6
   see also Copenhagen
Derbyshire 99
detonation 'guns' for atomic bombs 233–4, 237, 242, 244, 246, 249, 252, 255
deuterium 38, 39
   see also heavy water
Deutsche Physik ('German Physics') 16, 25, 26, 32, 34, 38, 49, 63
diamonds 103, 104

419

INDEX

Diebner, Kurt 120, 121, 137, 138, 174, 269, 310
   Alsos Mission hunt for/capture of 231, 232, 272
   Farm Hall internment 311, 314
diffusion membranes, U-235 enrichment 146, 160, 194, 235
'dirty' bombs, Allies' fear of 221
Disinfection, Operation 14
Donne, John 242
Döpel, Robert 172, 183
'dragon' experiment, Frisch 233, 244, 256–7, 316, 317–18, 319, 321
dry ice 172
Du Pont Company 199, 200
Dunning, John 82
'Dutch cloth' 142

East Germany 335–6
Einstein, Albert
   German campaigns against 24, 32, 34
   and Lenard 16
   letter to Roosevelt (August 1939) 64, 66, 78–9, 343, *p*32
   relativity theory 23–4
   relocation to United States 32–3
   and Uranium Advisory Committee 84
'Einstein debate', Bad Nauheim (1920) 24
Eldorado Mining Company 160, 186, 195, 204
electric detonators 260
electricity generation *see* nuclear energy/power
electrolysis 39
electromagnetic separation, U-235 enrichment 171, 181, 235–6, 237, 256
   see also calutron (modified cyclotron)
elements 9
Enola Gay, Hiroshima bombing 293–7, 298
enrichment of U-235 *see* U-235 enrichment (isotope separation)
Epsilon, Operation 273, 309
Esau, Abraham 136

Ettrick, SS 107, 108, 144
explosive lenses 259, 260, 274, 279, 322

Falkenhorst, General von 207
fallout, radioactive 262
Falmouth Bay 104
Farm Hall, Godmanchester, *Uranverein* detainees in 310–15, 344
Farrell, Thomas 284, 302, 303
Fat Man (plutonium bomb) 246–7, 249, 257–61
   arming of 302–3
   design of 240, *259*
   gun detonators 233–4, 237, 242, 244, 246, 249, 252, 255
   implosion device 238–40, 244, 252, 257–8, 322
   initiator 249–50, 260–1, 322
   internal structure *259*
   lens implosion 259, 260, 274, 279, 322
   mock-up bomb test 301
   Nagasaki bombing 301–4, *p*28
   naming of 240–1
   test drops of dummy bombs 291–2
   US Air Force trials 242
Fat Man and Little Boy (film) 318–19
Feather, Norman 114–15, 127–8
Fermi, Enrico xxii, 27, 43–7, 58
   'CP-1 pile' 195, 197–9
   and embargo on publication of nuclear fission research 81, 83
   Met Lab 192
   military use of atomic bomb 276
   'pile' collaboration with Szilard 76–7, 84, 106, 126, 181, 186
   relocation to United States 53–4
   Trinity test 284
Fermi, Laura 192, 198–9
Feynman, Arline 224–5, 275
Feynman, Richard 224–5, 233, 246, 248, 275
First World War 12–15, 16, 30
Fischer, Emil 12
Fleischmann, Rudolf 251
Foley, Frank 133, 134, 149–50, 178
France
   German invasion/occupation of 101–5, 129

INDEX

safeguarding of scientific assets from Nazi control 100–5
transportation of heavy water from Norsk Hydro plant 86–8
Franck, James 276
Franck Report 276
Freshman, Operation 189, 190
Frisch, Otto Robert xxii–xxiii, 40–3, 52, *p*6
   American Medal of Freedom presentation *p*31
   autobiography of 340–1
   Cambridge collaborations 115
   'Category C' alien status 91
   Copenhagen conference (1937) *p*7
   death of 341
   'dragon' experiment 233, 244, 256–7, 316, 317–18, 319, 321
   escape from Germany 67
   farewell party at Los Alamos 308
   as 'father of the atomic bomb' 341
   Frisch–Peierls memorandum 93–5, 100, 324
      text of 351–7
   Hiroshima bombing, reaction to 299
   Manhattan Project 216, 217, 225, 248
   nuclear chain reaction theory 59
   nuclear fission 54–7, 62, 68, 91–2, 99–100, 108–9
   and Peierls 70, 107
   piano-playing *p*21
   post-war activities 339–41
   and Rotblat 109
   and Second World War outbreak 65–6
   security risk, perceived as 97
   Trinity test 281, 283, 285
   U-235 enrichment 171
Front National 232
Fuchs, Klaus xxiii, 107, 143–4, 148, *p*15
   cinematic portrayal of 319
   contribution to atomic bomb development 321, 322
   death of 336
   departure from Los Alamos 309
   espionage activities 248, 335–6
   and the Peierlses 338
   relocation to East Germany 335–6
   trial and imprisonment of 335

Fuller Lodge, Los Alamos 307
Furman, Robert 212, 221

gadget (implosion device), plutonium bomb detonation 238–40, 244, 252, 257–8, 322
Gadget Group 239–40
gas centrifugation, U-235 enrichment 181
gaseous diffusion, U-235 enrichment 110, 111, 194, 201, 235, 256, 317
Geiger, Hans 19, 137, 139
General Election, UK (1945) 278, 288
Gentner, Wolfgang 119–21, 128, 129, 176
George Washington University, Washington DC 58
Gerlach, Walther 269, 270, 272, 309, 311, 312
German Army, Ordnance Department 136, 137, 138
German atomic bomb, feasibility/threat of 62–4, 66, 75, 76, 94, 117, 131–2, 135–40
   'alternative history' of 314–15
   and Groves 212, 213
   Heisenberg 172–5
   Hitler's axing of atomic bomb project 184–5
   and Manhattan Project 192
   Nazi apathy towards atomic bomb 315
   Perrin memo 169, 170
Germany
   anti-Einstein campaign 24, 32, 34
   'bibliocaust' 33–4
   chemical warfare 14, 17
   escape of Jewish scientists from 51
   Kristallnacht (9–10 November 1938) 53
   Nazism 23–5, 32, 33–4
   nuclear programme 170–5, 213, 221
   racialisation of science 38
   scientific pre-eminence 11
   uranium exports 66
   villages of Haigerloch, Hechingen and Bisingen **xxxvi**
   see also Deutsche Physik ('German Physic'); Second World War

421

INDEX

Gestapo 232
Goebbels, Josef 34
Goering, Hermann 174
Goudsmit, Samuel 230–1, 232, 251
Gowing, Margaret 323, 327–8
graphite (moderator) 77, 139, 172, 195
Groningen, University of 52
Grouse, Operation 189
Groves, Leslie R. xxiii, *p*18, *p*30
  and Akers 196
  anti-British bias 193, 203, 319, 320
  atomic bomb target selection 288
  atomic bomb use 290
  British animosity towards 197
  and British contribution to atomic bomb 349
  budget management 201, 254
  and Chadwick 215–16, 243, 320
  Director of Manhattan Project appointment 189
  disaster planning, Trinity test 274–5
  financial cost of Manhattan Project 201, 254
  and German atomic bomb threat 212, 213
  Hanford (Site W) 199–200
  Hiroshima bombing report 299
  historical narrative of Manhattan Project 318
  Los Alamos (Site Y)
  Oak Ridge (Site X) 235, 249
  personality of 193, 200, 319
  and purpose of atomic bomb 220
  and Russian atomic bomb programme 268
  Scientific Intelligence (Alsos) Mission 220–1
  security imperative 193, 194–5, 224, 306–7
  Smyth Report 264
  Trinity test 274–5, 282, 286, 287
Guéron, Jules 249
gun detonators for atomic bombs 233–4, 237, 242, 244, 246, 249, 252, 255
Gunnerside, Operation 205–7, 213

Haagen, Eugen 251
Haakon VII, King of Norway 95

Haber, Fritz 11, 17
Hahn, Otto xxiii, 11–12, 14, 15, 41, 65, 310, *p*1
  Alsos Mission capture of 273
  barium anomaly 54
  Farm Hall internment 311, 312, 314
  and German atomic bomb 63
  Hiroshima bombing, reaction to 311
  and Meitner 45–6, 52
  Nobel Prize awarded to 313–14
  nuclear fission discovery 61
  relocation to Black Forest 269
  Schumann Conference attendance 137
Haigerloch **xxxvi**, 270–2, 346
  B-VIII reactor 271, *272*, *p*22, *p*23
Halban, Hans von 61–2, 73, 85, *p*8
  American treatment of 204
  clandestine transportation of heavy water from France to England 103
  escape from Paris 102
  Kowarski collaboration 62, 113–16, 127, 147
  relocation to Canada 195–6
  Tube Alloys tour of US 180
  viewed as security risk by Americans 249
Halifax bombers 190–1
Hammett, Dashiell 240
Hanford (Site W), Washington 199–200
  plutonium production 247, 258
  working conditions 247–8
Hankey, Lord 75, 142, 144, 155, 159, 183, 187
Hard, Fred 117, 118, 344, 345
Harrison, George L. 286, 287–8, 299
Harteck, Paul 39, 134, 136, 169, 170, 171, 269, 309–10
Harwell, RAF 330, 339, 340
Hawaii 300
Hawkins, David 317–18
Haworth, Norman 106
heavy water
  chemical composition and production of 38–9
  clandestine transportation from Norsk Hydro plant, Vemork 86–8, 150

422

INDEX

German access to/production of 96, 138, 150, 170, 172, 213, 214, 222
ICI investigations 226
neutron moderation 76–7
sabotage of Vemork plant 177
transportation of from France to England 102, 103, 104, 105
US production of 147
Hechingen **xxxvi**, 271
Heidelburg University 33, 270
Heisenberg, Werner xxiv, 26–7, 35, 63, 132, 134, *p3*
Alsos Mission capture of 272–3
American trip (1939) 65
Bohr meetings in Copenhagen (September 1941) 148–9, 211
Farm Hall internment 311, 312–13
and German atomic bomb 140, 170–5
kidnapping plan 252
L-IV reactor 172, 183, 184
Meitner's assessment of 149, 211
MI6 file on 133
Peierls's assessment of 69
presentation to Milch and Speer 184
Schumann Conference attendance 137, 138
Stark's campaign against 50
Uranverein leadership 169
as weakest link in *Uranverein* chain 315
Zurich trip 252–3
Helberg, Claus 207
Hersey, John 341–2
Hertz, Gustav 111
'hesperium' ('element 94') 44
Hess, Rudolf 150
Hevesy, George 96
hex *see* uranium hexafluoride
hibakusha (survivors of the atomic bombs) 342
Hill, Archibald (A.V.) 71, 94–5, 122
Himmler, Heinrich 49, 50, 174, 175
Hinton, Joan 281, 284
Hirohito, Emperor 301, 304, 305
Hiroshima
atomic bombing of 293–7, 341–2, *p27*
destructive impact of atomic bomb 295–7
eyewitness testimony of atomic bombing 296, 297
ground zero plaque 342
Peace Memorial Museum 342
target selection 266, 289–90
Hitler, Adolf 25, 33, 41, 53, 59, 75, 119
assassination attempt 230
axing of German atomic bomb project 184–5
Horsa glider mission (Freshman) 190–1
Hovde, Frederick 164, 167
Howard, Charles (Jack) 100–1, 103–5, 117–18, 344–5
Hubbard, Jack 281, 282
Hutton, R.S. 135
Hyde Park Aide-Mémoire 250–1
hydrodynamics 239, 245
hydrogen
alpha-particle experiments 21
neutron moderation 46
radioactive isotopes of 39
hydrogen (H-) bomb 331

ICI
and Akers 157–8, 204, 332
Chadwick's collaborations with 105–6, 157
diffusion membranes 235
heavy-water sources 226
and nuclear energy 147, 156
and Perrin 157–8, 332
and Tube Alloys 203–4, 321
U-235 enrichment 146, 157
IG Farben 86, 87, 138
implosion device, plutonium bomb detonation 238–40, 244, 252, 257–8, 322
implosion dynamics 239–40
Indianapolis, USS 292
initiator (device igniting an explosive chain reaction) 234–5, 249–50, 260–1, 274, 322
Institute of Theoretical Physics, Copenhagen 27
Interim Committee 275–7
isotopes 9–10, 29–30

423

INDEX

Italy
  Alsos Mission in 221
  Fascist regime 43, 44, 52–3
Iwo Jima 291

Japan
  decision to use atomic bomb against 250
  propaganda leaflets dropped over 301
  refusal to respond to surrender demands from US 301
  surrender of 305
  Tokyo firebombing 265
  ultimatum to surrender or face 'utter devastation' 288
  US invasion plan 265, 323
Jensen, Hans 185
Joliot, Frédéric xxiv, 22, 23, 29, 35–6, 37, 46, 175–7, p8, p11
  Alsos Mission interrogation of 231–2
  atomic bomb patent 62
  'boiler' research 85–6, 97, 113–16, 126
  in Clairvivre 105, 119
  escape from Paris 101–2
  Gentner collaboration 120–1, 128, 176
  Meitner–Frisch security risk perception 97
  publication of uranium research 61, 63
  Schumann interrogation 120
  securing of heavy water from Norsk Hydro plant 86, 87, 88
Joliot-Curie, Irène 14, 15, 22, 23, 29, 35–6, 37, 176
  'element 93' 46
  escape from Paris 102
  marriage to Frédéric Joliot 85
  tuberculosis treatment 105, 119
Joos, Georg 135–6
Jungk, Robert 344

Kaiser Wilhelm Institutes (KWI) 11, 14, 134, 137, 170, 313
Keitel, Wilhelm 174, 175
Kemmer, Nicholas 128

Kistiakowsky, George 239, 252, 260, 281, 284
Klaproth, Martin Heinrich 1
Kokura, plutonium bomb primary target 302, 303
Korschung, Horst 269, 310, 311, 312
Kowarski, Lew 85, p8
  escape from Paris 102
  Halban collaboration 62, 113–16, 127, 147, 195
  Tube Alloys tour of US 180
Kristallnacht (9–10 November 1938) 53
KWI *see* Kaiser Wilhelm Institutes
Kyoto, atomic bomb target veto 275

L-I reactor 172
L-IV reactor 172, 183, 184
La Sapienza University, Rome 43
Langevin, Paul 129
Laue, Max von 26, 34, 95–6, 310
  Alsos Mission capture of 273
  Farm Hall internment 312, 314
  relocation to Black Forest 269
Laurence, William 336
Lauritsen, Charles 152, 163
Lawrence, Ernest xxiv, 153, 154, 162, 167
  and approval for American atomic bomb 180
  at Berkeley 48, 83, 181, 182, 216, 235, 243, p16
  cyclotron invention and development 181
  evangelist for atomic bomb 163
  military use of atomic bomb 276
  U-235 separation, electromagnetic 235
Lenard, Philipp 15–16, 24, 25, 33, 34, 38, 49
lens implosion 259, 260, 274, 279, 322
Lessing, Theodor 32
Lewis, Gilbert 39
Lindemann, Frederick (Lord Cherwell) 73, 76, 98, 110, 111, 167, 227
  Anglo-American military and scientific collaboration 123
  Bohr–Churchill meeting 228
  and Britain's commercial nuclear priorities 204, 208

INDEX

and British super-bomb 164, 165, 183, 187
Conant meeting 151–2
and M.A.U.D. Committee 112
Peierls meeting 112–13
plutonium put down 142
scepticism towards atomic bomb 278
support for atomic bomb 146
liquid thermal diffusion, U-235 enrichment 171, 236–7, 256, 257
lithium atom, splitting of by Cockcroft and Walton 31–2
Little Boy (uranium bomb)
  altitude of detonation 294
  construction of 274, 292
  critical mass of U-235 321
  design of 241, 255
  destructive power of 252
  fireball description 295
  flash and blast damage to people 296
  gun detonator 242, 246, 249, 252, 255
  Hiroshima bombing 293–7, p27
  internal structure 255
  loading pit on Tinian p26
  mushroom cloud 295
  propellant charge loading 292, 293
  target selection 275
  test drops of dummy bombs 291–2
  US Air Force trials 242
Littler, Derrik 284
Liverpool
  Blitz 108–9, 125, 343
  Chadwick's research at University of Liverpool 48, 343
Locker-Lampson, Oliver 33
Los Alamos (Site Y), New Mexico **xxxv**, 202, 217, 223–6
  Anchor Ranch 234
  farewell party 307–8, p29
  Groves selects site 202
  Hiroshima bombing, reactions to 299
  Los Alamos Primer 224
  Nagasaki bombing, reactions to 306
  social and leisure activities 225, 307–8
  Technical Area 223
  town amenities 223
  trouble-shooting sessions 224
  see also Trinity (test plutonium bomb)

McMillan, Edwin (Ed) 82, 83, 116, 126, 128
Macmillan, Harold 105
magnetron, cavity 74, 89, 125, 151
Manhattan Project
  accident, fatal 257
  atomic bomb target selection 266
  'Atomic Cities' **xxxiv**, 200–2
  British animosity towards Groves 197
  British Mission 209, 215–20, 223, 225, 243–6, 248, 263, 306–9
  contributions to atomic bomb development 320–3, 326–9
  farewell party 307–8
  members of 358–9
  post-war activities of members 333
  British perceived as security risk by Americans 204
  cinematic accounts of 318–19
  'compartmentalisation' of 193
  espionage activities 227, 248, 335–6
  establishment of 325
  exclusion of British from 203
  financial cost of 201, 233, 254
  and German atomic bomb threat 192
  historical narratives of 317–18
  and Interim Committee 275–6
  Manhattan Project: Official History and Documents 317
  minimisation of British contribution to by Americans 316–19, 326–7
  naming of 188
  pessimistic attitudes towards 254
  security imperative, Groves's 193, 194–5, 224, 306–7
  Smyth Report 264, 316–17, 319, 349–50
  and Tube Alloys 209
  workforce employed by 233
  see also Oppenheimer, J. Robert; Trinity (test plutonium bomb)

425

INDEX

'Manifesto of Rome' 52–3
maps
    Europe before Second World War **xxxi**
    Haigerloch, Hechingen and Tailfingen **xxxvi**
    Los Alamos, New Mexico **xxxv**
    Manhattan Project, 'Atomic Cities' **xxxiv**
    Tube Alloys, research and supply centres **xxxii**
    Uranverein, research and supply centres **xxxiii**
Marie Curie Radium Fund 22
Marsden, Beryl 101, 104, 117, 118, 344, 345
Marshall, Leona 197, 199
Massey, Harrie 320
Mattauch, Josef 134, 135
M.A.U.D. Committee 97–9, 105–6, 113–14, 130, 141–2, 143, 165, 334
    barring of Frisch and Peierls from 100
    final report of 144–7, 151, 152, 154, 155–6, 163, 324, 328
    nuclear fission research 116
    Policy Committee 116
    Technical Committee 112, 116, 144, 152, 157
    see also Tube Alloys
MAYSON (proposed unification of British and US atomic bomb programmes) 164, 165, 179
Meitner, Lise xxiv–xxv, 12, 15, 40–1, 42, *p*1
    'atomic histories' interview 340
    barium anomaly 54
    death of 340
    escape from Germany 51–2, 96, 134–5
    on Heisenberg 149, 211
    nuclear fission, explanation of 54–7
    Rosbaud friendship 134–5
    security risk, perceived as 97
    telegram to Physical Society, London 99
    and 'transuranic elements' 45–6
membrane diffusion, U-235 enrichment 110, 111, 194, 201, 235, 256, 317

Mendeleyev, Dmitri 2
Met Lab (Metallurgy Laboratory), University of Chicago 180, 182, 186, 192
    petition to stop atomic bomb's use against people 290
    plutonium production 188
Metropolitan-Vickers 142, 146, 160, 161
MI5 141, 221, 338
MI6 132–3, 139, 149, 150, 177, 310
Milch, Erhard 184
military intelligence
    Abwehr 86, 87, 88–9, 129, 138
    British intelligence sharing with Americans 185, 192–3, 212
    Farm Hall covert recordings of *Uranverein* detainees 311–14
    see also MI5; MI6
moderators, for nuclear fission
    graphite 77, 139, 172, 195
    heavy water 76–7
    hydrogen 46
Moon, Philip 89
Morrison, Philip 290, 291
Murphree, Eger 180
mushroom cloud 295
Mussolini, Benito 43, 44, 53

Nagasaki, atomic bombing of 301–4, 342, *p*28
napalm 265
National Academy of Science (NAS) 162, 166, 324, 325
National Defense Research Committee (NDRC) 84, 126, 153, 162
    London bureau 125, 130, 144, 151
Nature (journal) 58
Naturwissenschaften (journal) 54, 134
Naval Gun Factory, Washington 234, 237
Nazism 23–5, 33–4, 35
Neddermeyer, Seth 238, 239, 322
neptunium, naming of 128
Nernst, Walther 20
Netherlands
    German invasion/occupation of 98
    University of Groningen 52
Neumann, John von 239
neutrons
    discovery of 29–30

426

INDEX

irradiation of elements to create novel radioisotopes 43–7
nuclear chain reaction 60, 77, 90
slow and fast neutrons in nuclear fission 90
New York, Columbia University 76, 84, 181
New York Times 82, 300
Next of Kin, The (film) 192
nickel, sintered, diffusion membranes 235, 317, 321
Niepce de Saint-Victor, Claude 2
Nier, Albert 65, 81–2, 143
Nobel Institute for Physics, Stockholm 52
Nobel Prizes
  Becquerel 4
  Bohr 20
  Chadwick 37
  Cockcroft and Walton 331
  Curie, Marie 4–5, 13
  Curie, Pierre 4–5
  Einstein 24
  Fermi 53
  Haber 17
  Hahn 313–14
  Heisenberg 35
  Joliot and Joliot-Curie 36
  von Laue 26
  Lawrence 154
  Planck 16–17
  Soddy 21
  Stark 17
Noddack, Ida 44–5
Normandy landings (D-Day) 229, 230
Norsk Hydro ferry, sinking of 222
Norsk Hydro heavy-water plant, Vemork 38, 39, 77, 346
  clandestine transportation of heavy water to France 86–8
  German control of 96, 138, 150, 169–70, 213
  plan to relocate plant to Germany 222
  sabotage activities/missions 177, 189–91, 205–7
  US Eighth Air Force bombing of 214
Northern Mariana Islands 290–1

Norway
  Allied sabotage missions against Norsk Hydro heavy-water plant 189–91, 205–7
  Eric Welsh 150–1
  German invasion/occupation of 86, 95–6
Norwegian High Command 177, 205
Now It Can Be Told (Groves) 318, 349
Nuclear Age, birth of 197, 198
nuclear chain reaction 59–62, 64, 73, 77
  and neutron speed 77
  slow and fast neutrons 90
  speed of 145
  Thomson–Moon research 89
  in uranium (U-235) 60
  uranium bomb 234
nuclear disarmament campaign 331
nuclear energy/power
  Fermi's nuclear reactors 76–7, 106, 126, 181, 186, 195, 197–9
  German nuclear programme 138–9, 170–5, 183, 213, 271, *272*, 346, *p22*, *p23*
  Halban–Kowarski research 62, 113–16, 127, 147
  Rutherford's scepticism about 49
nuclear fission
  Bretscher–Feather collaboration 127–8
  energy released by 62
  fast neutrons 90, 116
  Meitner–Frisch theoretical explanation of 54–7
  publication of research 61, 63, 81–4
  secondary neutron production 77–8
  in uranium (U-235) 57
nuclear physics
  artificial radioisotopes 43–7
  atomic number 21, 30
  atomic structure 8, *9*, 20
  atomic weight 21, 30
  'splitting of the atom' 30–2
  structure of nucleus 29–30

Oak Ridge (Site X), Tennessee 200–1, 217, 321
  gaseous diffusion plant 235
  racetracks (calutron arrays) 236

INDEX

'S-50' liquid thermal diffusion plant 249, 256, *257*
   workforce employed by 235
   X-10 reactor 247
   Y-12 electromagnetic U-235 separation plant p19
Office for Scientific Research and Development (OSRD) 162
Office for Strategic Services (OSS) 252
Okinawa 304
Oliphant, Marcus (Mark) xxv, 28, 31, 39, 67, 70, 74–5, 89, *p4*
   Anglo-American military and scientific collaboration 167–8
   Berkeley trips 153–4, 163, 166, 236, 320
   Birmingham University research group 236, 244, 320, *p20*
   Britain's nuclear programme 263
   calutron work 244
   contribution to atomic bomb development 320–1
   Copenhagen conference (1937) *p7*
   death of 338
   and Frisch–Peierls memorandum 95, 100
   and M.A.U.D. Committee final report 163
   post-war activities 337–8
   and Tube Alloys 159
   U-235 enrichment 236
   and US volte-face on atomic bomb programme 325–6
   and Wilkins 160
Operation Disinfection 14
Operation Epsilon 273, 309
Operation Freshman 189, 190
Operation Grouse 189
Operation Gunnerside 205–7, 213
Operation Overlord 229
Operation Peppermint 222
Operation Swallow 206–7, 222
Oppenheimer (film) 319
Oppenheimer, J. Robert xxv, 27, 182, 225, *p18*
   'Army Navy Excellence' award 308
   'Coordinator of Rapid Rupture' appointment 188
   educational role at Los Alamos 224
   Fat Man (plutonium bomb) photographs 302
   implosion device ('gadget') 238, 239, 242
   initiator for plutonium bomb 260
   military use of atomic bomb 276
   Scientific Director of Manhattan Project appointment 202
   social life of 219
   Trinity test 281, 282, 283, 284, 285
OSRD *see* Office for Scientific Research and Development
OSS *see* Office for Strategic Services
Overlord, Operation 229
Oxford, Clarendon Laboratory 111, 142

Paris
   Allied liberation of 231
   German assault on 102–3
   Radium Institute 22, 35, 85, 176
   see also Collège de France, Paris
Parsons, William 290, 293, 298
Pash, Boris 220, 230, 231, 232, 251, 270, 271
patents on nuclear technology 62, 159, 258
Peace Memorial Museum, Hiroshima 342
Pearl Harbor, Japanese attack on 167
Peenemünde 178, 184
Pegram, George 165–6
Peierls, Genia Nikolaevna 69, 107, 216, 217, 218–19, 264, 338, 339
   departure from Los Alamos 309, *p29*
Peierls, Rudolf (Rudi) xxv–xxvi, 69–70, 91, 99, *p9*
   American Medal of Freedom presentation *p31*
   autobiography of 339
   and Bretscher 142
   Cambridge collaborations 115
   contribution to atomic bomb development 321, 322
   Copenhagen conference (1937) *p7*
   death of 339
   departure from Los Alamos 309
   farewell party at Los Alamos 307

INDEX

fire-fighting duties in Birmingham Blitz 107, 108
Frisch collaboration 70
Frisch–Peierls memorandum 93–5, 100, 324
   text of 351–7
   and Fuchs 143, 148, 321–2, 338
   German atomic bomb feasibility 140
   Hiroshima bombing, reaction to 299
   implosion device 258
   implosion dynamics 239–40
   Lindemann meeting 112–13
   Manhattan Project 194–5, 216, 217, 218–19, 225, 226
   MI5 file on 338
   Nagasaki bombing, reaction to 306
   and Oppenheimer 182
   post-war career 338–9
   security risk, perceived as 100
   Trinity test 281, 285
   Tube Alloys tour of US 180
   U-235 enrichment 109–12
   and von Neumann 239
Penney, Adele 250, 266
Penney, William (Bill) 245–6, 250, *p*31
   atomic bomb target selection 266, 289–90
   post-war career 331–2
Peppermint, Operation 222
periodic table 9–10, 44
Perrin, Francis 62, 70
Perrin, Michael 157–8, 166, 169, 170, 186–7
   and Alsos Mission 231
   Bohr's visit 211
   diary of 332–3
   and ICI 157–8
   post-war career 332–3
   Smyth Report 317
Perutz, Max 107
Pétain, Marshal Philippe 103–4, 119
Petites Curies ('radiological cars') 13, 14
Philadelphia Naval Shipyard 236
Philipp Lenard Institute, Heidelberg 38
photoelectric effect 16
Physical Review (journal) 65, 77, 81, 82–3, 116–17, 131
Physical Society, London 99

'pile' (prototype nuclear reactor)
   CP-1 (Chicago Pile-1) 195, 197–9
   Fermi–Szilard collaboration 76–7, 84, 106, 126, 181, 186
pitchblende 1, 4, 73, 347
Placzek, George 65, 81
Planck, Max 26, 32, 38, 69
plutonium 83, 162, 170, 182
   critical mass 258
   fissionability of 131, 142, 173
   naming of 128
   plutonium salts 188, 258
   predetonation problem 237–8
   production of 188, 199, 247–8
   Pu-239 isotope 237
   Pu-240 isotope 238
   radioactive decay of 237
   see also atomic bomb; Fat Man (plutonium bomb)
poison gas 14, 17
polonium 4, 6
polonium–beryllium initiators 260–1
positrons 36
Potsdam Conference (July 1945) 276, 277–8, 286–8
Potsdam Declaration 288, 299
Poulsson, Jens 213
Project A (Alberta) 262, 290–5
   see also Fat Man (plutonium bomb); Little Boy (uranium bomb)
'Project Larson' 252
Project TR *see* Trinity (test plutonium bomb)
Project Y: The Los Alamos Story (Hawkins) 317–18
protactinium 15, 45
protons, etymology 21
Pugwash Conferences 331, 338, 339

Quadrant Conference (August 1943) 209
quantum mechanics 26
Quebec Agreement (August 1943) 209, 277

racetracks (calutron arrays) 236
racism 16
   see also anti-Semitism
radar 72, 73, 74–5, 122–3, 125

429

INDEX

radiation
  components of ionising radiation 6
  Curies' research on 4–5
  discovery of 2
  Rutherford's research on 6–7
radio altimeters 241, 255
radioactive decay 7
radioactive fallout 262
radioactivity
  Becquerel's research 5–6
  etymology 4
  Hahn and Meitner's research 12
  Joliot-Curies' research 35–6
radioisotopes, artificial 36, 37, 43–7
radium 4, 5–6, 7, 10, 22, 102
  medical applications for 13
  'milking' of radon from 99
Radium Institute, Paris 22, 35, 85, 176
radon 7, 99
Ragazzi di Via Panisperna 43
Randall, John 74, 89, 151
Rascher, Sigmund 251
Reich University of Strasbourg 251
relativity theory, Einstein 23–4, 34, 35
rhenium 44
Rhodes, Richard 325–6
Rhydymwyn, 'Valley' (Special Products factory) 160–1, 227, 344
Rittner, Thomas 310, 311, 314
Rockefeller Foundation 41
Roentgen, Konrad 2
Rome
  Alsos Mission in 221
  La Sapienza University 43
  'Manifesto of Rome' 52–3
Rønneberg, Joachim 205, 206–7, 213, 346
Roosevelt, Franklin D. 166–7, 196
  Anglo-American military and scientific collaboration 122
  approval for American atomic bomb 163, 179–80
  and Churchill 121, 179, 185–6, 208–9, 250
  death of 267
  Einstein's letter to (August 1939) 64, 66, 78–9, 343, *p*32
  Hyde Park Aide-Mémoire 250–1
  letter to Churchill (October 1941) 163–4, 165, 167
  Uranium Committee 84
  Yalta Conference 267
Rosbaud, Hilde 134
Rosbaud, Paul xxvi, 133–5, 139, 148, 149, 169, 184–5, *p*14
  espionage activities 178, 212, 270
  Uranverein relocation 222
Rosenfeld, Léon 57, 58
Rotblat, Joseph 89–90, 98, 109
  British Mission to the Manhattan Project 219–20
  departure from Los Alamos 253
  revulsion for atomic bomb 248
  spying allegation against 253
Rotblat, Tola 109
Royal Engineers 190
Royal Society 113, 114, 313, 332
'Rufus' (plutonium hemispheres for Fat Man device) 300
Russia
  Anglo-Russian Treaty 196
  atomic bomb secrets, sharing with 196, 219, 228, 248, 253, 268, 335–6
  declaration of war against Japan 304
  Groves's bragging about nuclear intimidation of 220
  nuclear programme 227, 268, 335
Rust, Bernhard 135, 136, 174, 175
Rutherford, Ernest 6–8, 12, 13, 15, 19, 28, 31, 60
  atomic energy scepticism 49
  Cavendish Laboratory, Cambridge 20–1
  and Chadwick 47–8
  cyclotron rants 48
  death of 48–9
  and Fermi 43
  Lenard book review 33
  peerage award 47
  Rutherfordisms 49

S-1 Committee *see* Uranium Committee
'S-50' liquid thermal diffusion plant, Oak Ridge 249, 256, *257*

INDEX

Sachs, Alexander 64, 78–9, 80
St Joachimsthal 346–7
St John the Baptist, Charlton 344, 345
Saipan 301
Salomon, Jacques 129, 176
Sapienza University of Rome 43
Scherrer, Paul 252, 253
Schintlmeister, Josef 173
Schumann, Erich 119, 120, 121, 136–7, 173, 175
   Alsos Mission hunt for 231
   Uranverein report 174
Schumann Conferences 137
science journals, embargo on publication of nuclear fission research 81–4
Scientific Intelligence Mission *see* Alsos Mission
Scientists Against Time (Baxter) 317
Seaborg, Glenn 131, 142, 181–2, 187–8, 275–6
Second World War
   1943 as turning point in 220
   Europe before Second World War **xxxi**
   French armistice with Germany 119
   German air-raids over Britain 108–9
   German invasion of Belgium and the Netherlands 98
   German invasion of Denmark and Norway 95–6
   German invasion of France 101–5
   Hiroshima, atomic bombing of 266, 293–7, 341–2, p27
   modern warfare and civilian deaths 265–6
   Nagasaki, atomic bombing of 301–4, 342, p28
   Normandy landings (D-Day) 229, 230
   outbreak of 65–6
   Pearl Harbor attack 167
   Tokyo, firebombing of 265
   Victory in Europe Day 265
   see also Potsdam Conference (July 1945); Yalta Conference (February 1945)
Senate Appropriations Committee 237
Sengier, Edgar 73, 226

separation of U-235 *see* U-235 enrichment (isotope separation)
Serber, Robert 224, 240
Seuss, Hans 185
Sherwin, Martin 318
'Shetland Bus' (Norwegian fishing boats) 177
shockwaves
   modelling of 245, 246, 322
   Trinity test 283
'Sigurd' (Norwegian engineering student and spy) 178
Simon, Charlotte 111
Simon, Franz (Francis) xxvi, 110–13, 194, p13
   knighthood awarded to 334
   post-war activities 334
   Tube Alloys tour of US 180
   U-235 enrichment 126–7, 160–1
Skyrme, Tony 321
Smyth, Henry D. 316, p30
Smyth Report 264, 316–17, 319, 349–50
   Statements Relating to the Atomic Bomb appendix 317
Soddy, Frederick 7, 8–10, 17–18, 21
SOE *see* Special Operations Executive
Solvay Physics Conference, Brussels (1927) 26
Soviet Union *see* Russia
Special Operations Executive (SOE) 151, 189
Special Training Schools 205–6
Speer, Albert 184, 269
Springer, Ferdinand 133
Springer Verlag 133
Stalin, Joseph
   Anglo-Russian Treaty 196
   'news' about atomic bomb 288, 335
   Potsdam Conference 277, 278, 286, 287, 288
   Yalta Conference 268
Stark, Johannes 25, 26, 33, 34, 35, 38, 49–50
Stern, Otto 41
Stimson, Henry 208, 230, 256, 267
   Hiroshima bombing 299
   Interim Committee 275

431

INDEX

Potsdam Conference 277, 278, 286, 287
Senate Appropriations Committee, appearance before 237
US invasion of Japanese mainland 265
Stockholm 96, 210
Strasbourg, Reich University 251
Strassmann, Fritz 45–6, 54, 61, 63
STS *see* Special Operations Executive, Special Training Schools
'Substance 38' (uranium oxide) 136, 139, 172
Sudetenland 53
'Swallow' team 206–7, 222
Sweden
  Bohr's sojourn in 209–10
  Meitner's escape to 96
Sweeney, Charles 302, 303, 304
Szasz, Ferenc 323, 328–9
Szilard, Leo xxvi, 62–4, 66, 80, 81, 82
  and birth of Nuclear Age 198
  Copenhagen conference (1937) *p*7
  and Einstein's letter to Roosevelt (August 1939) 343, *p*32
  nuclear chain reaction theory 59–61
  objection to use of atomic bomb against civilians 275–6, 290
  'pile' collaboration with Fermi 76–7, 84, 106, 126, 181, 186

TA *see* Tube Alloys
Tailfingen **xxxvi**, 271
tamper (heavy shell surrounding nuclear explosive) 234, 255
Tatlock, Jean 242
Taylor, Geoffrey 245, 262, 281
Teller, Edward 63, 224
thermal diffusion, U-235 enrichment 68, 94, 109, 171, 201, 236–7, 256, *257*
Thomson, George xxvii, 73, 74, 89, 116, *p*12
  Anglo-American military and scientific collaboration 130
  and Frisch–Peierls memorandum 95
  Manhattan Project 193
  M.A.U.D. Committee 97–9, 105, 144, 152, 163, 334

Tube Alloys 159
Thomson, J.J. 6
thorium 4, 9
Tibbets, Paul 291, 293, 295
Tinian 290–1, 292, 300, 301
Titterton, Ernest 216, 217, 244
  contribution to atomic bomb development 322
  departure from Los Alamos 309
  electronc detonation 260, 283
  post-war career 333
Tizard, Henry 71–5, 94, 95, *p*10
  Anglo-American military and scientific collaboration (Tizard Mission) 122–6
  and German atomic bomb 139–40
  'super-bomb' feasibility 98
  uranium research projects 73–4
Tokyo, firebombing of 265
Tolman, Richard *p*30
'transuranic elements' 44, 45–6, 47, 58, 82, 83, 114–15
'trinitite' 289
Trinity (test plutonium bomb) 242, 245–6, 261–3, 278–85
  Arming Party 282
  Base Camp 261, 284
  blast wave 282
  bomb assembly and positioning 280
  Control Station 261
  countdown procedure 282
  date set for 276
  detonation procedure 282–3
  emergency evacuation plans 274–5
  explosion and fireball 283–4, 289, *p*25
  explosive yield of 289
  instrument monitoring 262–3
  newspaper report of 289
  plutonium core delivery 280, *p*24
  Point Zero 261, 289
  rehearsals for 274, 278–9
  shockwave 283
  site preparation 261
  success of 286–7
  test firing of TNT ('100-ton test') 263
  VIP observers of 262, 281
tritium 39

INDEX

Tronstad, Leif 150–1, 177, 189, 205, 212–13, 214
Truman, Harry S.
  becomes President 267
  Hiroshima bombing statement 299–300, 316
  Interim Committee 275, 276
  Potsdam Conference 277, 286, 287, 288
Tube Alloys (TA) 158–61, 167
  and American nuclear programme 186, 187
  animosity towards Groves 197
  and ICI 203–4, 321
  and Manhattan Project 193, 204, 209
  tour of United States 180–2
  uranium supply 195
  see also M.A.U.D. Committee
Tube Alloys Consultative Council 158, 183, 263
Tube Alloys Directorate 158, 159
Tube Alloys Intelligence Committee 269
Tube Alloys Technical Committee xxix, 158, 243
  research and supply centres **xxxii**
Tuck, James 244–5, 285, 308, 322, 333
Turner, Louis 82, 116
Tuve, Merl 90, 126, 143
Tyndall, A.M. 74
typhus vaccine 251

U-235 enrichment (isotope separation) 98, 109–12, 126–7
  Clarendon Laboratory, Oxford 142
  electromagnetic separation 171, 181, 235–6, 237, 256, *p*19
  gas centrifugation 181
  gaseous diffusion 110, 111, 194, 201, 235, 256, 317
  German nuclear programme 171
  and ICI 146, 157
  plan for US separation plant 156, 161
  Rhydymwyn separation plant 160–1, 227, 344
  thermal diffusion 68, 94, 109, 171, 201, 236–7, 256, *257*
  ultracentrifugation 110, 171, 181

zinc–silver alloy 166
  see also Oak Ridge (Site X), Tennessee
U-boats 13
ultracentrifugation, U-235 enrichment 110, 171, 181
unexploded bomb (UXB) team 117–18, 344–5
Union Minière Company, Brussels 73, 74, 232
United Kingdom *see* Britain
United States (US)
  decision to use atomic bomb against Japan 275
  invasion of Japanese mainland, planning for 265, 323
  nuclear supremacy, post-war 267–8
  scepticism/indifference to concept of atomic bomb 122, 126, 151–4, 161–2, 323–5
  see also Anglo-American military and scientific collaboration; Manhattan Project
Uranbrenner ('uranium burner')
  nuclear reactors 138–9, 172, 173, 175, 213
  B-VIII reactor 271, *272*, 346, *p*22, *p*23
  L-IV reactor 183
'uranic rays' 2, 3
uranium
  critical mass 62, 70, 74, 91–2, 94, 100, 174
  Curies' research on 4
  energy potential of 10
  fissionability of U-235 82, 143
  neutron irradiation experiments 44–5, *83*
  nuclear chain reaction *60*, 61
  nuclear fission *57*
  ore deposits 2, 66, 136, 142, 226
  properties of 2
  radiation from 6
  radioactive decay 7
  sintered uranium 172
  structure of nucleus 29
  'transuranic elements' 44, 45–6, 47
  uranium oxide 136, 160, 232
  uranium-235 isotope 47

INDEX

uranium-238 isotope 82, *83*, 116
see also atomic bomb; Little Man (uranium bomb); U-235 enrichment (isotope separation)
Uranium Committee (S-1 Committee) 79, 84–5, 126, 130, 144, 153, 154, 166
  approval for American atomic bomb 179–80
  first meeting of (October 1939) 80–1
  negativity towards US atomic bomb 324
uranium hexafluoride (hex) 98, 110, 161, 171, 201, 236–7
uranium mines 73, 160, 226
Uranverein ('Uranium Club') 66, 121, 136, 137–9, 148, 169, 173, 174–5
  Alsos Mission pursuit and capture of scientists and uranium 230–2, 269–73, 309–10
  espionage against 178
  Farm Hall internment 310–15, 344
  German atomic bomb, alternative history of 314–15
  Heisenberg's pitch to Reich High Command 183–4
  relocation from Berlin to Black Forest 222, 251–2, 269
  research and supply centres **xxxiii**
  Russian pursuit of scientists 268
  scientific misapprehensions of 313
  withdrawal of support for 175
'urchin' (plutonium bomb initiator) 260–1, 274, 322
Urey, Harold 38, 84, 147, 165–6, 167, 201
  and approval for American atomic bomb 180
  objection to use of atomic bomb against civilians 275–6
  U-235 enrichment 181
US Air Force
  atomic bomb trials 241–2
  Tokyo firebombing 265
  Vemork bombing 214

US Army Corps of Engineering 188, 200, 201
'UzM' papers (Stern) 41

V-1 flying bombs 250
'Valley' (Special Products factory), Rhydymwyn 160–1, 227, 344
Vemork plant *see* Norsk Hydro heavy water plant, Vemork
Villager, The (community newspaper at Hanford) 300
Virus House (*Uranbrenner* reactor facility) 137, 138, 139, 170, 221

Wales, 'Valley' (Special Products factory at Rhydymwyn) 160–1, 227, 344
Walke, Harold 91, 154
Walton, Ernest 30–1, 331
Watson-Watt, Robert 72
Weisskopf, Viktor 211–12
Weizsäcker, Carl von 50, 132, 134, 148
  'alternative version' of Nazi nuclear programme 314
  capture by Alsos Mission 271–2
  Farm Hall internment 312, 344
  and German atomic bomb 63, 64, 173, 314
Wellcome Trust 332
Wells, H.G. 10, 17
Welsh, Eric ('Theodor') 150, 177–8, 185, 192, *p*23
  and Alsos Mission 231
  atomic bomb disinformation 221
  and Bohr 209, 210, 211
  failure to share military intelligence with Americans 212
Wendover Air Base, Utah 291, 301
whale blubber 91
What Little I Remember (Frisch) 340–1
Wheeler, John 58, 59, 65, 81
Wigner, Eugene 80, 84, 198
Wilkins, Maurice 159–60, 300
Wimperis, Harry 71, 72
Wirtz, Karl 134, 138, 150, 169, 170, 269, 309–10
  Farm Hall internment 314
  Uranbrenner reactor design 172

434

INDEX

World Set Free, The (Wells) 10, 17

X-10 reactor, Oak Ridge (Site X) 247
X-rays 2, 3, 13, 14

Y-12 electromagnetic U-235 separation plant, Oak Ridge *p*19

Yalta Conference (February 1945) 267, 268

zinc–silver alloy 166
Zinn, Walter 61, 77
Zurich, Heisenberg lecturing in 252–3